Maritime Accidents and Environmental Pollution – The X-Press Pearl Disaster

This book discusses in detail the facts and findings related to the X-Press Pearl container vessel accident that occurred in May 2021 off the coast of Colombo, Sri Lanka. The ship was carrying a large consignment of chemicals and diverse hazardous materials that caused a disastrous and vast environmental and social catastrophe in the region. Through many case studies, accumulated knowledge, and experiences, the authors discuss the accident response, risk mitigation, investigation, and damage assessment activities from the very onset of the accident. It helps researchers and regulators understand the facts of this unique marine chemical accident and formulate necessary future regulations as well as develop robust safety and sustainability management systems and safety cultures.

Features:

- Written by authorities who led the team involved in accident response and damage assessment
- Focuses on identifying plausible root causes, pitfalls in accident response, and weaknesses in current regulatory and management protocols
- Delivers in-depth understanding of a unique marine chemical accident to help formulate necessary future policies and regulations related to such disasters
- Includes many case studies related to the accident illustrated with photos and figures that are true evidence of the disaster, the response, and the mitigation
- Explains and discusses key research findings in a streamlined manner understandable for a wide audience

A valuable resource for readers in environmental management and policy creation, as well as for researchers, professionals, academics, and students involved in environmental science, chemical engineering, technical safety, sustainability management, maritime, polymer, and ocean sciences. Countries where maritime disasters are a concern will also find this book to be an important guide for taking a responsible approach when handling similar situations in the future; not least to avert such events from occurring.

Emergent Environmental Pollution
Series Editor: Prof. Jörg Rinklebe, University of Wuppertal, Germany

The nature of pollutants in the environment depends largely on their interactions in the soil, water, air, and organisms. Understanding the fate and transport of pollutants in the environment is essential for humans and animals, plants, water bodies, air, soils, and sediments. This series provides timely information on the biological, chemical, and physical processes governing the interactions of pollutants in the environment. Such knowledge is needed for minimizing the negative impact of pollutants on human health and the health of ecosystems. Books from this series are of value to professionals and researchers, students, industrial and mining engineers, government regulators, authorities, and consultants in the environmental arena.

Vanadium in Soils and Plants
Edited by Jörg Rinklebe

Nitrate Handbook: Environmental, Agricultural, and Health Effects
Edited by Christos Tsadilas

Nickel in Soils and Plants
Edited by Christos Tsadilas, Jörg Rinklebe, and Magdi Selim

Trace Elements in Waterlogged Soils and Sediments
Edited by Jörg Rinklebe, Anna Sophia Knox, and Michael Paller

Phosphate in Soils: Interaction with Micronutrients, Radionuclides and Heavy Metals
Edited by H. Magdi Selim

Permeable Reactive Barrier: Sustainable Groundwater Remediation
Edited by Ravi Naidu and Volker Birke

Competitive Sorption and Transport of Heavy Metals in Soils and Geological Media
Edited by H. Magdi Selim

For more information about this series please visit: https://www.routledge.com/Advancesin-Trace-Elements-in-the-Environment/book-series/CRCADVTRAELE

Maritime Accidents and Environmental Pollution – The X-Press Pearl Disaster

Causes, Consequences, and Lessons Learned

Edited by
Meththika Vithanage, Ajith Priyal de Alwis,
and Deshai Botheju

CRC Press
Taylor & Francis Group
Boca Raton London New York

CRC Press is an imprint of the
Taylor & Francis Group, an **informa** business

Designed cover image: The cover image is a composite image. "Pellets" and "Ship fire" images are from Mr. Indika, Marine Environment Protection Agency. The image of the turtle is from Mr. Ranil Nanayakkara. All images have been used with permission.

First edition published 2024
by CRC Press
2385 Executive Center Drive, Suite 320, Boca Raton, FL 33431

and by CRC Press
4 Park Square, Milton Park, Abingdon, Oxon, OX14 4RN

CRC Press is an imprint of Taylor & Francis Group, LLC

© 2024 selection and editorial matter, Meththika Vithanage, Ajith Priyal de Alwis, and Deshai Botheju; individual chapters, the contributors.

Reasonable efforts have been made to publish reliable data and information, but the author and publisher cannot assume responsibility for the validity of all materials or the consequences of their use. The authors and publishers have attempted to trace the copyright holders of all material reproduced in this publication and apologize to copyright holders if permission to publish in this form has not been obtained. If any copyright material has not been acknowledged please write and let us know so we may rectify in any future reprint.

Except as permitted under U.S. Copyright Law, no part of this book may be reprinted, reproduced, transmitted, or utilized in any form by any electronic, mechanical, or other means, now known or hereafter invented, including photocopying, microfilming, and recording, or in any information storage or retrieval system, without written permission from the publishers.

For permission to photocopy or use material electronically from this work, access www.copyright.com or contact the Copyright Clearance Center, Inc. (CCC), 222 Rosewood Drive, Danvers, MA 01923, 978-750-8400. For works that are not available on CCC please contact mpkbookspermissions@tandf.co.uk

Trademark notice: Product or corporate names may be trademarks or registered trademarks and are used only for identification and explanation without intent to infringe.

Library of Congress Cataloging-in-Publication Data
Names: Vithanage, Meththika, editor. | De Alwis, Ajith, editor. | Botheju, Deshai, editor.
Title: Maritime accidents and environmental pollution - the X-Press Pearl
disaster: causes, consequences, and lessons learned / edited by
Meththika Vithanage, Ajith Priyal de Alwis, Deshai Botheju.
Other titles: Causes, consequences, and lessons learned
Description: 1st edition. | Boca Raton : CRC Press, 2024. |
Includes bibliographical references and index. |
Summary: "This book discusses in detail the facts and findings related to the X-Press Pearl container vessel accident that occurred in May 2021 off the coast of Colombo, Sri Lanka. The ship was carrying a large consignment of chemicals and diverse hazardous materials that caused a disastrous and vast environmental and social catastrophe in the region. Through many case studies, accumulated knowledge, and experiences, the authors discuss the accident response, risk mitigation, investigation, and damage assessment activities from the very onset of the accident. It helps researchers and regulators understand the facts of this unique marine chemical accident and to formulate necessary future regulations as well as to develop robust safety and sustainability management systems and safety cultures"– Provided by publisher.
Identifiers: LCCN 2023022415 (print) | LCCN 2023022416 (ebook) |
ISBN 9781032315270 (hbk) | ISBN 9781032322919 (pbk) | ISBN 9781003314301 (ebk)
Subjects: LCSH: X-Press Pearl (Container ship) | Marine accidents–Environmental aspects–Sri Lanka. |
Environmental disasters–Sri Lanka. | Environmental disasters–Prevention. |
Hazardous substances–Accidents–Prevention. |
Hazardous substances–Transportation–Safety measures.
Classification: LCC VK1288.S65 M37 2024 (print) |
LCC VK1288.S65 (ebook) | DDC 363.12/3095493–dc23/eng/20230906
LC record available at https://lccn.loc.gov/2023022415
LC ebook record available at https://lccn.loc.gov/2023022416

ISBN: 978-1-032-31527-0 (hbk)
ISBN: 978-1-032-32291-9 (pbk)
ISBN: 978-1-003-31430-1 (ebk)

DOI: 10.1201/9781003314301

Typeset in Times New Roman
by Newgen Publishing UK

Contents

Editor Biographies ix
List of Contributors xi

Chapter 1 X-Press Pearl: The Fateful Journey 1
Ajith Priyal de Alwis

Chapter 2 Recent Maritime Accidents: Chemicals and Plastic Spills................. 15
R.S.M. Samarasekara, A.H.M.S. Siriwardana, and D.R. Anthony

Chapter 3 X-Press Pearl: The Cargo and Its Complexity in Impact 40
Thejanie Jayasekara, Sajani Piyatilleke, Dilantha Thushara, Shantha Egodage, and Ajith Priyal de Alwis

Chapter 4 The Fate of the Pearl 56
Deshai Botheju

Chapter 5 Onshore Responses during the X-Press Pearl Disaster 67
K.P.G.K.P. Guruge, T.S. Ranasinghe, I.P. Amaranayake, K.R. Gamage, T. Maes, and P.B.T.P. Kumara

Chapter 6 Societal and Media Response... 121
Hemantha Withanage and Chalani Rubesinghe

Chapter 7 Plastics, Nurdles, and Pyrogenic Microplastics in the Coastal Marine Environment: Implications of the *X-Press Pearl* Maritime Disaster 134
Madushika Sewwandi, Kalani Imalka Perera, Christopher M. Reddy, Bryan D. James, A.A.D. Amarathunga, Indika Hema Kumara Wijerathna, and Meththika Vithanage

Chapter 8 Sources and Fate of Plastics into Microplastics: Degradation and Remediation Methods... 155
G.M.S.S. Gunawardhana, U.L.H.P. Perera, and Amila Sandaruwan Ratnayake

Chapter 9 X-Press Pearl Ship Impact on Sea Turtles..........173

                *E.M. Lalith Ekanayake, U.S.P.K. Liyanage, and
M.S.O.M. Amararathna*

Chapter 10 Impacts of the MV X-Press Pearl Ship Disaster on Marine Mammals in Sri Lankan Marine Waters..........190

                *U.S.P.K. Liyanage, P.R.T. Cumaranatunga,
E.M. Lalith Ekanayaka, and G.A. Tharaka Prasad*

Chapter 11 Impact of MV X-Press Pearl Disaster on Coastal and Marine Fish and Fisheries..........215

                *P.R.T. Cumaranatunga, K.R. Dalpathadu,
H.M.U. Ayeshya, K.A.W.S. Weerasekera, S.S.K. Haputhantri,
W.N.C. Priyadarshani, G.K.A.W. Fernando,
L.D. Gayathry, and H.B.U.G.M. Wimalasiri*

Chapter 12 Impact of the X-Press Pearl Disaster on Coastal and Marine Birds..........273

                Sampath S. Seneviratne and Jude Janith Niroshan

Chapter 13 Beach and Marine Microplastics: Physiochemical Removal Techniques Targeting Marine Disasters..........297

                *Kalani Imalka Perera, Madushika Sewwandi,
Indika Hema Kumara Wijerathna,
Sujitra Vassanadumrongdee, and Meththika Vithanage*

Chapter 14 Navigating Container Ship Impacts..........315

                Xiaokai Zhang and Mona Wells

Chapter 15 Management and Policy Recommendations: MV X-Press Pearl Ship Accident..........338

                Shakeela S. Bandara

Chapter 16 X-Press Pearl Maritime Disaster: A Framework for Valuation of Environmental Damages..........367

                *U.A.D.P. Gunawardena, J.M.M. Udugama, and
G.A.N.D. Ganepola*

Chapter 17 Lessons from the Pearl 387
  Deshai Botheju

Index 399

Editor Biographies

Meththika Suharshini Vithanage's contribution to science has been recognized by The World Academy of Sciences (TWAS), presenting the Fayzah M. Al-Kharafi Award in 2020. She was listed as a Highly Cited Researcher in 2021 by Clarivate. She was selected as one of the Early Career Women Scientists by the Organization for Women Scientists in Developing Countries, Italy. Additionally, she received the Best Graduate Researcher Award for Natural Hazards from the American Geophysical Union in 2010 for her doctoral research work on the effect of tsunami on groundwater. She has published more than 200 *SCI* journal articles and 10 edited books. Further, she was listed as Top 2% of the most cited scientists across various disciplines globally in from 2017 to 2023. She served as an expert panel member for the X-Press Pearl environmental damage assessment team established by the Marine Environmental Protection Agency of Sri Lanka.

Ajith Priyal de Alwis is Senior Professor at the Department of Chemical and Process Engineering at the University of Moratuwa, Sri Lanka, and is also the Director of the University of Moratuwa. He completed his PhD at the University of Cambridge, UK. He is the recipient of numerous awards in recognition of his work, most notable being the Senior Moulton Medal of Institution of Chemical Engineers (UK), Danckwerts-Maxwell Award from University of Cambridge, Department of Chemical Engineering (Best PhD thesis), and University of Moratuwa Research Awards from 1997 to 2007. He co-chaired the expert panel of the X-Press Pearl damage assessment committee established by the Marine Environment Protection Authorities of Sri Lanka. He has published many research articles and book chapters in various international and local journals. He was also the science team leader for Sri Lanka Institute of Nanotechnology (SLINTEC) from its inception in 2008 to 2011 and Chairman of the National Nanotechnology Committee at National Science Foundation, Sri Lanka from 2011 to 2012.

Deshai Botheju is currently consulting for the Norwegian petroleum and marine industry within his field of expertise in Safety Engineering and Sustainability (Environmental) Management. For more than 19 years he has been involved in numerous large-scale marine and offshore projects as well as in various research and development projects. He received his PhD from Norwegian University of Science and Technology (NTNU) in 2010. Since then, he has been working closely with industry as well as academia as a Senior Engineering consultant, Assistant Professor, and a visiting lecturer. He served as an expert panel member of the X-Press Pearl damage assessment committee established by the Marine Environmental Protection Authority of Sri Lanka.

Contributors

I.P. Amaranayake
Marine Environment Protection
    Authority
Colombo, Sri Lanka

M.S.O.M. Amararathna
Department of Wildlife
    Conservation (DWC)
Battaramulla, Sri Lanka

A.A.D. Amarathunga
National Aquatic Resources Research
    and Development Agency
Crow Island, Colombo, Sri Lanka

D.R. Anthony
Department of Civil Engineering
General Sir John Kotelawala Defence
    University
Colombo, Srilanka

H.M.U. Ayeshya
Marine Biological Research
    Division
National Aquatic Resources Research
    and Development Agency
Crow Island, Colombo, Sri Lanka

Shakeela S. Bandara
Central Environmental Authority
Parisara Piyasa, Denzil Kobbekaduwa
    Mawatha
Battaramulla, Sri Lanka

Deshai Botheju
Managing Director – Safety &
    Sustainability
Academy of Safety and
    Environmental Studies AS
Sandefjord, Norway

P.R.T. Cumaranatunga
Department of Fisheries and Aquaculture
Faculty of Fisheries, Marine Sciences
    and Technology
University of Ruhuna
Matara, Sri Lanka

K.R. Dalpathadu
Marine Biological Research Division
National Aquatic Resources Research &
    Development Agency
Crow Island, Sri Lanka

Ajith Priyal de Alwis
Department of Chemical and Process
    Engineering
University of Moratuwa
Moratuwa, Sri Lanka

Shantha Egodage
Department of Chemical and Process
    Engineering
University of Moratuwa
Moratuwa, Sri Lanka

E.M. Lalith Ekanayake
Bio Conservation Society (BCSL)
Kandy, Sri Lanka

G.K.A.W. Fernando
Marine Biological Research Division
National Aquatic Resources Research
    and Development Agency
Crow Island, Colombo, Sri Lanka

K.R. Gamage
Department of Fisheries and Aquaculture
Faculty of Fisheries and Marine
    Sciences & Technology
University of Ruhuna
Matara, Sri Lanka

G.A.N.D. Ganepola
Department of Agribusiness Management
Faculty of Agriculture and Plantation Management
Wayamba University of Sri Lanka
Makandura, Gonawila (NWP), Sri Lanka

L.D. Gayathry
Department of Fisheries and Aquaculture
Faculty of Fisheries, Marine Sciences and Technology
University of Ruhuna
Wellamadama, Matara, Sri Lanka

U.A.D.P. Gunawardena
Resource Economics Unit, Department of Forestry and Environmental Science
Faculty of Applied Sciences
University of Sri Jayewardenepura
Nugegoda, Sri Lanka

G.M.S.S. Gunawardhana
Department of Applied Earth Sciences
Faculty of Applied Sciences
Uva Wellassa University
Passara Road, Badulla, Sri Lanka

K.P.G.K.P Guruge
Department of Animal Science
Faculty of Animal Science and Export Agriculture
Uva Wellassa University
Sri Lanka

S.S.K. Haputhantri
Marine Biological Research Division
National Aquatic Resources Research and Development Agency
Crow Island, Colombo, Sri Lanka

Bryan D. James
Woods Hole Oceanographic Institution
Falmouth, Massachusetts, USA

Thejanie Jayasekara
Department of Chemical and Process Engineering
University of Moratuwa
Moratuwa, Sri Lanka

P.B.T.P. Kumara
Department of Oceanography and Marine Geology
Faculty of Fisheries, Marine Sciences and Technology
University of Ruhuna
Wellamadama, Matara, Sri Lanka

U.S.P.K. Liyanage
National Institute of Oceanography and Marine Sciences
National Aquatic Resources Research and Development Agency (NARA)
Crow Island, Colombo, Sri Lanka

T. Maes
GRID-Arendal
Arendal, Norway

Jude Janith Niroshan
Avian Sciences and Conservation (ASC)
Department of Zoology and Environment Sciences
Faculty of Science
University of Colombo
Colombo, Sri Lanka

Kalani Imalka Perera
Graduate School
Chulalongkorn University
Pathumwan, Bangkok, Thailand

U.L.H.P. Perera
Department of Applied Earth Sciences
Faculty of Applied Sciences
Uva Wellassa University
Badulla, Sri Lanka

Contributors

Sajani Piyatilleke
Department of Chemical and Process Engineering
University of Moratuwa
Moratuwa, Sri Lanka

G.A. Tharaka Prasad
Director Wildlife Health
Department of Wildlife Conservation
Battaramulla, Sri Lanka

W.N.C. Priyadarshani
National Institute of Oceanography and Marine Sciences
National Aquatic Resources Research and Development Agency
Crow Island, Colombo, Sri Lanka

T.S. Ranasinghe
Sri Lanka Turtle Conservation Project
Panadura, Sri Lanka

Amila Sandaruwan Ratnayake
Department of Applied Earth Sciences
Faculty of Applied Sciences
Uva Wellassa University
Passara Road, Badulla, Sri Lanka

Christopher M. Reddy
Woods Hole Oceanographic Institution
Falmouth, Massachusetts, USA

Chalani Rubesinghe
Centre for Environmental Justice
Colombo, Sri Lanka

R.S.M. Samarasekara
Department of Civil Engineering
Faculty of Engineering
University of Sri Jayewardenepura
Nugegoda, Sri Lanka

Sampath S. Seneviratne
Avian Sciences and Conservation (ASC)
Department of Zoology and Environment Sciences
Faculty of Science
University of Colombo
Colombo, Sri Lanka

Madushika Sewwandi
Ecosphere Resilience Research Centre
Faculty of Applied Sciences
University of Sri Jayewardenepura
Nugegoda, Sri Lanka

A.H.M.S. Siriwardana
Department of Civil Engineering
Faculty of Engineering
University of Sri Jayewardenepura
Nugegoda, Sri Lanka

Dilantha Thushara
Department of Chemical and Process Engineering
University of Moratuwa
Moratuwa, Sri Lanka

J.M.M. Udugama
Department of Agribusiness Management
Faculty of Agriculture and Plantation Management
Wayamba University of Sri Lanka
Makandura, Gonawila (NWP), Sri Lanka

Sujitra Vassanadumrongdee
Environmental Research Institute
Ühulalongkorn University
Wangmai Pathumwan, Bangkok, Thailand

Meththika Vithanage
Ecosphere Resilience Research Centre
Faculty of Applied Sciences
University of Sri Jayewardenepura
Nugegoda, Sri Lanka

K.A.W.S. Weerasekera
Environmental Studies Division
National Aquatic Resources Research
 and Development Agency
Crow Island, Colombo, Sri Lanka

Mona Wells
The Meadows Center for Water and the
 Environment
Texas State University
San Marcos, Texas, USA
and
Natural Sciences
Ronin Institute
Montclair, New Jersey, USA

Indika Hema Kumara Wijerathna
Marine Environment Protection
 Authority
Colombo, Sri Lanka

H.B.U.G.M. Wimalasiri
National Institute of Oceanography and
 Marine Sciences
National Aquatic Resources Research
 and Development Agency
Crow Island, Colombo, Sri Lanka

Hemantha Withanage
Centre for Environmental Justice
Colombo, Sri Lanka

Xiaokai Zhang
Institute of Environmental Processes
 and Pollution Control
School of Environmental and Civil
 Engineering
Jiangnan University
Wuxi, China

1 X-Press Pearl
The Fateful Journey

Ajith Priyal de Alwis

1.1 INTRODUCTION

X-Press Pearl ship which caught fire on 21 May 2021 made headline news across the world as it battled the fire for more than 12 days. It was carrying a significant quantity of chemicals and plastics on board. She was positioned approximately 9.5 nautical miles off the Port of Colombo, just off the capital of Sri Lanka. Finally, the ship sank just 850 m away from its original anchor position as it was being towed into the deep seas as she was found to be sinking after the fire was extinguished. The wreck removal which started in November 2021, is expected to be completed only in April 2023 as monsoon weather is hampering the process by salvors.

This maritime accident is now considered to be the worst maritime accident in the world where chemicals and plastics are concerned. With the cargo completely destroyed along with the ship, it has led to an environmental catastrophe. This chapter covers the final fateful journey of the ship which was a feeder vessel belonging to one of the largest feeder vessel companies in the world – X-Press Feeders. It was on its third voyage taking its assigned route and had previously visited Colombo twice.

The X-Press Pearl was effectively a new ship coming out of a Chinese shipyard in 2021. Details of the ship are given in Table 1.1.

1.2 X-PRESS PEARL – THE FINAL VOYAGE

X-Press Pearl as a container feeder vessel operated covering several ports facilitating the swift movement of cargo. On this particular voyage, she was carrying 1486 containers. It had all types of cargo on board from general cargo, reefer cargo to dangerous goods. As per the manifest and other information available, the ship had on board 7 out of the 9 classes of dangerous goods. The two categories that were not present were explosives and radioactive cargo. The initial departure was from the Port of Hamad in Qatar on 9th May and the ship first called over at Jabel Ali in Dubai. Then the ship returned to Hamad Port again on 10th May. From Hamad, it sailed to Port of Hazira, Gujarat in India. She left Hazira on 15th May destined for Colombo, Sri Lanka. From Colombo, the ship was to sail to Tanjung Pelepas, and Port Klang, Malaysia before heading back to the Persian Gulf, more or less on the same route

TABLE 1.1
X-Press Pearl Specifications and Details

Ship Owner – EOS RO Pvt Ltd
Class and Type – Super Eco 2900 Container ship
Vessel Operator – X-Press feeders
Port of Registration – Singapore
Ship Builder – Zhoushan Changhong International Shipyard, China
Ship Launch and Completion – 28 September 2020, February 2021
Identification – IMO No 9875343
Tonnage – 36,149 DWT, 31,629 GT
Displacement – 48,848 T
Length and Breadth – 186 m, 34.8 m
Depth – 17.9 m
Capacity – 2,756 TEUs
Crew – 26

which she had followed twice before. The ship however was never to set sail from Colombo again while ending in its grave off the coast of Colombo after causing one of the worst maritime disasters in the history. Figures 1.1a and 1.1b provide the final route of the ship and its final position.

1.3 CARGO AND THE CONSEQUENCES

On the X-Press Pearl, the containers on deck amounted to 691 and the number below deck was 796. This is important considering the way the vessel suffered from fires and explosions. There was a wide scatter of containers on top whereas what was below the deck went down with the ship completely. The overall outcome over the contents had to be ascertained alongside the fire that took place. On-deck, the fire swept across while below-deck contents were not all affected by the fire directly.

The fire on board which ultimately led to the demise of the ship has been attributed to one single container with dangerous cargo (DG) – mainly a nitric acid. Multiple evidences clearly indicate that this assertion is accurate. There is however a long history for the event as it did not happen all of a sudden. There were lots of observations and interactions that clearly demonstrate that the final conclusion could have been well avoided saving much heartache to many which included multiple eco-systems as well.

Out of 1486 containers, 512 were to be discharged at the Port of Colombo. Of them, 22 containers were DG cargo.

International Tanker Owners Pollution Federation Limited (ITOPF) issued a quick report and it carried the initial fire to have happened at Cargo Hold 2, Bay 11 at the position 110582. This corresponds to the position of nitric acid stowage. The 25 MT nitric acid container had come on board at Jabel Alli. Almost immediately (upon sailing from Jabel Ali), a leakage had been detected. The initial observation has been recorded on 10 May. As per ship records on 12 May, heavy corrosion has

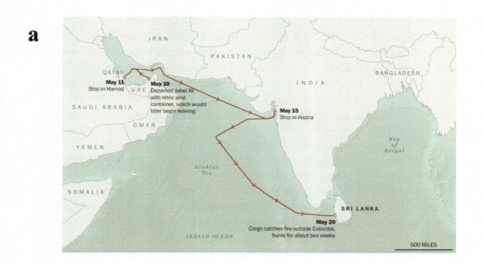

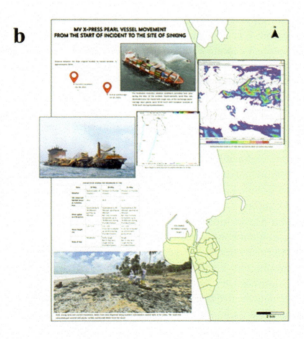

FIGURE 1.1 (a) The ship route. (b) The ship journey to the final location. Credits: MEPA.

been noted on the edge of the hatch cover which is within 12 hours of the initial loading. However, the ship had not returned to Jabel All for off-loading the leaky container. Instead, it had preceded to Hamadi, a port in Qatar. Due to the limited facilities at Hamadi, off-loading of the leaky container had not taken place and the vessel had been allowed to sail away from that port. The vessel was supposed to have then returned to Jabel Ali where the leaky container was loaded but for unknown reasons, the container had not been attended to there too and the vessel was allowed to sail to Hazaria, a port in Northwestern India which was vessels scheduled next port of call. At Hazaria too, the leaky container was not off-loaded sighting the same reasons as Hamadi. However, analyzing the port information and the claims by the respective port operators, it is generally felt that this lack of ability to off-load a container and support a ship in difficulty should never be accepted. Anyhow, this will never get highlighted during the investigation into the accident. The perceived acid leakage has been continuously observed and discussed throughout the journey of the vessel. From Hazaria too, the advice had been to move the vessel to the next port, Colombo.

Hamad refusal to discharge the container had been on the ground that they only handle direct delivery of the container. Apparently, it is due to cost considerations that X-Press Feeders has decided not to make a second call to Jabel Ali but to continue to Hazira.

The ship that sailed from Hazaria in Gujarat cannot be considered as seaworthy in the technical sense as it was not complying with Reg 6 of Chapter XII of SOLAS at that time. A comment that had been recorded at the Port of Hazaria says that the leak is 'massive' and therefore it is difficult to be handled there. On the contrary, however, the X-Press Pearl Captain Tyutkalo Vitali, has reported that he has not observed a leak from the container at the Indian Port and that is the reason why he did not report it to the Sri Lankan Authorities while approaching Colombo.

If however there was a leak at the time departing Hazaria, whether observed or not, the situation may have been aggravated by the fact that the vessel encountered turbulent weather with heavy rain on-route to Colombo which had delayed the vessel reaching Colombo by a whole day.

Different leakage rates have been reported from 0.5 L/hr to 1.2 L/hr. The journey from Hazaria to Colombo had been affected by bad weather that prevailed. As a result, the ship even had got delayed by one day to enter Sri Lankan waters. Consequently, the ship calling into Colombo did not request direct passage to berth nor communicate to the Sri Lankan Authorities that there is an issue on board to the Sri Lankan port authorities. The evidence indicates that the shipping agent had been aware, but the agent had failed to raise attention or request support from the port authorities due to his failure to communicate. The ship, on arrival late at night on 19 May (at 23:48 hrs), anchored off Colombo.

Various statements and records made by the ship captain and other officers clearly indicate that there had been some awareness of the issue since departing Jabel Alli. What appears is that they did not comprehend fully the potential danger of it. The leak had been estimated to be around 1 litre per hour and this had been communicated to their Head Office in Singapore. Singapore office in turn failed to advise what action to be taken. Effectively, the leakage continued for over 10 days. In the investigation, the crew has not been questioned on the adherence to the CTU Code (which is regarding the securing of dangerous goods in shipping containers), though this is not a standard procedure. As per Hapaz Lloyd, it is noncompliance in transporting restricted commodities and dangerous goods that has been identified to be the most relevant root cause of major shipboard fires at present.

The particular nitric acid container has a story of its own though complete verification is not possible. The particular container had been picked up in Jabel Ali and the container had arrived there from Iran on MV Ronika. In fact, MV Ronika had carried two nitric acid containers destined for Malaysia and should have been loaded on the X-Press Pearl. One of the containers however, had shown signs of a leak while at Jabel Ali and the port had arranged for that container to be reworked (reloaded) to another container. That reworked container, however, had not found its way to X-Press Pearl. The container that was loaded on the X-Press Pearl was detected to be also leaking only after the X-Press Pearl had departed Jabel Ali.

It is likely that the 25 MT of nitric acid was loaded in a 20-foot container in 18 intermediate bulk containers (IBCs). According to the evidence given by the captain of the ship at the enquiry, they have heard significant movement of cargo from within the container which signifies that there seems to have been a serious failure in securing the IBCs carrying DG cargo. IBCs carrying DGs have strict storing and securing guidelines. If there was evidence of one container showing signs of leakage, common sense suggests that the other container also should have been examined prior to loading.

Legally, the responsibility in accepting and assuring the safe stowage of cargo on board a ship lies with the Chief Officer but how practical that is in the container trade where most times over thousands of containers are handled in a matter of a few hours to verify how each of those containers have been stuffed and secured inside needs to be reexamined by the relevant Authorities of the global shipping trade.

As per the stowage plan, the particular location where the leaky nitric acid container was stowed (Figure 1.2), may have caused the now potent leaked material because of being diluted by rainwater, to have eroded the hatch cover and its packing (seal) and leaked through the hatch into the cargo hold below. No specific technical details including the container planning had been communicated to the Sri Lankan authorities nor to the environmental damage assessment team despite requesting such information for analysis.

In a piece of investigative journalism tracking open resources, the nitric acid container had been tracked from Iran to Malaysia (Figure 1.3). This was initially reported in a Sri Lankan daily, *Ceylon Today*.

The ship had observed trouble as it approached Colombo. Almost immediately after anchoring the ship, it had to deploy its fixed firefighting system (CO_2 based) against a potential fire in the Cargo Hold No. 2. Activation was due to a fire alarm emanating from the No. 2 Hold. It is important to understand the position of cargo in understanding the situation that unfolded. Twenty-five members were on board – Indians, Chinese, Filipinos and Russians – on the ship at the time of the accident. Some senior officers were actually on relief as a changeover had taken place at Hazira port.

On deck closer to the nitric acid container, a Li-ion battery container was present. This is quite likely to be the first instance where a complete battery consignment got lost through fire and sinking. Lithium has the ability to catch fire. If a battery gets in touch with nitric acid, there would be leaching and the leached lithium can easily catch fire. ITOPF's initial communication covering the incident reveals that the fire was first observed in Cargo Hold 2, Bay 11, position 110582 (bay plan), which points

6 Maritime Accidents and Environmental Pollution

FIGURE 1.2 The location of the nitric acid container as per the stowage plan.

Source: SLAir Force.

X-Press Pearl – The Fateful Journey

FIGURE 1.3 The source and the final destination of the nitric acid cargo.

to the nitric acid container. There is a question over the identification of fire and fume. Nitric acid by itself cannot catch fire as it is only a strong oxidizing chemical and for a fire to initiate, it must come into contact with an organic substance. By itself, there is no opportunity for a fire. Yet the ITOPF report is explicit over the fire in the specific position.

It is evident from the statements given at the inquiry that an issue was brewing on-board for a few days. The crew had reported orange smoke emanating from the container for almost 10 days. However, the assessment had been that it is emanating only a chemical smell and that there is no evidence of a fire. Smell of chemicals as fumes had been noted though this has not been recognized as an issue as there had not been any smoke. It appears that there have been significant lapses in understanding the consequences of chemical spills. There are reports of the crew using sawdust in an attempt to prevent the spreading of the spill. It seems definite that yellow smoke had emanated for over 2–3 days as the ship approached Colombo (Figure 1.4).

The event as it unfolded since anchoring the vessel is given in chronological order as that is the best way to present the situation that developed from then on.

1.4 CHRONOLOGY OF EVENTS

The following sequence is the chronology of events after the ship anchored off Colombo.

- **19 May 2021** – Two days after the scheduled arrival, X-Press Pearl reached Colombo at 23:48 hrs. The ship did not request for priority berthing sighting an emergency but opted to wait for normal berthing on arrival basis.
- **20 May 2021** – At 2:00 hrs, the smoke alarm was activated. Captain had alerted the Port Control, Colombo. It has been indicated that at 10.19 hrs on this day, the ship's local agent, Sea Consortium Lanka has sent SLPA a mail informing about the acid leak. Port Control had requested temperature readings of Cargo Holds 1, 2 and 3 every hour. At 12:00 hrs, fire had started inside Cargo Hold 2 which was reported to Port Control. The crew had initiated boundary cooling on deck with sea water using ship's firefighting system. At 12:23 hrs, carbon dioxide firefighting system for cargo hold had been activated. The Chief Engineer has reported initially about a burning rubber beading on Cargo Hold 2. At about 2

FIGURE 1.4 X-Press Pearl approaching Colombo (the emanating vapour/smoke is visible). Image: Author.

a.m., a strong fume has been indicated but still the evidence has not led to any concerted action. The decision has been there is nothing abnormal. Evidence recorded however indicates that the Chief Engineer had noted a heated red colour metal surface on a container on the second tier from the bottom. Another observation had been rust bubbles on the surface of a container but the location is unclear. At around 16:30 hrs, SLPA had sent an inspection team who had returned for a follow-up inspection around 18:30 hrs. At 23:05 hrs, a fire erupted on deck at Bay 10. It had been reported as a huge fire that was difficult to manage.

- **21 May 2021** – At 01:20 hrs, the SLPA tug boat arrived on the scene but they were unable to fight the fire. At 06:00 hrs, two more tug boats arrived on the scene but failed in controlling the fire. At 13:35 hrs, Sri Lanka Air Force (SLAF) helicopters used dry powder to control the fire. Despite swift action and early success, the fire was not completely extinguished – this is probably because firefighting action could not reach the root of the fire and was only extinguishing the visible flames, which was a case of getting rid of the visible fire but not addressing the cause. This is a scenario possible with fighting chemical-induced fires. The firefighting efforts were mostly using water cannons deployed from both Sri Lanka Ports Authority (SLPA) Tugboats and

X-Press Pearl – The Fateful Journey

FIGURE 1.5 MOD Social media feed.

Sri Lanka Navy Vessels (Figure 1.5). MEPA requested the ship owner, local agent and the captain to move the ship to a different location about 50 nautical miles from the anchor position. There had also been a request to take the ship to the Eastern coast of Sri Lanka. This has been refused. Then the request had been to proceed to deeper waters.

- **22 May 2021** – As a precautionary measure, fuel on board the vessel was transferred from the No. 1 tanks that were close to the seat of fire to the No. 2 tanks which were relatively further. The task was completed on 23 May around 05:30 hrs. At around 14.35, 12 firefighters from the Netherlands boarded the vessel. The team had tried to change the direction of the vessel to avoid spreading of the fire from bow to stern due to strong wind that prevailed at the time but the manoeuvre did not help and the fire had spread unabated.
- **23 May 2021** – At 16.04 and 16.45, more vessels with firefighting capabilities joined the fight including a couple of offshore vessels with greater firefighting capabilities. Unfortunately, none of those efforts were successful. By this time, the entire Bay 10 was on fire. Twelve members of the crew were removed from the vessel as a safety precaution.
- **24 May 2021** – Members of the Dutch firefighting team confirmed stability and satisfactory conditions on Cargo Holds 3–5. However, on Cargo Hold 2, they have indicated difficulty in checking due to presence of ammonia.
- **25 May 2021** – Five days after the fire commenced, a violent explosion was reported on-board the vessel. The explosion that was reported around 3:00 a.m. had come from the engine room. At that point, the crew of the vessel was evacuated with two crew members suffering injuries. One of the injured also

tested COVID-19 positive. Only eight containers were initially reported to be thrown overboard as a result of the aforementioned explosion. However, later evidence (based on debris washed ashore) pointed towards many more cargo containers being dislodged from the ship and ending up in the sea, as a result of this initial explosion or due to the ongoing fire. Within 24 hours after the first explosion, debris started to wash up along the coastline from Colombo to Negombo. The initial debris that washed ashore included a whole shipping container, deformed clearly due to the explosion, many sacks of low-density polyethylene beads (nurdles/pellets), some rolls of yarn, food items, etc. Soon the shoreline was covered with polymer beads and a gummy textured dark substance, which is suspected to be self-polymerized resins. MEPA again requested the movement of the ship to a safer location as on 21st May.

- **26 May 2021** – Despite continued firefighting efforts, the fire intensified and escalated to engulf the entire vessel from the forward to aft. Indian Coast Guard sent in additional resources to join in to support the firefighting operation. The firefighting vessels had faced intense heat of the fire and at times had to retreat as well (Figure 1.6).
- **27 May 2021** – The fire had largely consumed much of the flammable substances aboard the ship and the intensity of the fire began to subside. The firefighting efforts were resumed with water cannons, as well as with air drops of dry chemical packs. By this time, more and more plastic beads were found to be washing ashore, even South of Colombo than initially anticipated, in locations as far as Hikkaduwa and Galle. A number of carcasses of dead fish,

FIGURE 1.6 The ship subjected to boundary cooling at the initial stages of fire.
Source: SL Navy.

sea turtles and other marine organisms were also found along the coastline. The fire had largely depleted and had been reduced to a small fire deep within the interior of the Vessel.

- **28 May 2021** – The fire had been mostly extinguished and only smoke plumes were visible. On the same day, at around 23:00 hrs, a second explosion was reported on the ship, but no additional intensification of fire occurred.
- **1 June 2021** – The fire had been extinguished, and the ship was sufficiently cooled, allowing salvors to board the Vessel and commence preliminary investigations. This was salvor's first inspection on board and they reported smoke still coming out from Cargo Holds 1, 2 and 3 intermittently (Figure 1.7).
- **2 June 2021** – As the vessel was found sinking, attempts were made to tow her to deeper waters, and as the towing commenced, a large section of the hull virtually disintegrated, massive flooding occurred in the engine room. Within 850 m of towing, the ship's stern struck ground preventing any further towing. By the end of the day, the aft section of vessel had settled on the seabed at a depth of about 21 m, while the forward section was still slowly sinking. The midsection was completely submerged in water while several of the now derelict containers were still visible above the water level. The superstructure (accommodation) located at the aft of the vessel continued to be visible above the surface due to its height. The vessel was now a shipwreck located within the territorial waters of Sri Lanka that required further action. On this day, the X-Press feeders issued a statement 'regrets to report that despite salvors successfully boarding the vessel and attaching a tow wire, efforts to move the ship to deeper waters have failed'.

FIGURE 1.7 Stages of the X-Press Pearl disaster. Image: Author.

FIGURE 1.8 X-Press Pearl resting on the sea bed. Image: Author.

- **8 June 2021** – An oil slick emanating from the ship was sighted by satellite. The slick continued to be visible extending up to 4 km during the ensuing days. The ship had been using intermediate fuel oil (IFO), a blend of 95% heavy fuel oil (HFO) and 5% gasoline.
- **10 June 2021** – With the ship almost fully resting on the bottom, the smoke emanating from the wreck completely stopped.
- **June 17 2021** – The entire vessel had settled on the seabed at a depth of about 21 m (Figure 1.8).

1.5 CONCLUSIONS

X-Press Pearl was a brand-new ship with new and modern features as per the ship details offered. Unfortunately, human neglect was to be her undoing. It is quite clear that professional negligence had a direct contribution to this disaster. The saga of X-Press Pearl is definitely going to be an international maritime case study. Issues with claiming Place of Refuge, the effectiveness of the CTU Code, Conformity of Sea Worthiness, the effect of ISM Code and the absolute Authority ship's command, etc., are matters that should definitely come under scrutiny. The lack of availability of email communications between the ship and the Colombo Port also increased tension in the investigations.

A leaking nitric acid container appears to be the initiator of this whole disaster. The issue is that the leakage had been known for quite some time but the vessel continued to sail with no proper action to remedy the matter. It appears that there had been a serious underestimation of the magnitude of the potential danger. Dangers associated with leaking nitric acid should have been ascertained and identified much earlier. The continuance of the voyage with a leaky container on board appears to be the singular root cause for the incident. Did this happen because of the lack of emphasis on the risks associated with dangerous goods on container or general cargo vessels, unlike those on oil or gas tanker vessels or chemical carriers? For the latter vessels, specific training is mandatory.

XPP is leaving behind multiple lessons for the future. The accident with Stolt Rotterdam was caused by nitric acid leaking out of a stainless-steel tank into the hull and damaging the structural steel. The ship had caught fire and sank at the terminal. The investigation came out with the conclusion that the leak had started at least 8 hours prior to the accident and the mechanism for the fire had been flowing acid reacting with a wooden structure. Nitric acid is a strong oxidizer and is not flammable. Hence, this is a reaction necessary to initiate a fire. This accident is a clear lesson in the danger of leaking nitric acid in a ship. The other example is the fire observed on another X-Press feeder ship. The captain observed the smoke emanating from the lithium battery container and diverted the ship to the harbour and had the container duly removed and action taken. The ship safely proceeded after the intervention and the whole operation took only a couple of hours. It was fortunate that the smoke was quickly noticed on deck as the incident happened during the daytime. However, both these cases indicate what did not happen with X-Press Pearl leading the journey of XPP to be its final journey.

X-Press Pearl's final casualty site is in close proximity to Colombo at North 7.081564 and East 79.778778. The depth at the site is 21 m. The location is completely exposed to elements from North, South and West with the land providing shelter from the East. A simple cargo loading neglect or a violation has led to a significant maritime disaster with ecological consequences for many years if not decades to come.

ACKNOWLEDGEMENT

Special acknowledgements to Master Mariner Capt Lasitha Cumaranathunge for the exchange of insights and continuous discussions on the X-Press Pearl.

BIBLIOGRAPHY

https://news.mongabay.com/2022/06/a-year-since-x-press-pearl-sinking-sri-lanka-is-still-waiting-for-compensation/
https://en.wikipedia.org/wiki/X-Press_Pearl
https://openfacto.fr/2021/10/31/pourquoi-la-proliferation-est-mauvaise-pour-les-tortues/
https://twitter.com/slpauthority/status/1395594354069151744
https://twitter.com/IndiaCoastGuard/status/1397518948384395265
www.adaniports.com/Ports-and-Terminals/Hazira-Port
www.dpworld.com/en/uae/ports-and-terminals/jebel-ali-port
www.mwani.com.qa/English/Ports/HamadPort/Pages/default.aspx
www.x-presspearl-informationcenter.com

The Island (2021). https://island.lk/place-of-refuge-the-international-debate/
www.washingtonpost.com/world/interactive/2021/sri-lanka-cargo-ship-fire-pollution/
https://dailyexpress.lk/glocal/8770/
Zunkel A, Tiebe C and Schlischka J (2014), 'Stolt Rotterdam' the sinking of an acid freighter, Engineering Failure Analysis http://dx.doi.org/10.1016/j.engfailanal.2014.03.002

2 Recent Maritime Accidents
Chemicals and Plastic Spills

R.S.M. Samarasekara, A.H.M.S. Siriwardana, and D.R. Anthony

2.1 INTRODUCTION

2.1.1 MARITIME ACCIDENTS

Over 70% of global trade by economic value is carried through maritime transportation and thus it has a high economic significance (UNCTAD, 2018). However, maritime transportation has long been considered a complex activity, and the associated risk has impacts on human lives, the economy, the marine environment, and marine life. Transport and handling of hazardous chemicals, crude oil, and other types of cargo through shipping have considerably increased over the recent decades owing to the rapid industrial development and thus increasing the risk of heavy marine environmental pollution during marine accidents (Häkkinen and Posti, 2013). The term 'Marine Accident' has been defined by the legislation of different countries in slightly different ways [Sri Lanka Marine Pollution Prevention Act No. 35 of 2008, 2008; The Merchant Shipping (Accident Reporting and Investigation) Regulations No. 1743 of the United Kingdom, 2012]. However, in general, a 'Marine Accident' can be defined as a marine casualty or a marine incident, which is an event or a sequence of events that has resulted in damage to people (injury or loss of life), damage, loss or presumed loss, or abandonment of a ship or a fleet or a marine facility or damage or pollution of the marine environment. The fire on the MV X-Press Pearl ship and the fire on MT New Diamond ship in Sri Lankan territorial sea and MV Wakashio ship oil leakage on the Mauritius coast are a few examples of such maritime accidents that caused devastating impacts on the marine environment (Jeong, Capelouto, and Chavez, 2020; Ellis-Petersen, 2021; Jayasinghe, 2021). General Cargo ships are the most accident-prone ship type followed by bulk carriers, tankers, and container carriers respectively. Furthermore, the sea areas around the United Kingdom, Denmark, Southeast Asia, East Asia, and Singapore are considered as regions with the highest frequencies of maritime accidents (Y. Zhang *et al.*, 2021). However, in the present study, the maritime accidents associated with environmental impacts caused by chemical and plastic spills are considered. Oil spills, hazardous and noxious substances (HNS)

DOI: 10.1201/9781003314301-2

spills, and microplastic spills resulting from such marine accidents have devastating effects with long recovery periods on the marine environment and marine flora and fauna (Guzzetti *et al.*, 2018; Walker *et al.*, 2018; Sinanaj, 2020). Once discharged into the ocean caused by a maritime accident, these oil patches and microplastics undergo various physical and chemical processes under environmental conditions and wave action causing them to adapt complex transport mechanisms. Oil patches can be broken into smaller droplets and are incorporated into the water column under the action of waves, subsequently, they can cluster with other mineral particles and sink, being able to interact with various forms of marine life (Walker *et al.*, 2018; Khatmullina and Chubarenko, 2019). Furthermore, these harmful pollutants can be transported to other regions from the point of accident occurrence through motions induced by oceanic currents and have adverse effects over wide areas of the ocean. These microplastics and oils cause several lethal and sublethal effects threatening the survival of planktonic, nektonic, and benthic marine organisms (Buskey, White, and Esbaugh, 2016; Li *et al.*, 2021). Hazardous and Noxious Substances (HNS) spills exhibit a considerably wider range of behaviours when compared to oil and microplastic spills and different levels of toxicities to marine organisms depending on the nature of the given HNS (Cunha, Moreira, and Santos, 2015). Therefore, the behaviour of those substances released into the ocean caused by maritime accidents needs to be comprehensively studied together with their impacts to provide effective prevention and mitigation mechanisms.

2.1.2 COMMON CAUSES OF MARITIME ACCIDENTS

Figure 2.1 shows maritime accidents by type of accidents based on the nearly 1400 accidents (Häkkinen and Posti, 2013). The most common accident type is grounding

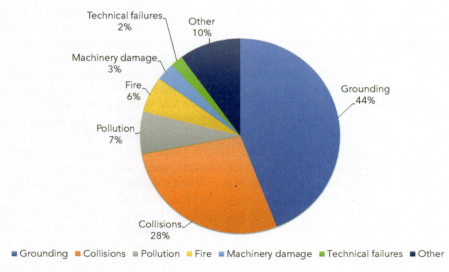

FIGURE 2.1 Maritime accidents by type of accidents based on the nearly 1,400 accidents.

and the next type is collisions. Over the years, a huge effort has been made to investigate the root causes of maritime accidents and eliminate them to improve maritime safety. Typically, maritime accidents are caused by the failure of one or more components which are required to function accurately for the successful operation of a ship and when a ship is taken as a system, these individual components may be mechanical or human (Soares and Teixeira, 2001). However, most studies to date have shown that 75% to 96% of maritime calamities are associated with human errors or failures (Lützhöft, Grech, and Porathe, 2011; Faturachman and Mustafa, 2012; Özdemir and Güneroğlu, 2015; Uğurlu, Yildirim, and Başar, 2015). Maritime accidents can be classified into different categories, such as fires or explosions, collisions, grounding, foundering, mechanical problems, heavy weather, and hull problems. At the same time, collisions and fires are at the forefront, respectively (IAEA, 2001; Soares and Teixeira, 2001; Panagiotidis *et al.*, 2021). Hanzu-Pazara *et al.* (2008) showed that 89% to 96% of collisions and 75% of fires and explosions occur due to human errors. It should be noted that the majority of those human errors are linked to inadequate procedures (e.g. having obsolete contingency plans, inadequate training of the workforce, inadequate work preparation) or deviation from the standard operating procedure (SOP) (Coraddu *et al.*, 2020). However, SOPs developed by regulatory frameworks of different countries for the maritime sector are not followed on certain occasions by crew members of ships due to various reasons, such as impracticality of some SOPs and lack of clarity in SOP guidelines (Kurt *et al.*, 2016). The fire on the MV X-Press Pearl ship is an example of such maritime accidents influenced by human errors and it's believed that the fire was caused by improperly packed nitric acid stocks and the crew and the shipowners are heavily blamed for their negligence (Claudio Bozzi, 2021; Farzan, 2021). However, poor design of technological equipment to match the human performance and control console ergonomics have also caused maritime accidents. Technological developments have influenced maritime transport to adopt automated systems such as Integrated Bridge Systems (IBS), Integrated Navigation Systems (INS), Central Alert Management Human Machine Interface (CAM-HMI), Electronic Chart Display Integrated Systems (ECDIS), etc. Though these complex automated systems minimize human involvement, thus minimizing the possibility of human errors, they also possess a high level of autonomy and make it a challenge for the crew to monitor, integrate, and interpret the information provided by these systems (Znanstveni *et al.*, 2017). The maritime accident that happened due to heavy contact of Sirena Seaways ferry with berth 3 of Harwich International Port in 2013 is a good example of such maritime accidents caused by Control console ergonomics (MAIB, 2014). Climatic or environmental factors may also cause shipping accidents (Sampson *et al.*, 2018). Significant wave height (H_s), mean wave period (T), and mean wave direction are three distinctive parameters that are used to describe the behaviour of oceanic waves. Studies have shown that most maritime accidents have happened when H_s is in between 0 and 3 m, and this range is not generally recognized as a rough sea state. However, the swell wave energy within this range exceeds 50%, which indicates swells may threaten ships in relatively low oceanic states. More than two-thirds of the maritime accidents occurred when the difference of wave periods of swell waves and wind waves is very small (<3 seconds) and T is less than 7 seconds. Furthermore, a majority of ship accidents have been taken place when the difference between mean wave directions of swell waves and wind waves is less than 30° even though optimum coupling of swell and wind waves occurs when this difference remains between 30° and 40° and the difference of wave periods is much smaller (~1.8 seconds) (Tamura, Waseda, and Miyazawa, 2009; Zhang and Li, 2017). The behaviour of oceanic waves is complex and cannot be directly attributed to maritime disasters, however, they exhibit certain characteristic behavioural patterns in relation to maritime accidents.

2.1.3 Recent Maritime Accidents

In terms of recent maritime accidents, the X-Press Pearl accident is a notorious and serious environmental disaster caused in Sri Lankan coastal waters in 2021. At the time of the catastrophe, the ship was carrying 1,486 containers with a cargo of 25 tonnes of nitric acid, various chemicals, and low-density polyethylene pellets (LDPE) (MTI Network, 2022). According to the reports, the ship caught fire on 20 May, 9.5 nautical miles off the Colombo Port. The Sri Lankan Navy, Airforce, Coast Guard, and the Indian Coast Guard were able to successfully control the fire and allow salvors to board the ship to attach a tow wire. On 2 June, the stern of the X-Press pearl hit the ocean bottom amidst the efforts of towing to the deeper waters. Aside from the oil spill from the estimated 378 tonnes of oil on board at the time, The containers that were on board the X-Press Pearl fell overboard and the cargo was released into the ocean. The marine environment was contaminated due to the addition of nitric acid and other chemicals as well as the LDPE pellets which are known as nurdles. Multiple sightings of dead turtles, whales, and dolphins washed up along the coastline of North-Western, Western, and Southern Sri Lanka which is suspected as a result of the environmental damage caused by the X-Press Pearl accident.

In October 2020, nurdles were washing up ashore at Cape Beach, South Africa. The liability for the incident is still under investigation as the nurdles are suspected to have come from the container ship CSAV Trancura. The ship was reported to be berthed with a collapsed container stow at the Ngqura container terminal, Eastern Cape in August 2020. Around 30.18 tonnes of nurdles have been collected from the beach after the incident by June 2021 (Spear, 2021). Plastic pellets accounting for over 10 tonnes of weight, were lost into the Sea from the container ship MV Trans Carrier when a container broke open due to being moved as it was sailing through rough seas at the German Bight on 23 February 2020 (KIMO, 2020). The vessel was operated under Seatrans Ship Management. Twenty-six tonnes of polypropylene pellets were included in the cargo in its voyage from Rotterdam, Netherlands to Tananger, Norway. Even though the incident was reported, no action was taken by the Norwegian Coastal Administration stating that the spill was too small to be concerned. Large quantities of plastic pellets have been reported to be washed up on the shores of Sweden, Denmark, and Norway since mid-March. A few other incidents to be mentioned are, the Hong Kong plastic Nurdle Spill in 2012, the Cosco Yokohama fly swatter spill in 2012, where 29 containers fell overboard due to the dangerous weather in the Gulf of Alaska. The Atlantic Lego Spill-Tokyo Express in 1997, the mid-pacific rubber duck spill in 1992, and the Nike sneaker spill in 1990 were all examples of plastic spills resulting from maritime accidents (Ebbesmeyer and Ingraham, 1992; Abeynayaka, 2021). Figure 2.2

Recent Maritime Accidents: Chemicals and Plastic Spills

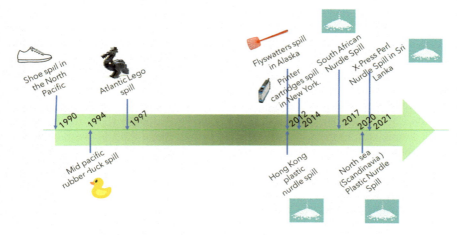

FIGURE 2.2 Chronology of recent maritime plastic spills.

shows the chronology of recent maritime plastic spills including the above-mentioned incidents. It starts from the shoe spill in the North Pacific in 1990 and ends with the X-Press Perl incident in Sri Lanka.

There has been a considerable increase in the transportation and handling of chemicals and chemical products during the past two decades which increased the potential of pollution accidents (Häkkinen and Posti, 2013). According to the HELCOM (Helsinki Commission), approximately 1,400 ship accidents have taken place in the Baltic Sea alone during 1989–2010. Out of the above-mentioned accidents, 210 accidents involving tankers (including crude oil tankers, chemical tankers, and other liquid bulk cargo carriers) have taken place. But only 28 tanker accidents have led to pollution and out of the said accidents, almost all were a form of the oil spill, and only one chemical spill (orthoxylene) was reported. But it has been found that the possibility of a chemical tanker accident is very rare (Häkkinen and Posti, 2013). However, the above statistics might not comply with the worldwide statistics due to marine accidental risks being different and varying in different oceanic regions. Considering the world data, 13 accidents have occurred between the years 1991 and 2000 involving Hazardous and Noxious Substances (HNS). After the year 2000, only five accidents have occurred considering the bulk transported HNS (Galieriková et al., 2021). However, since 2013, the number of global maritime accidents has been on a decline. General cargo ships, bulk ships, oil tankers, fishing boats, and container ships are the most accident-prone types of marine vessels (Y. Zhang et al., 2021). This book chapter aimed to give an overview of the recent global maritime accidents concerning chemicals and plastics.

2.2 METHODS AND DATA OF THE STUDY

A Socio-Economic System (SES) is formed (e.g. marine ecosystem) when an ecological system is intricately linked with and affected by one or more social systems

(Anderies, Janssen, and Ostrom, 2004). Due to the high environmental and socio-economic sensitivity of such SESs, comprehensive environmental assessment and adaptive management frameworks are required to overcome the existing and emerging problems (Gari, Newton, and Icely, 2015). For this purpose, various frameworks have been introduced such as the outcome approach in Integrated Coastal Management (ICM) (Olsen, 2003), Millennium Ecosystem Assessment (World Resources Institute, 2005), Systems Approach for Sustainable Development (Newton, 2012), and Driver-Pressure-State-Impact-Response (DPSIR). However, DPSIR Framework has been widely adopted to understand links between anthropogenic pressures and relevant state changes in marine and coastal ecosystems. The DPSIR Framework was first developed by the Organization of Economic Corporation and Development (OECD) (Gari, Newton, and Icely, 2015). In the DPSIR Framework, Drivers, Pressures, States, and Impacts can be defined as follows with respect to the marine and coastal SES; Drivers refers to the socio-economic and socio-cultural forces (e.g. need for food, need for transportation, need for recreation) driving human activities; Pressures refers to the human activities that directly affect the marine and coastal environment resulting from drivers (e.g. abrasion from trawling, disposal of inorganic and organic waste, release of harmful materials to the ocean through marine accidents); States refer to the observable changes of the marine and coastal environment (e.g. changes in water quality); Impacts refer to the effects of a changed marine and coastal environment (e.g. reduction of fish harvest, reduction of marine biodiversity); and Responses refer to the efforts of the society to solve the problem or minimize the adverse effects (e.g. increasing maritime safety, introduction of policies to minimize marine and coastal environmental pollution, containing and cleaning of the harmful chemicals during maritime accidents) (Patrício et al., 2016). Up to the present, several studies and projects in the marine and coastal environment have used the DPSIR Framework for their respective assessments (Atkins et al., 2011; Liu et al., 2018; Dzoga et al., 2020). The DPSIR framework can be applied to a wide range of environmental problems as a tool for risk assessment and risk management and thus it can be considered as a wide-ranging tool (Fock, Kloppmann and Stelzenmüller, 2011; Elliott et al., 2016). The usage of the DPSIR Framework is effective in identifying the lacking areas of understanding and prioritizing monitoring, research, and investments in the identified areas. Furthermore, this framework has been used by various stakeholders such as academics and researchers, environmental organizations, policy, and decision-makers. It allows easy integration and communication among various stakeholders. However, there are certain disadvantages and anomalies in this framework too. The framework is used in both social sciences and natural sciences, while social science gives more emphasis to the Drivers and Pressures and natural science gives more emphasis to States, Impacts, and Responses. However, it's important to cover both aspects of science for a successful and comprehensive analysis (Patrício et al., 2016). Furthermore, the first three elements of this framework that are Drivers, Pressures, and States are strongly interrelated, interlinked and not mutually exclusive even though the framework considers them as separate elements. This can be considered as an anomaly of the selected framework. Therefore, the application of

Recent Maritime Accidents: Chemicals and Plastic Spills

this framework requires a deeper understanding of the inter-relation of the elements. However, the authors of the present study put adequate attention to all the elements of the framework and comprehensively address the inter-relatability of the elements when defining them to obtain a precise analysis of chemical and plastic spills.

A comprehensive literature review was conducted, with over 300 journal articles, government reports, newspaper articles, and non-government organizations' (NGOs) reports related to oil spills and their environmental consequences, with emphasis on socio-economic impacts. The literature consists largely of case studies across different geographic areas and practical domains. Several previous oil and chemical spills which are prominent in the literature are focused on in this study. They are the 1989 Exxon Valdez oil spill in Prince William Sound in Alaska, 2010 Deepwater Horizon oil spill, 2002 Prestige oil spill that occurred off the coast of Galicia in Spain, 2020 MV Wakashio oil spill in Mauritius, 2011 Bohai Bay oil spill in China, MT Hebei Spirit oil spill in South Korea in 2007, X-Press Pearl in 2021 in Sri Lanka, rubber ducks (Moby-Duck) in North Pacific in 1992, MV Trans plastic pellets spill in the German Bight in 2020, Hansa Carrier (loss of Nike shoes) South of the Alaska Peninsula in 1990, 2012 Gulf of Alaska COSCO Yokohama flyswatter spill, and Legos spill by Tokio Express at United Kingdom's Southwestern coast in 1997. Figure 2.3 shows the spatial extent of reported maritime accidents caused by chemical and plastic spills that are discussed in this book chapter. Each literature was sorted into five categories based on their research focus. These five categories are (1) drivers of chemical and plastic spills; (2) pressure on chemical and plastic spills; (3) status of chemical and plastic spills; (4) impacts of chemical and plastic spills; and (6) responses to chemical and plastic spills.

FIGURE 2.3 Spatial extent of reported maritime accidents caused by chemical and plastic spills.

2.3 CHEMICALS AND PLASTIC SPILLS CAUSED BY MARITIME ACCIDENTS

2.3.1 Drivers of Chemical and Plastic Spills

Maritime transport is one of the main activities of the blue economy and it accounts for around 80% of worldwide trade (Fratila *et al.*, 2021). Global seaborne trade surpassed 2019 levels by 0.5% to reach 12 billion tonnes in 2021, representing an annual growth rate of 4.2% (2021) even with COVID-19 (Cullinane and Haralambides, 2021). Ninety percent of the world's trade is at the present day carried by sea (Harlaftis and Theotokas, 2020). Until the last third of the 20th century, Europe dominated the world fleet to be gradually replaced by the Asian fleets in the 21st century. Maritime accidents are more likely to occur around the United Kingdom, Denmark, Singapore, and Shanghai of China. This may be due to the large cargo volume and the high density of routes (Y. Zhang *et al.*, 2021). Therefore, future maritime chemical and plastic spills could more often occur with the increased number of fleets. Marine transportation still generates negative impacts on the marine environment, including air pollution, greenhouse gas emissions, releases of ballast water containing aquatic invasive species, historical use of antifoulants, oil and chemical spills, dry bulk cargo releases, garbage, underwater noise pollution; ship strikes on marine megafauna, risk of ship grounding or sinkings, and widespread sediment contamination of ports during transhipment or shipbreaking activities (Walker *et al.*, 2019). The risk of future chemical and oil spills should be mitigated using adaptive management as the source (i.e. increased number of fleets) is also beneficial to mankind. Adaptive management of oil spills has become common in some places (Webler and Lord, 2010; Navrud, Lindhjem, and Magnussen, 2017) and more models are required to get more accurate decisions.

There are lots of causes for maritime accidents. Many of them are unavoidable like severe or extreme weather conditions. Long hours, lack of sleep leading to fatigue, inexperience, and lack of training, long voyages, extended time at sea, personal relationships aboard the vessel, reckless behaviour, including abuse of drugs and alcohol, poor decision making and/or negligence, pressures and stress of job duties are some causes of human errors that have led to maritime accidents (Chen *et al.*, 2013; Puisa *et al.*, 2018; Coraddu *et al.*, 2020). The weak links in interactions between ship operators and equipment manufacturers are considered as the main issue of maritime safety control as accidents take place in a complex socio-technical context (Awal and Hasegawa, 2017). Among all the above factors, it was identified that working within a multidisciplinary team is very fruitful to enhance maritime safety, though it is challenging (Nuutinen and Norros, 2009). The main focus of research in maritime accidents has shifted over the past 50 years from naval architecture to human error and may continue to expand into socio-economic factors (LUO and SHIN, 2019). The so-called 'Fourth Industrial Revolution', also termed 'Industry 4.0' minimize issues due to human errors with autonomous ships such as Yara Birkeland and the Autonomous Spaceport Drone Ship (ASDS) (Ichimura *et al.*, 2022).

Expected demand for offshore oil and gas drilling gradually increases and transforms into a more attractive business. This demand is a driver for future chemical and oil spills in coastal waters. Deepwater Horizon is the largest marine oil spill in history which took place on 20 April 2010. Its hydrocarbon distribution was widely spread than previously predicted or reported (Spier *et al.*, 2013). This incident highlights that the existing knowledge is not sufficient in the chemical spills. However, the oil spillage is reduced by 1,003 to 24,000 tons from 1970' to 2000' (Chen *et al.*, 2019) by major oil spill accidents in different Periods with the support of research.

In the last 10 years, the use of ocean renewable energy sources has experienced significant growth globally (Shadman *et al.*, 2019). The main sources of ocean energy are tidal streams, ocean currents, tidal range (rise and fall), waves, ocean thermal energy, and salinity gradients. Chemical impacts (or accidents) in the construction, installation, maintenance, and operational phases of power plants can also cause spills. The development of large-scale power plants in coastal waters could impose anomalies in nutrients and microconstituents (e.g. copper, manganese, iron), carbon dioxide, total organic carbon, pH, redox potential, alkalinity and dissolved oxygen (Cunningham, Magdol, and Kinner, 2010; Rivera, Felix, and Mendoza, 2020). Therefore, the growth of ocean renewable energy is one of the drivers of chemical spills in coastal waters.

2.3.2 THE PRESSURE ON CHEMICAL AND PLASTIC SPILLS

Considerable uncertainty regarding the nature of chemical and plastic spills is a pressure on the mitigation of chemical and plastic spills. Accurate prediction of chemical dispersion has been challenging due to its highly deformed/dispersed interface as well as small-scale oil-water mixing. Some researchers successfully reduced the uncertainty and errors associated with numerical simulations. Small-scale oil-water mixing is modelled with the incorporation of sub-particle-scale turbulence models and diffusion equations (Shimizu *et al.*, 2020). Numerical models can predict the evolution and the behaviour of oil spilt at sea, independently of the atmospheric conditions (Carmo, Pinho, and Vieira, 2010). Some studies are carried out to compare the results of different models (Elizaryev *et al.*, 2018). Generally, the changes are influenced by the wind, tide, and weather in the context of oil and chemical spills (Zhilnikova *et al.*, 2019). Microplastic distribution and pathways were also modelled using numerical models (Ebbesmeyer and Ingraham Jr., 1994; Hardesty *et al.*, 2017; Onink *et al.*, 2021). Some of these models include the effects of currents, waves, and wind along with a series of processes that influence how particles interact with ocean currents, with fragmentation and degradation. Availability of archived wave, current, bathymetry, sediment, and biota data are key factors to successful mathematical/numerical modelling. Even though some sophisticated models are available, a lack of essential baseline data cases and considerable uncertainty affects accurately forecasting the dispersion (and spread) of chemical spills.

Underreporting of maritime accidents is a problem for authorities trying to improve maritime safety through regulation and risk management companies and other entities who use maritime casualty statistics in risk and accident analysis. Oil spills have become more reported in the last decades. Usually, spillages of millions of tons are reported as spills (Oliveira *et al.*, 2021). Most container losses are unreported and undocumented because there is no obligation for lost cargo to be declared unless it is dangerous and possible to pose an immediate hazard to the environment. Therefore, the evidence of cargo of distinctive plastic items, from lost containers is usually restricted to the public and noticed primarily by regular beachcombers (Peters and Steinberg, 2019; Turner, Williams, and Pitchford, 2021). A mysterious oil spill along Brazil's Northeast and Southeast seaboard (2019–2020) is another case that the causes are unclear (or unknown)(Lourenço et al., 2020). The spilt oil underwent minimal weathering and light hydrocarbons were present in the stranded oil. Therefore, investigators had to pay extra effort to identify the chemical and physical properties of spilt oil. The reporting to the International Maritime Organization (IMO) of marine casualties and incidents, together with marine safety investigations, is founded on the provisions of several regulations. Adhering to the guidelines of IMO is required to mitigate hazards. One good example for adhering to the guideline is the MV Wakashio grounding incident in Mauritius in 2020. IMO has adopted the 2020 Guidelines to cutting sulphur oxide emissions to improve air quality, preserve the environment and protect human health. MV Wakashio is the world's first major spillage of Very Low Sulphur Fuel Oil (VLSFO). The aromatic content in the VLSFO was relatively low, signifying that the potential ecosystem harm arising from exposure to toxic components was less than with traditional fuel oil spills (Scarlett et al., 2021).

2.3.3 The Status of Chemical and Plastic Spills

Both HNS and oil are considered to be dangerous. Weathering processes affect the behaviour of oil spilt at sea. Most studies are focused primarily on liquid substances, but also data on packaged goods with dangerous properties are presented by different researchers (Galieriková et al., 2021). Generally, sorbents are most often used to remove final traces of oil, or in areas that cannot be reached by skimmers in small spills. Natural, graphemic, nano, polymeric, and waste materials are the main used sorbents worldwide (Oliveira et al., 2021). Natural-based modified materials such as rice husk, coir fibre, and jute can also use lignocellulosic sorbents for oil spill clean-up is driven by their abundance, inexpensiveness, non-toxicity, reusability, and biodegradability. Drawbacks of these materials are low hydrophobicity, compromised oil sorption performance, and buoyancy properties (Zamparas et al., 2020). Oil spill clean-up using combined sorbents (i.e. add rice straw to synthetic hydrophobic sorbent) can also be successfully used as oil sorbents (Tayeb et al., 2020). The other common clean-up method is the use of the dispersant, in situ burning, and use of skimmers (mechanical recovery) (Karlsson et al., 2018).

Plastic spills cannot be ignored because of the hazardous nature of pellets. The main pollution is local but long-range transport may also be possible (Karlsson et al., 2018). Once a plastic spill has occurred, unless removed with the longevity of plastic, it may remain in the environment for hundreds of years. Specially designed booms and capturing devices are helpful to collect plastics (Abeynayaka, 2021). The three-tiered approach (snapshot assessment, pollution hotspot assessment, long-term environmental monitoring) was considered more practical and useful against the mitigation of main pollutants (Hassan Partow et al., 2021) in the case of the X-Press Perl incident. Those main pollutants were (i) various types of plastic pellets (linear low-density polyethylene (LLDPE), LDPE, and high-density polyethylene (HDPE); (ii) burnt plastic fragments of various sizes (micro < 5 mm to macro > 5 mm); (iii) other debris and cargo; (iv) foam and sludge of unknown composition. The identified debris was mostly confirmed as low-density polyethylene, epoxy resins, olefin copolymers, aromatic polyamides, natural rubber, and polyethylene terephthalate. A blue treatment facility is recommended as effective in the physical separation of microplastics from the sand (Sewwandi et al., 2022). Nearly, 53,000 bags of unburnt nurdles, burnt plastic, and other debris were cleaned from Sri Lankan beaches. The added complexity was caused by the fire and subsequent burning of plastic nurdles (de Vos et al., 2021). Still, plastic pellets are not classified as hazardous material in maritime transportation regulations. Environmental Investigation Agency prepared a petition to IMO to Stop Plastic Pellet Pollution at Sea (EIA, 2021).

Under international law, flag states, port states, and coastal states have jurisdiction over the prevention of pollution caused by maritime accidents. The IMO is the Institution of the United Nations, which is responsible for the safety and security of maritime affairs and the prevention of maritime and air pollution by ships. The International Convention for the Prevention of Pollution from Ships (MARPOL) is the key international convention covering the prevention of pollution of the marine environment by ships from operational or accidental causes by IMO since 1973. The flag of the merchant ship represents information such as whether it has been registered or licensed and also it is considered as the nationality of the ship. The flag states to certify the following of the international environmental law such as MARPOL. The Merchant Shipping Secretariat is the flagship state of Sri Lanka under the merchant shipping Act No 52 of 1971. Port state control is the inspection of foreign ships in national ports to ensure that the condition of the ship and its equipment conform with the requirements of international regulations and that the ship is operated following payment laws. The Merchant Shipping Secretariat, the Sri Lanka Ports Authority, and the Sri Lanka Customs are primarily engaged in port state control activities in Sri Lanka. The coastal state is the state containing a coastal line that has jurisdiction over the activities of its territorial waters along with its implications. The coastal and maritime boundaries of a country are governed by the legal system of that country. Marine Environment Protection Authority (Marine Pollution Prevention Act No. 35 of 2008), Department of Fisheries and Aquatic Resources (Act No. 02 of 1996 Fisheries and Aquatic Resources), Department of Coast Conservation and Coastal Resources Management (the Coastal Conservation Act Number 57 of 1981 [Amendment No. 49 of 2011)], Sri Lanka Coast Guard (Coast Guard Act No. 41 2009), and Department of Wildlife Conservation [Fauna and Flora Perfection Ordinance No. 22 of 2009 (Amended)] play main roles with the coastal state in Sri Lanka (NAOSL, 2020).

2.3.4 THE IMPACTS OF CHEMICAL AND PLASTIC SPILLS

Table 2.1 shows the human health effects, physical damages to properties, environmental threats, social impacts, and economic impacts of chemical and plastic spills presented by different researchers. The impacts of chemical and plastic spills are difficult to evaluate with precision, given the limitations in available data. A wide range

TABLE 2.1
The Human Health Effects, Physical Damages to Properties, Environmental Threats, Livelihood Impacts, and Economic Impacts of Chemical and Plastic Spills

Human Health Effects	Physical Damages to Properties	Environmental Threats	Social Impacts	Economic Impacts
Allergic symptoms (Rubesinghe et al., 2022) **[X-Press Pearl]**	Destroyed nets, decreased fish catch (Rubesinghe et al., 2022) **[X-Press Pearl]**	Spilt nearly 1680 tons of spherical pieces of plastic or 'nurdles' (de Vos et al., 2021); Toxins claimed the lives of 176 turtles, 20 dolphins, and four whales (Ovcina, 2021);	Affected around 7,000 families engaged in fishing. They lost their livelihoods (Rubesinghe et al., 2022) **[X-Press Pearl]**	Of the LKR 720 million compensation received by the government (Rabel, 2021) **[X-Press Pearl]**
There are (i) effects on mental health; (ii) physical/physiological effects; and (iii) genotoxic, immunotoxic, and endocrine toxicity from oil spills (Major and Wang, 2012; Laffon, Pásaro, and Valdiglesias, 2016) **[Exxon Valdez, Prestige, and Gulf oil spills]**	Clean-up operations (e.g. use of hot water and chemicals) may cause additional damage beyond that caused by the oil spill (Seveso et al., 2021) **[Mauritius Oil Spill]**	Fish found (*Acanthopagrus berda*) with plastic pellets in opercle (Hassan Partow et al., 2021) **[X-Press Pearl]**	Captain and first officer get 20-month jail terms (Dahir, 2020);	Compensation regimes play a role in ameliorating the economic impacts of oil spills (Chang et al., 2014) **[Vancouver oil spill]**
		Affect a young mangrove propagule as well as the mangrove roots (Seveso et al., 2021) **[Mauritius Oil Spill]**	A march (nearly 100,000 protesters march) was organized to denounce government handling of the oil spill disaster (Raghoo, 2021) **[Mauritius Oil Spill]**	Estimated damage over 10 billion USD from the oil spill of Mauritius (Degnarain, 2020) **[Mauritius Oil Spill]**
Explosions, during the tank cleanings and sank 43 died (Kozanhan, 2019) **[Golar Patricia]**	The frequencies of ship damage and cargo damage are 430 and 133, respectively (Zhang et al., 2019)	Indicative recovery periods after oiling, for various habitats. Mangrove 10 years. Salt marshes 5 years. Sand beaches 2 years (Akten, 2006) **[Turkish Seas]**	The economic loss (fisheries and tourism industries) of marine plastic pollution is between 6 and 19 billion USD (data from 86 countries) (Diggle and Walker, 2022)	

of human health impacts was observed in many incidents. The remains from X-Press Pearl posed health risks, with allergic reactions including skin conditions, breathing difficulties, and other long-term health concerns as well because of the content of the chemicals aboard the vessel (Hassan Partow et al., 2021; MTI Network, 2022; Rubesinghe et al., 2022). Not only physical health but also mental health problems were posed due to oil spills such as Exxon Valdez, Prestige, and Gulf oil spills (Major and Wang, 2012; Laffon, Pásaro, and Valdiglesias, 2016). Golar Patricia marine accident caused 43 fatalities. Surface fishing nets were damaged by debris or dissolved by the chemicals in X-Press Pearl (Hassan Partow et al., 2021). Consumption of seafood with harmful heavy metals and other chemicals can lead to various ailments such as cancer and nervous system-related diseases in humans. Some toxic chemicals released into the ocean as chemical waste are insoluble in water and sink to the seabed. Very small aquatic organisms feed on these and then travel to the bodies of other animals over the food chain. Mercury pollution in Minamata Bay in Japan and lead pollution in India are examples of such cases (Harada, 1995; Kumar et al., 2020).

Environmental impacts of chemical and plastic spills are widely discussed by many researchers. The biological impact of oil and chemical spills is critical to sensitive ecosystems such as coral reefs and mangrove forests. Sometimes restoration or clean-up operations may cause additional damage beyond that caused by the oil spill such as the use of hot water and chemicals to reach a fast oil removal (Seveso et al., 2021). Indicative recovery periods for Mangroves, Salt marshes, and Sand beaches are, respectively, 10 years, 5 years, and 2 years (Akten, 2006). Humpback dolphins could serve as a biomonitoring species for microplastic pollution based on stomach microplastic pollution (Zhang et al., 2021). Therefore, plastic spills could break the food chains in some environments and could destroy an ecosystem. The plastic can go between 500 and 1000 years to decompose and is possible to be carried by ocean currents to coasts hundreds of kilometres away from the shipwreck (Pattiaratchi and Wijeratne, 2021). Some chemicals that are washed into the ocean don't decay and stay there for years. They use oxygen to decompose. As a result of this, the oxygen content of the water decreases and marine life such as whales, turtles, and sharks do not get the necessary oxygen to breathe. Improper exposure to sunlight due to substances such as oil floating on the ocean can distract photosynthesis and thus destroy corals and associated organisms; increasing the sea acidity also causes coral reef destruction. It can also make the regulation of temperature hard, leading to the death of aquatic life. The oil spills on the gills of fish and the feathers of marine birds, making it difficult for them to navigate and disrupting the feeding of young. Figure 2.4 shows the major impacts of oil, chemical, and plastic spills caused by maritime accidents. Coral and mangrove destruction, food poisoning via food chains, loss of income of fishermen, and human health effects are illustrated here.

X-Press Pearl affected around 7,000 families engaged in fishing by losing their livelihoods (Rubesinghe et al., 2022) and the global economic loss (fisheries and tourism industries) of marine plastic pollution is between 6 and 19 billion USD (data of 86 countries) (Diggle and Walker, 2022). Different cases go to different social dimensions as well. As an example, a march (nearly 100,000 protestors march) was organized to denounce the government's handling of the oil spill disaster of Mauritius (Raghoo, 2021). Captain and first officer of the damage caused ship got 20-month

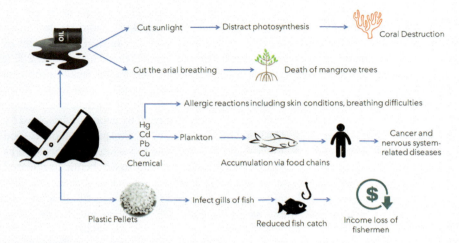

FIGURE 2.4 Major impacts of oil, chemical, and plastic spills caused by maritime accidents.

jail terms as they were found guilty on the charges of endangering safe navigation (Dahir, 2020). Compensation regimes for financing victims play a vital role to reduce distress after the disaster (Chang et al., 2014). The Sri Lankan government received 3.54 million USD compensation (Rabel, 2021) and Sri Lanka claimed 40 million USD (TME, 2021) for the X-Press Pearl disaster. The damage is estimated at over 10 billion USD from the oil spill of the Mauritius Disaster (Degnarain, 2020). Therefore, compensation through financial schemes plays a vital role to minimize the impacts of chemical and plastic spills.

2.3.5 The Responses to Chemical and Plastic Spills

The maritime industry has been occupied in a continuous evolutionary process through automation and digitalization. Automation can be beneficial to operators of complex systems in terms of a reduction in workload or the release of resources to perform other onboard duties (Hanzu-Pazara et al., 2008). Minimizing vessel collision is a benefit of autonomous ship navigation. Evolutionary algorithms, fuzzy logic, expert systems, and neural networks (NN), and a combination of them (hybrid system) are used to get critical decisions related to collision avoidance (Statheros, Howells, and Maier, 2008; Mizythras et al., 2021). Another benefit of using AI and machine learning techniques is that it allows sailors to predict sea conditions, letting ship captains alter their fuel consumption. Autonomous shipping can reduce fuel consumption by increasing sustainability (Tsou and Cheng, 2013; Burmeister et al., 2014; Simonsen, Walnum, and Gössling, 2018). Advanced analytics are used to extract useful information from a variety of data sources, and thus ensure that judgements are originated on data-backed principles in automated ships. Automation can also detect threats and other malicious activities. Therefore, the occurrence of maritime accidents can be reduced by AI and automation.

Maritime Law enforcement has fallen into the hands of Governments. The first international convention on oil pollution was adopted in 1926 by the International Maritime Conference in Washington. Shipping has to comply with the provisions of the International Convention for the Prevention of Pollution From Ships, 1973 as modified by the Protocol of 1978 (MARPOL) (Anyanova, 2012). Researchers have highlighted the strict enforcement of relevant laws concerning marine pollution based on the experiences of various countries such as Nigeria and Taiwan (Chiau, 2005; Elenwo and Akankali, 2015; Chitrakar et al., 2019). Maritime disputes in the potential flashpoints between countries with overlapping claims are one reason for poor law enforcement (Østhagen, 2020). Inconsistent actions make dangerous situations even worse. It can erode community trust in governments. X Press Pearl disaster is an example because the crew sought a priority berth for shelter while the ship was immersed in flames at Colombo port. It was difficult to enforce the International Maritime Dangerous Goods regulations, which control their handling and storage, to provide better-trained workers who apply these regulations, and to issue robust sanctions by states where cargoes originate, and by shipping companies (Claudio Bozzi, 2021).

Clean-up methods for chemical and plastic spills play a vital role in restoration processes. The first option to consider for cleaning oil spills is dispersant use. The conventional type dispersants are a mixture of non-aromatic hydrocarbon solvents. A concentrated dispersant (mixture of oxygenates like glycol and non-aromatic hydrocarbon) is more popular due to easy care (Anish Wankhede, 2019). The second method to clean the spilt oil is in situ burning. The third option is conducting mechanical recovery using skimmers (Prabowo and Bae, 2019). Hot water and high-pressure washing are mainly used in situations where the oil is inaccessible to mechanical removal methods such as via booms and skimmers. Extremophilic microorganisms are used for the treatment of toxic pollutants in the environment during chemical spills due to their toughness, adaptability, and strong resistance to extreme conditions (Jeong and Choi, 2020). Volunteers of all ages have manually removed the nurdle over 1,500 kg from beaches after the X-Press Pearl disaster (Rodrigo, 2021). Blue removal methods such as manual removal are also suggested by the chemical analysis of pollutants (nurdles) (Sewwandi et al., 2022). Scooping plastic out of the ocean is the most popular method of removing plastic from the ocean waters. Solar-powered vessel with conveyor belts collects floating debris, the remote-controlled robot collects garbage from waterways, and an automated bucket uses a pump to capture floating debris, including plastics can also be used to clean up operations of macro and microplastics (Schmaltz et al., 2020).

Compensation for clean-up costs and damages caused by chemical and plastic spills has clear rules about who pays for the direct response activities, the cost of assessing environmental damages and implementing the necessary restoration in the case of the United States (NOAA, 2015). Only 11% of claims were approved for compensation for the Hebei Spirit oil spill on the West coast of Korea (Kim et al., 2014). There are many similar cases and sometimes the compensation process is also distressing (Ritchie, Gill, and Long, 2018). The 2011 Bohai Sea oil spill has led to successful compensation of 0.48 billion USD and some of the disappointed fishermen have thus been seeking legal

30   Maritime Accidents and Environmental Pollution

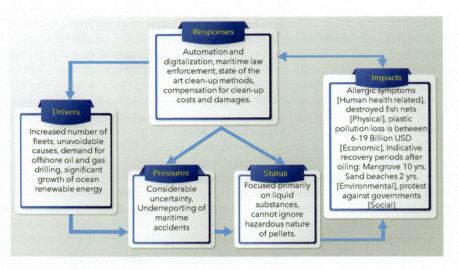

FIGURE 2.5 DPSIR framework on chemical and plastic spills caused by maritime accidents.

support in U.S. courts (Feng and Tu, 2014). All these cases show that the compensation cannot satisfy all stakeholders and the right pay still matters. However, compensation is a factor especially to enhance the livelihoods of victims.

Figure 2.5 shows a summary of the items described in the result section. Increased number of fleets, unavoidable causes, demand for offshore oil and gas drilling, and significant growth of ocean renewable energy plants are recognized as the drivers of maritime accidents caused by oil and plastic spills. The considerable uncertainty regarding the nature of chemical and plastic spills, and underreporting of maritime accidents are the main pressure to mitigate the damage of chemical and plastic spills. The studies on maritime spills primarily focused on liquid substances as well as they highlight the hazardous nature of pollution caused by pellets. Impacts of chemical and plastic spills can be categorized into human health-related, physical, environmental, social, and economic subdomains to accurate elaboration. Various types of incidents were recorded in many cases in different countries such as allergic symptoms on beach users, destroyed fishnets, loss of 6 to 19 billion USD, and protest against governments. Automation and digitalization of maritime shipping, maritime law enforcement, state-of-the-art clean-up methods, compensation for clean-up costs, and damages are the primary responses to mitigate oil and plastic spills.

2.4 SUMMARY AND CONCLUSIONS

Although maritime transportation is vital to world trade, the risk of environmental pollution and damages to marine life has been increased with the wide range of transported cargo. In this study, maritime accidents which caused chemical and plastic spills are considered where the various chemical reactions and transportation under the oceanic currents ultimately lead to multiple polluting instances where multiple stakeholders are affected. It has been found that the majority of maritime accidents are associated with human errors or failures, even though there could be mechanical failures that can cause a malfunction of a successful operation of a ship. Collisions and fires are found to be the most common maritime accidents. The MV Wakashio and the MV X-Press Pearl are two of the recent maritime accidents which lead to massive environmental pollution. The statistics indicate that chemical and plastic spills are not frequent amongst the accident data records. Out of the multiple framework systems that have been introduced, the DPSIR framework is used to identify the pressures and state changes in marine and coastal ecosystems. A comprehensive literature review was carried out and each literature was sorted into five categories: Drivers, Pressures, Status, Impacts, and Responses related to chemical and plastic spills. The increasing demand for offshore oil and gas drilling and the growth of ocean renewable energy could be identified as drivers of chemical spills. Even though some models are available to predict the dispersion of chemical and plastic spills, still considerable uncertainty is considered as pressure on the mitigation of chemical and plastic spills. The underreporting of maritime accidents was also noted during the study where most container losses go unreported, while spillages in large amounts are reported as oil spills. Usually, sorbents are used to remove final traces of oil spills. In-situ burning and the use of skimmers are some other common clean-up methods. Plastic spills can do long-lasting damage to the environment due to their longevity. A wide range of human health impacts was observed in many incidents where chemical and plastic spills had occurred. Not only physical health but also mental health was affected according to some studies. The accumulation of heavy metals and other chemicals through the food chain can lead to various diseases in the human population. Furthermore, physical damages to properties, environmental threats, and social and economic impacts occurred due to chemical and plastic spills of maritime accidents. The environmental impacts could be disastrous as some sensitive marine ecosystems have a very long recovery period. Automation and digitalization of marine transportation systems and ship navigation systems are in play to reduce the possibilities of having human error in the process. Since the first international convention on oil pollution in 1926, there are several provisions ships have to comply with within international shipping. The compensation payments are important while the assessment of the environmental damage and sometimes the compensation process could get distressing. While the compensation can't satisfy all stakeholders, it is vital to enhance and recover the damaged livelihood of victims. Maritime accidents caused by oil spills are thoroughly studied, but not chemical and plastic spills as their occurrences are not frequent. Environmental impacts are discussed in greater depth by various researchers but human health, socio-economic, policy, and physical impacts are not acutely discussed due to limited baseline information. There is a need to identify the potential impacts and better solutions via research with the expertise of different disciplines. The focus of marine accident mitigation shifted from marine architecture to human and social factors with the development of developments in AI and computational power. The hazardous nature of pellets cannot be ignored though they are not classified as hazardous materials. There is an increasing emphasis on the mathematical modelling of spills, safety enhancement of ships, and ship operations while sustainability and economic perspectives are somehow neglected in the research.

REFERENCES

Abeynayaka, A. (2021) 'Plastic Spills in Maritime Transportation and Consequences', *The Green Guardian*, 3(2). Available at: www.researchgate.net/publication/352449640_Plastic_Spills_in_Maritime_Transportation_and_Consequences

Akten, N. (2006) 'Shipping accidents: a serious threat for marine environment', *Journal of the Black sea/Mediterranean environment*, 12(3), pp. 269–304.

Anderies, J. M., Janssen, M. A. and Ostrom, E. (2004) 'A Framework to Analyze the Robustness of Social-ecological Systems from an Institutional Perspective', *Ecology and Society*, 9(1). doi: 10.5751/es-00610-090118

Anish Wankhede (2019) *Different Types of Dispersants Used in an Oil Spill, Marine Insight*. Available at: www.marineinsight.com/environment/different-types-of-dispersants-used-in-an-oil-spill/ (Accessed 5 March 2022).

Anyanova, E. (2012) 'Oil Pollution and International Marine Environmental Law', in Curkovic, S. (ed.) *Sustainable Development*. Rijeka: IntechOpen. doi: 10.5772/37399

Atkins, J. P. et al. (2011) 'Management of the marine environment: Integrating ecosystem services and societal benefits with the DPSIR framework in a systems approach', *Marine Pollution Bulletin*, 62(2), pp. 215–226. doi: 10.1016/j.marpolbul.2010.12.012

Awal, Z. I. and Hasegawa, K. (2017) 'A Study on Accident Theories and Application to Maritime Accidents', *Procedia Engineering*, 194, pp. 298–306. doi: https://doi.org/10.1016/j.proeng.2017.08.149

Burmeister, H.-C. et al. (2014) 'Can unmanned ships improve navigational safety?', in *Proceedings of the Transport Research Arena, TRA 2014, 14–17 April 2014, Paris*.

Buskey, E. J., White, H. K. and Esbaugh, A. J. (2016) 'Impact of oil spills on marine life in the Gulf of Mexico: effects on plankton, nekton, and deep-sea benthos', *Oceanography*, 29(3), pp. 174–181. doi: 10.5670/oceanog.2016.81

Chang, S. E. et al. (2014) 'Consequences of oil spills: a review and framework for informing planning', *Ecology and Society*, 19(2), Article 26.

Chen, J. et al. (2019) 'Oil spills from global tankers: Status review and future governance', *Journal of Cleaner Production*, 227, pp. 20–32. doi: https://doi.org/10.1016/j.jclepro.2019.04.020

Chen, S.-T. et al. (2013) 'A Human and Organisational Factors (HOFs) analysis method for marine casualties using HFACS-Maritime Accidents (HFACS-MA)', *Safety Science*, 60, pp. 105–114. doi: https://doi.org/10.1016/j.ssci.2013.06.009

Chiau, W.-Y. (2005) 'Changes in the marine pollution management system in response to the Amorgos oil spill in Taiwan', *Marine Pollution Bulletin*, 51(8), pp. 1041–1047. doi: https://doi.org/10.1016/j.marpolbul.2005.02.048

Chitrakar, P. et al. (2019) 'Current status of marine pollution and mitigation strategies in arid region: a detailed review', *Ocean Science Journal*, 54(3), pp. 317–348. doi: https://doi.org/10.1007/s12601-019-0027-5

Claudio Bozzi (2021) *Could the X-Press Pearl Disaster Have Been Prevented?, The Maritime Executive*. Available at: www.maritime-executive.com/editorials/could-the-x-press-pearl-disaster-have-been-prevented (Accessed 5 March 2022).

Coraddu, A. et al. (2020) 'Determining the most influential human factors in maritime accidents: A data-driven approach', *Ocean Engineering*, 211, p. 107588. doi: https://doi.org/10.1016/j.oceaneng.2020.107588

Cullinane, K. and Haralambides, H. (2021) 'Global trends in maritime and port economics: the COVID-19 pandemic and beyond', *Maritime Economics & Logistics*, 23(3), pp. 369–380. doi: https://doi.org/10.1057/s41278-021-00196-5

Cunha, I., Moreira, S. and Santos, M. M. (2015) 'Review on hazardous and noxious substances (HNS) involved in marine spill incidents-An online database', *Journal of Hazardous Materials*, 285, pp. 509–516. doi: https://doi.org/10.1016/j.jhazmat.2014.11.005

Cunningham, J. J., Magdol, Z. E. and Kinner, N. E. (2010) 'Ocean thermal energy conversion: assessing potential physical, chemical, and biological impacts and risks', Coastal Response Research Center, University of New Hampshire, Durham, NH, 33.

Dahir, A. L. (2020) 2 Ship Officers Plead Guilty to Light Charges in Mauritius Oil Spill, *New York Times*. Available at: www.nytimes.com/2021/12/21/world/africa/mauritius-oil-spill.html (Accessed 3 March 2022).

de Vos, A. *et al.* (2021) 'The M/V X-Press Pearl Nurdle Spill: contamination of burnt plastic and unburnt nurdles along Sri Lanka's Beaches', *ACS Environmental Au*. doi: https://doi.org/10.1021/acsenvironau.1c00031

Degnarain, N. (2020) Legal Loophole Opens $10 Billion Compensation Claim For Mauritius Oil Spill, Forbes. Available at: www.forbes.com/sites/nishandegnarain/2020/11/24/legal-loophole-opens-10-billion-compensation-claim-for-mauritius-oil-spill/?sh=10901ea21ab8 (Accessed 4 March 2022).

Diggle, A. and Walker, T. R. (2022) 'Environmental and economic impacts of mismanaged plastics and measures for mitigation', *Environments – MDPI*, 9(2). doi: 10.3390/environments9020015

do Carmo, J. A., Pinho, J. L. and Vieira, J. P. (2010) 'Oil spills in coastal zones: predicting slick transport and weathering processes', *The Open Ocean Engineering Journal*, 3, pp. 129–142.

Dzoga, M. *et al.* (2020) 'Application of the DPSIR framework to coastal and marine fisheries management in Kenya', *Ocean Science Journal*, 55(2), pp. 193–201. doi: 10.1007/s12601-020-0013-y

Ebbesmeyer, C. C. and Ingraham Jr., W. J. (1994) 'Pacific toy spill fuels ocean current pathways research', *Eos, Transactions American Geophysical Union*, 75(37), pp. 425–430. doi: https://doi.org/10.1029/94EO01056

Ebbesmeyer, C. C. and Ingraham W. James, J. (1992) 'Shoe spill in the North Pacific', *Eos, Transactions American Geophysical Union*, 73(34), pp. 361–365. doi: https://doi.org/10.1029/91EO10273

EIA (2021) *EIA supports Sri Lanka Gov't call for plastic pellets to be reclassified as a hazardous substance, Environmental Investigation Agency UK*. Available at: https://eia-international.org/news/eia-supports-sri-lanka-govt-call-for-plastic-pellets-to-be-reclassified-as-a-hazardous-substance/ (Accessed 2 March 2022).

Elenwo, E. I. and Akankali, J. A. (2015) 'The effects of marine pollution on Nigerian coastal resources', *Journal of Sustainable Development Studies*, 8(1), pp. 209–224.

Elizaryev, A. *et al.* (2018) 'Numerical simulation of oil spills based on the GNOME and ADIOS', *International Journal of Engineering and Technology (UAE)*, 7(2), p. 24.

Elliott, M. *et al.* (2016) 'Ecoengineering with ecohydrology: successes and failures in estuarine restoration', *Estuarine, Coastal and Shelf Science*, 176, pp. 12–35. doi: https://doi.org/10.1016/j.ecss.2016.04.003

Ellis-Petersen, H. (2021) Sri Lanka faces disaster as burning ship spills chemicals on beaches, Sri Lanka, *The Guardian*, www.theguardian.com

Faturachman, D. and Mustafa, S. (2012) 'Sea transportation accident analysis in Indonesia', *Procedia – Social and Behavioral Sciences*, 40, pp. 616–621. doi: https://doi.org/10.1016/j.sbspro.2012.03.239

Feng, J. and Tu, F. (2014) '19 oil spills and ocean pollution', in *Chinese Research Perspectives on the Environment, Volume 3*. Brill, pp. 271–277.

Fock, H. O., Kloppmann, M. and Stelzenmüller, V. (2011) 'Linking marine fisheries to environmental objectives: a case study on seafloor integrity under European maritime policies', *Environmental Science and Policy*, 14(3), pp. 289–300. doi: https://doi.org/10.1016/j.envsci.2010.11.005

Fratila, A. *et al.* (2021) 'The importance of maritime transport for economic growth in the European Union: a panel data analysis', *Sustainability (Switzerland)*, 13(14), pp. 1–23. doi: https://doi.org/10.3390/su13147961

Galieriková, A. *et al.* (2021) 'Study of maritime accidents with hazardous substances involved: comparison of HNS and oil behaviours in marine environment', *Transportation Research Procedia*, 55, pp. 1050–1064. doi: https://doi.org/10.1016/j.trpro.2021.07.182

Gari, S. R., Newton, A. and Icely, J. D. (2015) 'A review of the application and evolution of the DPSIR framework with an emphasis on coastal social-ecological systems', *Ocean and Coastal Management*, 103, pp. 63–77. doi: https://doi.org/10.1016/j.ocecoaman.2014.11.013

Guzzetti, E. *et al.* (2018) 'Microplastic in marine organism: environmental and toxicological effects', *Environmental Toxicology and Pharmacology*, 64, pp. 164–171. doi: https://doi.org/10.1016/j.etap.2018.10.009

Häkkinen, J. M. and Posti, A. I. (2013) 'Overview of maritime accidents involving chemicals worldwide and in the Baltic Sea', in Weintrit, A. and Neumann, T. (eds) *Marine Navigation and Safety of Sea Transportation Maritime Transport & Shipping*. 1st edn. Boca Raton, Florida: Taylor & Francis Group, pp. 20–21.

Hanzu-Pazara, R. *et al.* (2008) 'Reducing of maritime accidents caused by human factors using simulators in training process', *Journal of Maritime Research*, 5(1), pp. 3–18.

Harada, M. (1995) 'Minamata disease: methylmercury poisoning in Japan caused by environmental pollution', *Critical Reviews in Toxicology*, 25(1), pp. 1–24. doi: 10.3109/10408449509089885

Hardesty, B. D. *et al.* (2017) 'Using numerical model simulations to improve the understanding of micro-plastic distribution and pathways in the marine environment', *Frontiers in Marine Science*, 4. doi: https://doi.org/10.3389/fmars.2017.00030

Harlaftis, G. and Theotokas, I. (2020) 'Maritime business: a paradigm of global business', in *Oxford Research Encyclopedia of Business and Management*. doi: https://doi.org/10.1093/acrefore/9780190224851.013.184

Hassan Partow *et al.* (2021) *X-Press Pearl Maritime Disaster Sri Lanka*. Colombo, Sri Lanka. Available at: https://postconflict.unep.ch/Sri Lanka/X-Press_Sri Lanka_UNEP_27.07.2021_s.pdf

IAEA (2001) *IAEA-TECDOC-1231 – Severity, probability and risk of accidents during maritime transport of radioactive material*. International Atomic Energy Agency: Vienna, Austria.

Ichimura, Y. *et al.* (2022) 'Shipping in the era of digitalization: mapping the future strategic plans of major maritime commercial actors', *Digital Business*, 2(1), p. 100022. doi: https://doi.org/10.1016/j.digbus.2022.100022

Jayasinghe, C. (2021) MT New Diamond fire: Sri Lanka forwards Rs 3.4 bn claim to Greek shipping company, *EconomyNext*, www.economynext.com

Jeong, S., Capelouto, S. and Chavez, N. (2020) *A cargo ship leaking tons of oil off the Mauritius coast has split in two*, www.cnnphilippines.com

Jeong, S.-W. and Choi, Y. J. (2020) 'Extremophilic microorganisms for the treatment of toxic pollutants in the environment', *Molecules (Basel, Switzerland)*, 25(21), p. 4916. doi: 10.3390/molecules25214916

Karlsson, T. M. et al. (2018) 'The unaccountability case of plastic pellet pollution', *Marine Pollution Bulletin*, 129(1), pp. 52–60. doi: https://doi.org/10.1016/j.marpolbul.2018.01.041

Khatmullina, L. and Chubarenko, I. (2019) 'Transport of marine microplastic particles: why is it so difficult to predict?', *Anthropocene Coasts*, 2(1), pp. 293–305. doi: https://doi.org/10.1139/anc-2018-0024

Kim, D. et al. (2014) 'Social and ecological impacts of the Hebei Spirit oil spill on the west coast of Korea: implications for compensation and recovery', *Ocean & Coastal Management*, 102, pp. 533–544. doi: https://doi.org/10.1016/j.ocecoaman.2014.05.023

KIMO (2020) *Plastic pellets spill pollutes Danish, Norwegian, Swedish coastlines, KIMO International reacts to the loss of plastic pellets from MV Trans Carrier in the German Bight*. Available at: www.kimointernational.org/news/plastic-pellets-spill-pollutes-danish-norwegian-swedish-coastlines/ (Accessed 1 March 2022).

Kozanhan, M. K. (2019) 'Maritime tanker accidents and their impact on marine environment', *Scientific Bulletin 'Mircea cel Batran' Naval Academy*, 22(1), pp. 1–20.

Kumar, A. et al. (2020) 'Lead Toxicity: Health Hazards, Influence on Food Chain, and Sustainable Remediation Approaches', *International Journal of Environmental Research and Public Health*, 17(7), p. 2179. doi: https://doi.org/10.3390/ijerph17072179

Kurt, R. E. et al. (2016) 'SEAHORSE Procedure improvement system', *6th Conference on Design for Safety*. Glasgow, United Kingdom.

Laffon, B., Pásaro, E. and Valdiglesias, V. (2016) 'Effects of exposure to oil spills on human health: Updated review', *Journal of Toxicology and Environmental Health, Part B*, 19(3–4), pp. 105–128. doi: https://doi.org/10.1080/10937404.2016.1168730

Li, Y. et al. (2021) 'Research on the influence of microplastics on marine life', *IOP Conference Series: Earth and Environmental Science*, 631(1). doi: https://doi.org/10.1088/1755-1315/631/1/012006

Liu, X. et al. (2018) 'Evaluating the sustainability of marine industrial parks based on the DPSIR framework', *Journal of Cleaner Production*, 188, pp. 158–170. doi: https://doi.org/10.1016/j.jclepro.2018.03.271

Lourenço, R. A. et al. (2020) 'Mysterious oil spill along Brazil's northeast and southeast seaboard (2019–2020): Trying to find answers and filling data gaps', *Marine Pollution Bulletin*, 156, p. 111219. doi: https://doi.org/10.1016/j.marpolbul.2020.111219

Luo, M. and Shin, S.-H. (2019) 'Half-century research developments in maritime accidents: Future directions', *Accident Analysis & Prevention*, 123, pp. 448–460. doi: https://doi.org/10.1016/j.aap.2016.04.010

Lützhöft, M., Grech, M. R. and Porathe, T. (2011) 'Information environment, fatigue, and culture in the maritime domain', *Reviews of Human Factors and Ergonomics*, 7(1), pp. 280–322. doi: https://doi.org/10.1177/1557234X11410391

Major, D. N. and Wang, H. (2012) 'How public health impact is addressed: a retrospective view on three different oil spills', *Toxicological & Environmental Chemistry*, 94(3), pp. 442–467. doi: https://doi.org/10.1080/02772248.2012.654633

Marine Accident Investigation Branch (MAIB) (2014) *Accident Report No 6/2014*.

Mizythras, P. et al. (2021) 'A novel decision support methodology for oceangoing vessel collision avoidance', *Ocean Engineering*, 230, p. 109004. doi: https://doi.org/10.1016/j.oceaneng.2021.109004

MTI Network (2022) *Incident Overview, X-Press Pearl – Incident Information Centre*. Available at: www.x-presspearl-informationcentre.com/ (Accessed 1 March 2022).

NAOSL (2020) *Progress of Implementation of MARPOL Convention Executing in Sri Lanka for Preventing Marine Pollution by Ships*. Colombo, Sri Lanka. Available at: www.auditorgeneral.gov.lk/web/images/audit-reports/upload/2021/performance_2021/8-iv/Marpol-English-report.pdf

Navrud, S., Lindhjem, H. and Magnussen, K. (2017) 'Valuing marine ecosystem services loss from oil spills for use in cost–benefit analysis of preventive measures', in *Handbook on the Economics and Management of Sustainable Oceans*. Cheltenham, UK: Edward Elgar Publishing.

Newton, A. (2012) 'A systems approach for sustainable development in coastal zones', *Ecology and Society*, 17(3), pp. 3–4.

NOAA (2015) *Who Pays for Oil Spills?*, National Oceanic and Atmospheric Administration. Available at: https://response.restoration.noaa.gov/about/media/who-pays-oil-spills.html (Accessed 5 March 2022).

Nuutinen, M. and Norros, L. (2009) 'Core task analysis in accident investigation: analysis of maritime accidents in piloting situations', *Cognition, Technology & Work*, 11(2), pp. 129–150. doi: https://doi.org/10.1007/s10111-007-0104-x

Oliveira, L. M. T. M. *et al.* (2021) 'Sorption as a rapidly response for oil spill accidents: A material and mechanistic approach', *Journal of Hazardous Materials*, 407, p. 124842. doi: https://doi.org/10.1016/j.jhazmat.2020.124842

Olsen, S. B. (2003) 'Frameworks and indicators for assessing progress in integrated coastal management initiatives', *Ocean & Coastal Management*, 46(3), pp. 347–361. doi: https://doi.org/10.1016/S0964-5691(03)00012-7

Onink, V. *et al.* (2021) 'Global simulations of marine plastic transport show plastic trapping in coastal zones', *Environmental Research Letters*, 16(6), p. 64053. doi: 10.1088/1748-9326/abecbd

Østhagen, A. (2020) 'Maritime boundary disputes: what are they and why do they matter?', *Marine Policy*, 120, p. 104118. doi: https://doi.org/10.1016/j.marpol.2020.104118

Ovcina, J. (2021) Dead turtles, dolphins washing ashore amid impact from X-Press Pearl's sinking, Offshore Energy. Available at: www.offshore-energy.biz/report-dead-turtles-dolphins-washing-ashore-amid-impact-from-x-press-pearls-sinking/ (Accessed 2 March 2022).

Özdemir, Ü. and Güneroğlu, A. (2015) 'Strategic approach model for investigating the cause of maritime accidents', *PROMET – Traffic & Transportation*, 27(2), pp. 113–123. doi: https://doi.org/10.7307/ptt.v27i2.1461

Panagiotidis, P. *et al.* (2021) 'Shipping accidents dataset: data-driven directions for assessing accident's impact and improving safety onboard', *Data*, 6(12). doi: https://doi.org/10.3390/data6120129

Patrício, J. *et al.* (2016) 'DPSIR – Two decades of trying to develop a unifying framework for marine environmental management?', *Frontiers in Marine Science*, 3(Sep), pp. 1–14. doi: https://doi.org/10.3389/fmars.2016.00177

Pattiaratchi, C. B. and Wijeratne, E. M. S. (2021) X-Press Pearl Disaster: An Oceanographic Perspective, *Groundviews*. Available at: https://groundviews.org/2021/06/08/x-press-pearl-disaster-an-oceanographic-perspective/ (Accessed 13 March 2022).

Peters, K. and Steinberg, P. (2019) 'The ocean in excess: towards a more-than-wet ontology', *Dialogues in Human Geography*, 9(3), pp. 293–307. doi: 10.1177/2043820619872886

Prabowo, A. R. and Bae, D. M. (2019) 'Environmental risk of maritime territory subjected to accidental phenomena: correlation of oil spill and ship grounding in the Exxon Valdez's case', *Results in Engineering*, 4, p. 100035. doi: https://doi.org/10.1016/j.rineng.2019.100035

Puisa, R. *et al.* (2018) 'Unravelling causal factors of maritime incidents and accidents', *Safety Science*, 110, pp. 124–141. doi: https://doi.org/10.1016/j.ssci.2018.08.001

Rabel, G. (2021) Righting the Ship: The X-Press Pearl Disaster and the Imperative for Regional Maritime Cooperation, *Talking Economics*. Available at: www.ips.lk/talkingeconomics/2021/08/09/righting-the-ship-the-x-press-pearl-disaster-and-the-imperative-for-regional-maritime-cooperation/#:~:text=The X-Press Pearl fire,social inequality they already faced. (Accessed 3 March 2021).

Raghoo, P. (2021) 'Precautionary policy? The Wakashio oil spill in Mauritius: a hard lesson about coastal and marine management', *June 20, 2021. IPL Policy Brief Series*, 2.

Ritchie, L. A., Gill, D. A. and Long, M. A. (2018) 'Mitigating litigating: an examination of psychosocial impacts of compensation processes associated with the 2010 BP deepwater horizon oil spill', *Risk Analysis*, 38(8), pp. 1656–1671. doi: https://doi.org/10.1111/risa.12969

Rivera, G., Felix, A. and Mendoza, E. (2020) 'A review on environmental and social impacts of thermal gradient and tidal currents energy conversion and application to the case of Chiapas, Mexico', *International Journal of Environmental Research and Public Health*, 17(21), pp. 1–18. doi: 10.3390/ijerph17217791

Rodrigo, M. (2021) Burnt pellets complicate impact of plastic spill off Sri Lanka, study finds, Mongabay. Available at: https://news.mongabay.com/2021/12/burnt-pellets-complicate-impact-of-plastic-spill-off-sri-lanka-study-finds/ (Accessed 5 March 2022).

Rubesinghe, C. *et al.* (2022) X-Press Pearl, a 'new kind of oil spill' consisting of a toxic mix of plastics and invisible chemicals. Available at: https://ipen.org/sites/default/files/documents/ipen-sri-lanka-ship-fire-v1_2aw-en.pdf

Sampson, H. *et al.* (2018) *The causes of maritime accidents in the causes of maritime accidents in the period 2002–2016*. Seafarers International Research Centre (SIRC), Cardiff University.

Scarlett, A. G. *et al.* (2021) 'MV Wakashio grounding incident in Mauritius 2020: The world's first major spillage of Very Low Sulfur Fuel Oil', *Marine Pollution Bulletin*, 171, p. 112917. doi: https://doi.org/10.1016/j.marpolbul.2021.112917

Schmaltz, E. *et al.* (2020) 'Plastic pollution solutions: emerging technologies to prevent and collect marine plastic pollution', *Environment International*, 144, p. 106067. doi: https://doi.org/10.1016/j.envint.2020.106067

Seveso, D. *et al.* (2021) 'The Mauritius oil spill: what's next?', *Pollutants*, 1(1), pp. 18–28. doi: https://doi.org/10.3390/pollutants1010003

Sewwandi, M. *et al.* (2022) 'Unprecedented marine microplastic contamination from the Xpress Pearl container vessel disaster: mitigating efforts by the Blue Treatment Facility', *Science of The Total Environment. Available at SSRN 4002422*. Available at: http://dx.doi.org/10.2139/ssrn.4002422

Shadman, M. *et al.* (2019) 'Ocean renewable energy potential, technology, and deployments: a case study of Brazil', *Energies*, 12(19). doi: 10.3390/en12193658

Shimizu, Y. *et al.* (2020) 'An enhanced multiphase ISPH-based method for accurate modeling of oil spill', *Coastal Engineering Journal*, 62(4), pp. 625–646. doi: https://doi.org/10.1080/21664250.2020.1815362

Shkëlqim Sinanaj (2020) 'The impact of shipping accidents on marine environment in Albanian Seas', *Journal of Shipping and Ocean Engineering*, 10(1), pp. 9–23. doi: https://doi.org/10.17265/2159-5879/2020.01.005

Simonsen, M., Walnum, H. J. and Gössling, S. (2018) 'Model for estimation of fuel consumption of cruise ships', *Energies*, 11(5). doi: https://doi.org/10.3390/en11051059

Soares, C. G. and Teixeira, A. P. (2001) 'Risk assessment in maritime transportation', *Reliability Engineering and System Safety*, 74(3), pp. 299–309. doi: https://doi.org/10.1016/S0951-8320(01)00104-1

Spear, D. (2021) *Nurdle spills point to a bigger problem of too much plastic, Daily Maverick.* Available at: www.dailymaverick.co.za/article/2021-07-22-nurdle-spills-point-to-a-bigger-problem-of-too-much-plastic/ (Accessed 1 March 2022).

Spier, C. *et al.* (2013) 'Distribution of hydrocarbons released during the 2010 MC252 oil spill in deep offshore waters', *Environmental Pollution*, 173, pp. 224–230. doi: https://doi.org/10.1016/j.envpol.2012.10.019

Sri Lanka Marine Pollution Prevention Act No. 35 of 2008 (2008) *Marine Pollution Prevention Act, No. 35 of 2008.* Government Publications Bureau of the Democratic Socialist Republic of Sri Lanka.

Statheros, T., Howells, G. and Maier, K. M. (2008) 'Autonomous ship collision avoidance navigation concepts, technologies and techniques', *Journal of Navigation*, 61(1), pp. 129–142. doi: https://doi.org/10.1017/S037346330700447X

Tamura, H., Waseda, T. and Miyazawa, Y. (2009) 'Freakish sea state and swell-windsea coupling: numerical study of the Suwa-Maru incident', *Geophysical Research Letters*, 36(1), pp. 2–6. doi: 10.1029/2008GL036280

Tayeb, A. M. *et al.* (2020) 'Oil spill clean-up using combined sorbents: a comparative investigation and design aspects', *International Journal of Environmental Analytical Chemistry*, 100(3), pp. 311–323. doi: 10.1080/03067319.2019.1636976

The Merchant Shipping (Accident Reporting and Investigation) Regulations No. 1743 of the United Kingdom (2012). www.legislation.gov.uk.

TME (2021) Sri Lanka Files Initial $40M Claim Over X-Press Pearl Fire, The Maritime Executive. Available at: www.maritime-executive.com/article/sri-lanka-files-initial-40m-claim-over-x-press-pearl-fire (Accessed 4 March 2022).

Tsou, M.-C. and Cheng, H.-C. (2013) 'An Ant Colony Algorithm for efficient ship routing', *Polish Maritime Research*, 20(3), pp. 28–38.

Turner, A., Williams, T. and Pitchford, T. (2021) 'Transport, weathering and pollution of plastic from container losses at sea: observations from a spillage of inkjet cartridges in the North Atlantic Ocean', *Environmental Pollution*, 284, p. 117131. doi: https://doi.org/10.1016/j.envpol.2021.117131

Uğurlu, Ö., Yildirim, U. and Başar, E. (2015) 'Analysis of grounding accidents caused by human error', *Journal of Marine Science and Technology (Taiwan)*, 23(5), pp. 748–760. doi: https://doi.org/10.6119/JMST-015-0615-1

United Nations Conference on Trade and Development (UNCTAD) (2018) *United Nations Conference on Trade and Development. Review of Maritime Transport*, United Nations Conference on Trade and Development.

Walker, T. R. *et al.* (2018) 'Environmental effects of marine transportation', in *World Seas: An Environmental Evaluation Volume III: Ecological Issues and Environmental Impacts.* Second Ed. Elsevier Ltd., pp. 505–530. doi: https://doi.org/10.1016/B978-0-12-805052-1.00030-9

Walker, T. R. *et al.* (2019) 'Chapter 27 – Environmental effects of marine transportation', in Sheppard, C. (ed.) *World Seas: an Environmental Evaluation (Second Edition).* Academic Press, pp. 505–530. doi: https://doi.org/10.1016/B978-0-12-805052-1.00030-9

Webler, T. and Lord, F. (2010) 'Planning for the human dimensions of oil spills and spill response', *Environmental Management*, 45(4), pp. 723–738. doi: https://doi.org/10.1007/s00267-010-9447-9

World Resources Institute (2005) *Ecosystems and Human Well-Being: Biodiversity Synthesis, Millennium Ecosystem Assessment.* Washington, DC.

Zamparas, M. *et al.* (2020) 'Application of Sorbents for oil spill cleanup focusing on natural-based modified materials: a review', *Molecules (Basel, Switzerland)*, 25(19). doi: https://doi.org/10.3390/molecules25194522

Zhang, L. *et al.* (2019) 'Ship accident consequences and contributing factors analyses using ship accident investigation reports', *Proceedings of the Institution of Mechanical Engineers, Part O: Journal of Risk and Reliability*, 233(1), pp. 35–47. doi: https://doi.org/10.1177/1748006X18768917

Zhang, X. *et al.* (2021) 'Microplastics in the endangered Indo-Pacific humpback dolphins (Sousa chinensis) from the Pearl River Estuary, China', *Environmental Pollution*, 270, p. 116057. doi: https://doi.org/10.1016/j.envpol.2020.116057

Zhang, Y. *et al.* (2021) 'Spatial patterns and characteristics of global maritime accidents', *Reliability Engineering & System Safety*, 206, p. 107310. doi: https://doi.org/10.1016/j.ress.2020.107310

Zhang, Z. and Li, X.-M. (2017) 'Global ship accidents and ocean swell-related sea states', *Natural Hazards and Earth System Sciences*, 17, pp. 2041–2051.

Zhilnikova, N. A. *et al.* (2019) 'Special aspects of modeling on accidental oil spills in inland sea waters', in *IOP Conference Series: Materials Science and Engineering*. IOP Publishing, p. 62013.

Znanstveni, M. *et al.* (2017) 'Preventing marine accidents caused by technology-induced human error', *Scientific Journal of Maritime Research*, 31, pp. 33–37.

Zulfick Farzan (2021) *Investigations reveal that negligence was cause for X-Press Pearl fire*, www.newsfirst.lk

3 X-Press Pearl
The Cargo and Its Complexity in Impact

Thejanie Jayasekara, Sajani Piyatilleke, Dilantha Thushara, Shantha Egodage, and Ajith Priyal de Alwis

The MV X-Press Pearl, a Singaporean Cargo ship, caught fire and sank off Colombo at a distance of 4.5 nautical miles from the shore. The disaster had elements of fire, explosions and total destruction of cargo and the ship. At the time weather too was quite inclemental for some days and the South-West monsoon was in strength at the time. The cargo that X-Press Pearl carried on board created the basis for the impact of the disaster. This chapter identifies and discusses the cargo on board and the complexity of impact on the maritime ecosystem and livelihood of local fishing community.

3.1 CARGO ABOARD THE X-PRESS PEARL

The degree of damage caused by the X-Press Pearl disaster can be better elaborated by appraising the onboard cargo. Comprised of a variety of polymers and plastics, chemicals, metals, food items, electrical and electronic equipment and other items ranging from fabrics to pharmaceuticals, the X-Press Pearl carried approximately 1487 containers onboard. Of this lot, 699 of the containers were stowed on deck and 788 were stored in the holds. The Bill of Lading (BOL) has not been completely released and as per the details available from multiple sources, the underdeck containers consist of 509 – 20 foot containers and 279 – 40 foot containers. There are slight discrepancies in numbers between different sources; however, the most likely value is provided here. Majority of the components within the cargo containers were either flammable or had the propensity to negatively impact the environment and due to this disaster, this hazardous cargo was released into the environment. As Figure 3.1 shows, the fire which raged on for about 10 days would have contributed to significant hazardous air pollutants as the situation clearly indicates incomplete combustion.

Containers stored below deck carried the following dangerous goods as cargo:

Aluminium processing by-product (Aluminium dross)
Bright yellow sulphur
Caustic soda
Urea

The Cargo and Its Complexity in Impact 41

FIGURE 3.1 X-Press Pearl on fire with thick black smoke bellowing to the wind. Credit: MEPA.

Inorganic chemicals (there is listing of cargo with this heading without any specifics)
Sodium methylated solution in alcohol
and 192 containers of microplastic nurdles.

Containers stored on deck which carried hazardous cargo:

5 containers designated as chemical products – the exact details are not available
9 tank containers with methanol
1 container of nitric acid (on bay 02)
2 containers of vinyl acetate, stabilized

As indicated, there is no clarity with contents declared as chemical products in quite a few containers. Those were placed on deck (bay 05 and bay 07) and are likely to have been consumed in the flames.

The presence of dangerous cargo on board created the disaster as it unfolded to go viral across the world. The situation escalated to significant level till it culminated with the sinking of the ship. The detailed assessment of the cargo across major categories is presented below in Table 3.1. The major categories had been identified as Polymers, Chemicals, Metals, e-Waste, Foods and Other. Dividing across these categories enables better analysis as well as understanding interactions.

FIGURE 3.2 Cargo identified as acetyl cedrene is still a mystery with incomplete cargo information. Credit: SL Airforce.

Other hazardous materials on board included liquid epoxy resins, gear oil, brake fluid, lead, copper and lithium batteries. Some parts of manifest in detail are available and examination of this information indicated that though certain cargo is classified as personal items the contents are electrical and electronic items. The quantity was identified to be significant. Effectively all these Electrical and Electronic equipment became Waste Electrical and Electronic Equipment (WEEE) upon descending into the ocean. It was considered that labelling them as personnel is not the correct approach in an environmental analysis. The selected cargo was thus reclassified as e-Waste. This reclassification was carried out later as more information was made available. Initial communication made with cargo did not reflect this.

Accordingly, the main sources of concern can be identified as the hazardous and noxious substances (HNSs) on board and the electrical and electronic equipment that have the capacity to contaminate the water sources with long-term heavy metal pollutants.

Other than the cargo itself, the ship's onboard stock of hydrocarbons for its operations, reported to have been about 700 MT in total, provided a source of continued combustion and pollution by exacerbating the fire and spilling into the ocean. It is believed that with the fire burning continuously, reaching almost 1500°C in some instances, a majority of the ship's fuel stocks have burned out. But as the fuel stocks were stored in the lower levels of the ship there is a definite possibility of unburnt fuel remaining within the unharmed fuel tanks of the wreckage. This may require contingency planning during the salvaging process as fuel can seep into the ocean upon disturbing the shipwreck.

3.2 ROOT CAUSE OF FIRE

The instigation of the fire is mainly attributed to the nitric acid leak, but a possible combination of scenarios may have had a collaborative effect on the continued burning and consequent explosions.

The Cargo and Its Complexity in Impact

TABLE 3.1
Quantities of the Content Belong to Each Category in MT (Total)

Cargo Detail in Brief	Plastics and Polymers	Cargo Detail in Brief	Chemicals	Cargo Detail in Brief	Metals	Cargo Detail in Brief	Food Items	Cargo Detail in Brief	Other	Cargo Detail in Brief	Electronic and Electrical Equipment
Epoxy resins	9700.8	Urea (Dust, Prilled)	1843.3	Aluminium content	2542.2	Fish, dry fish, sprats	1275.9	Paper/wastepaper items	1705.9	Personal cargo, effects	310
Synthetic resins	177.3	Inorganic chemicals, nos	495.3	Copper content	474.6	Chicken	370.6	Used items (vehicle parts, goods, clothes)	263.9	Mobile and stationary accessories	26.7
HDPE	747.8	Caustic soda	1126.9	Iron and steel content	74.8	Shrimp	252	Cartons	14.3		
LLDPE	245.4	Chemical products, nos	160.2	Lead	187	Cuttlefish	33.6	Fabrics, wadding, thread	1297.1		
LDPE	574	Nitric acid	28.7	Metal scraps	142.7	Dates	194.6	Tyres (New)	22.8		
Packages of PS pellets	31.9	Perfumery products	8			Raisin	88	Furniture	11.1		
Bare foam pig	1.9	Assorted perfumes	16.8			Alcoholic beverages	34.4	Waterproof materials	23.4		
Packaging materials	22.8	Quicklime lumps	1196.4			Chocolates	24.5	Pharmaceutical packing	16.5		

(*Continued*)

TABLE 3.1 (Continued)
Quantities of the Content Belong to Each Category in MT (Total)

Cargo Detail in Brief	Plastics and Polymers	Cargo Detail in Brief	Chemicals	Cargo Detail in Brief	Metals	Cargo Detail in Brief	Food Items	Cargo Detail in Brief	Other	Cargo Detail in Brief	Electronic and Electrical Equipment
Polycarbonate	60.9	Sodium methylate solution, sodium methoxide	57.3			Honey	11.1	Carpets	46.2		
Plastic articles	81.3	Methanol	235.6			Fruit juice	29.1	Aseptic pack	29.5		
Polymeric beads	19.3	Bright yellow sulphur	562.4			Fresh orange	120.6	Scoured goat hair	22		
Polymers of propylene	57.3	Molybdic oxide	48			Jam and pistachio	48.9	Crushed stone	22.5		
PVC film	20.4	Env. hazardous subs.	10.2			Oats	26	Fire protective equipment	109.3		
Alkyd resin	19.5	Pharmaceutical stuffs	7.5			Other food stuffs	259.5	Automotive, vehicle parts (New)	126.3		
Unsat. Polyester resin	21	Paints	21.2					Wagon Models	28.9		
Vinyl copolymer	18.6	Colours for ceramic ware	15					Exhaust stack anchor bolts, nuts	16.1		
Vinyl Acetate	46.3	Shampoo	16.3					Consol Cargo	43.6		

The Cargo and Its Complexity in Impact

Polybutadiene	92.7	Cement conforming	1154.8				Empty	1114.3
		Modified asphalt	205.2					
		Silicon sealant	12.5					
		Engine coolant and grease	16.9					
		Liquid paraffin	130.2					
		Brake fluid	34.7					
		Lubricants	81.6					
		Base oil	47.6					
		Engine oil	102.9					
		Lubricating oil, lube oil, and pails of lubricating oil	440					
Total: Plastics	11939.2	**Total: Chemicals**	8075.5	**Total: Metals**	3421.3	**Total: Food Items**	2768.8 **Total: other**	4913.7 **Total: Electronic and Electrical Equipment**	336.7

Note: nos, not otherwise specified.

As nitric acid by itself is a non-combustible, it does not evoke fire or explosion upon contact with an ignition source [1]. But nitric acid has the predisposition to react exothermically with both organics and non-metals, and therefore it can be posited that the reactions with surrounding cargo components led to the initiation of the fire. Nitric acid acts as a strong oxidizer and can ignite combustibles such as wood, paper, oil, clothing, etc., even without the presence of an ignition source. Instead, the emission of heat from the exothermic reaction further propagates the reaction and if not physically remedied through cooling action turns into a runaway reaction ending in thermal explosion.

$$C + 2HNO_3 \rightarrow CO_2 + NO_2 + NO + H_2O \tag{3.1}$$

Reaction of organic and metallic matter with nitric acids leads to the generation of nitrogen dioxide (NO_2), a reddish-brown fume with toxic properties, bringing about air pollution, acid rain and harm to vegetation and the respiratory system of humans. Though NO_2 does not directly influence the propagation of fire, the negative impact on the environment as well as humans is significant. During the first two days of the fire, the emission of NO_2 was evident as a reddish-brown smoke that rose from the ship. Due to the surrounding wind patterns the released NO_2 travelled inland, reaching a significant portion of populated area at concentrations greater than National Ambient Air Quality 1 hour average.

On the other hand, the produced nitrous oxide (NO) is much more reactive, prompting further exothermic reaction through oxidative activity and thus fuelling the fires generated.

Upon contact with metallic surfaces, especially those of the container walls in this scenario, nitric acid reacts producing metal nitrates and hydrogen. With the leak flowing to lower areas of the ship it is possible that hydrogen thus produced in close confines may not be able to disperse easily and potentially create zones of combustion.

Further, nitric acid can decompose in the presence of fire, adding to the release of NO_2 while simultaneously supplying oxygen to the fire, allowing it to grow and in turn facilitate explosion.

$$4HNO_3 \rightarrow 4NO_2 + O_2 + 2H_2O \tag{3.2}$$

The decomposition of nitric was present before the fire in the X-Press Pearl leakage and thus the nitric acid leak can definitively be established as one of the major causes of the ensuing fire.

3.3 ONBOARD CARGO THAT INDUCED AND SUSTAINED THE FIRE

The containers adjacent to the nitric acid cargo contained epoxy resin and high-density polyethylene (HDPE), and other nearby containers on deck carried vinyl acetate, methanol and lithium-ion batteries. These are all prone to explosion and fire, thus creating an environment ripe for disaster when in contact with the leaking nitric acid.

The Cargo and Its Complexity in Impact

The epoxy resin stored closest to the nitric acid cargo is an inherently flammable and toxic substance, and when put in contact with nitric acid it readily decomposes exothermically to start fires. A cargo load of 9700.8 MT of epoxy resin stored in 349 containers was on board, providing a significant amount of source material for a fire. Additionally, through the combustion of this synthetic material a significant environmental impact occurred through the release of pollutant gases into the air and release of partially burnt epoxy resin into the water.

Methanol, stored on deck and within the cargo hold, can react with nitric acid, producing methyl nitrate, a highly explosive substance. This, along with the highly flammable vinyl acetate, aluminium re-smelting by-products and sodium methylate (in alcohol), could also have contributed to the occurrence of the first violent explosion that occurred on the 25th of June. The aluminium re-smelting by-products and sodium methylate, upon contact with water, produce highly flammable, toxic gases (such as hydrogen, acetylene, methane, ammonia and phosphine) and methanol, respectively. The methanol and aluminium re-smelting by-products can be identified as the major precursors of the first explosion. And due to the tightly packed nature of the cargo as well as because of the very effects of firefighting due to adding of water onto these highly reactive components, the explosion could have been intensified. Despite firefighting efforts, it was this explosion that caused the entire vessel to be engulfed in flames from the forward section to the aft.

Several spaces over from the nitric acid container a consignment of two types of lithium-ion batteries was stored as illustrated in Figure 3.3. They are highly flammable in nature, and it is notoriously difficult to extinguish fires formed from lithium-ion batteries as they tend to re-ignite even after fire extinguishing action has been taken. Moreover, upon contact with nitric acid, lithium (Li) reacts exothermally to

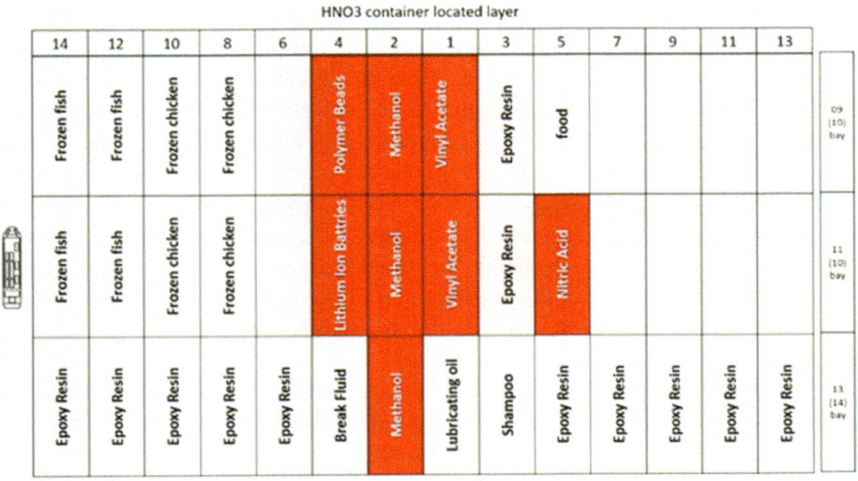

FIGURE 3.3 Cargo placement analysis – the cargo details close to the nitric acid storage.

FIGURE 3.4 A Tug boat of SL Navy on site as the event unfolds.
Source: SL Air Force.

form lithium nitrate and NO, of which both are oxidizers and can aggravate further exothermic reaction and thus the fire. As majority of the cargo on deck dislodged and fell overboard due to the shock of the explosion it is possible the unburnt Li batteries, which are located among uppermost deck containers as indicated by Figure 3.2, too fell into the ocean. This was ascertained by the Li found in gills of fish samples which were analyzed just after the incident and in the muscles of fish samples assessed after a certain time, showing the potential effects of Lithium bioaccumulation.

Other than the cargo on deck those below in the cargo hold could have directly affected the initiation as well as the magnification of the fire. As the nitric acid container was placed close to a hatch opening, it unfortunately allowed the acid leak to travel through the opening and into the cargo hold. An unfettered nitric acid leak can eat away at the rubber closures of cargo hold containers (built only to be waterproof carriers and not acid proof) and come in contact with a variety of chemicals and other consignments housed within the cargo hold which ultimately could have ended up in fire and explosion.

The fire was spread across the entire length of the ship and therefore would definitely have affected the deck structure due to weakening or distorting. Due to the melting and deforming on deck and hatches, some containers could have partially disappeared into the holds and as a result, the fire started on deck could have affected the containers in the hold as well. There was a period of high-intensity weather at the time of the initial fire while the vessel was still floating at anchor, which worsened the overall predicament.

The ship's hydrocarbons, used for its operations, consisted of approximately 300 MT of heavy fuel oil (HFO), of which 130 m^3 was held in storage tank #2 and 105 m^3

The Cargo and Its Complexity in Impact 49

in engine room tanks, <50 MT of marine diesel and ~52,300 L of different grades of marine lubricants, which were stored in the engine room. These hydrocarbons could have directly influenced the intensity of the fire. The bunker oil of the X-Press Pearl was an intermediate fuel oil (IFO380) which is a combination of 95% HFO and 5% gasoline. Though the IFO has a low inclination towards burning, evaporation of the gasoline component would increase its density and viscosity, creating a much higher persistence of the IFO in water upon spillage. This concern was realized when an oil slick of 3.23 km was observed on the 8th of June, 6 days after the sinking of the ship. By the 14th of June the oil sheen grew to 4.3 km and continued to spread in area and distance aided by the turbulence of the ocean waves. The spread was visible over a 5 km^2 area by August 20, 2021, but the overall extent could be much greater with the vertical dispersion of the oil through water columns under wave conditions, which in turn can transfer with wave currents.

After the first explosion, firefighting efforts over the duration of 25th to 27th of May managed to deplete the fire down to a pool fire deep within the interior of the vessel. But the second explosion still occurred on the 28th of May, and this could be due to the gas explosions caused by the extinguished pool fires in the ship's fuel tanks. It is a common occurrence where extinguished or partially extinguished ventilation-controlled pool fires often trigger explosions which may result in more fires.

3.4 ENVIRONMENTAL DAMAGE CAUSED BY THE CARGO THAT FELL INTO THE OCEAN AND SANK WITH THE SHIP

It is important to note the distribution of containers on deck and in the hold. Other than the cargo that got dislodged during the explosion and went overboard, all cargo that was on deck was subjected to the fire. It could be considered that almost all the containers stowed on deck were washed over the side. An imaging survey was carried out prior to the debris removal operation initiated by the ship owner's company. A multibeam survey was carried out with the help of the Indian Navy as well as a side scan sonar survey by National Aquatic Res and Research Agency's (NARA's) research ship Samudrika. A remotely operated vehicle (ROV) with camera, too, had been employed by the ship operator. The identification process revealed quite a lot of debris around the vessel within a radius of 1 km. Along with the cargo that directly fell overboard and polluted the sea, the cargo within the ship also leaked into the surrounding water as the ship sank.

3.4.1 COMPLEXITY ANALYSIS OF THE CHEMICAL COCKTAIL

The X-Press Pearl, at the moment of the unfortunate incident, was carrying over 8000 MT of a wide spectrum of chemicals where, most of which were designated as dangerous goods. Although the ship was carrying multiple chemicals, it can be assumed the best practices of segregation were implemented during the voyage. However, with nitric acid spillage, a gradual mixing of at least another chemical can be envisaged and should be identified as the prime cause for the fire as nitric acid forms explosive and exothermic reactions with many substances.

TABLE 3.2
2D Matrix of Impact of the Toxic Cocktail (an Extract)

	Water	Seawater	Nitric Acid	Vinyl Acetate	All Remaining By-products
Nitric acid	$HNO_3 + H_2O \rightarrow H_3O^+ + NO_3^-$ Exothermic. If spilt in huge amounts a localized decrease in pH and increase in temperature will occur (localized, surface). These are acute changes. High acidity and high temperature are hazardous to organisms in that location during that period. Due to dilution, later pH and temperature come back to normal levels.	$HNO_3 + H_2O \rightarrow H_3O^+ + NO_3^-$ Exothermic. If spilt in huge amounts a localized decrease in pH and increase in temperature will occur (localized, surface). These are acute changes. High acidity and high temperature are hazardous to organisms in that location during that period. Due to dilution, later pH and temperature come back to normal levels.	No reaction.	Violent explosive reaction; Can cause fire.	Al with conc. $HNO_3 - > NO_2^+ + H_2O$ $Al_2O_3 + H_2O$ Al_2O_3 acts as a protective layer and prevents continuation of the reaction.
Vinyl acetate	Form a slick on the surface. Evaporation continues from the upper surface and vapor over the water surface, slick would dissolve. React with water and air giving peroxides. Hazard to freshwater organisms.	Form a slick on the surface. Evaporation continues from the upper surface and vapor over the water surface, slick would dissolve. React with water and air giving peroxides. Hazard to freshwater organisms.	Rapidly volatile. On exposure to light spontaneous polymerization. Form peroxides by reacting with water and air that catalyze the exothermic polymerization reaction.	No reaction.	Violent explosive reaction; Can cause fire.

The Cargo and Its Complexity in Impact

| Water | No reaction | Miscible | $HNO_3 + H_2O \rightarrow H_3O^+ + NO_3^-$. Exothermic. If spilt in huge amounts a localized decrease in pH and increase in temperature will occur (localized, surface). These are acute changes. High acidity and high temperature are hazardous to organisms in that location during that period. Due to dilution, later pH and temperature come back to normal levels. | React in the presence of air to form Violent reaction. | Release flammable H_2. |

During the time when the ship was afloat and on fire, water that collected in the lower levels (due to firefighting and seawater leakage from damage to the hull) mixed in with the stored chemicals, creating a toxic cocktail. With the ship's unfortunate descent into the water, this mixture, carrying environmentally harmful chemicals including bioaccumulating organics, was released into the ocean. With the gradual introduction of plentiful seawater into the ship's deck, the chemical cocktail may have diluted away. However, the complex impact of the toxic cocktail which caused the accident sustained the fire and caused severe damage to marine animals and environment, clearly in three stages: formation of fire, sustainment of fire and peak of cocktail formation – needed to be assessed.

A 2D matrix of chemicals and its mixing effects was used as the base for the assessment as depicted in Table 3.2. The array of chemicals was repeatedly added into the x and y axis, and the mixing effect of sets of two chemicals i.e. (x_1,y_1), (x_2,y_3) or (x_3,y_1), and so on were evaluated on the respective cell with a colour code – green for no harmful reaction/no effect, orange for moderately harmful reaction and red for violent or explosive reactions.

Other than the toxic chemical mixture, evidence of other chemical materials in the cargo hold releasing to the ocean was visibly seen. A 562.4 MT stock of bright yellow sulphur, housed in the cargo hold, was suspected to be the source of the yellow substance floating near the ship on 2nd of June. This made it evident that the liquid cargo must have definitely dispersed within the ocean water. The spilled chemicals, such as sodium hydroxide, were expected to dilute and disperse along with the waves, minimizing the overall effect and limiting the severe consequence of aquatic toxicity closer to the ship's wreckage. Nevertheless, addition of highly alkaline chemicals like NaOH to water is exothermic, raising the water temperatures. There is a possibility that upon immediate dissolution with sea water the concentration of NaOH was higher in a singular location, causing harm to the marine life in the immediate vicinity by chemical burns as well as by the rise in temperature. Sea turtles, having a higher sensitivity to temperature changes, may have been affected to a greater degree [2]. As the total NaOH consignment on board was nearly 1127 MT in quantity the overall effect would have been severe.

The long-term effects of the spillage will no doubt affect the marine life and ecosystem for a while to come. The waste electrical and electronic equipment is one such concern as it will leach heavy metals such as cadmium, chromium, lead, mercury, nickel and thallium in the long term until they are salvaged from the wreckage. Other than the electrical and electronic waste, onboard containers carrying lead and molybdic oxide directly contributed to the increase in heavy metals of the surrounding waters of the shipwreck.

It should also be noted that the leaching of compounds such as urea, fertilizers and food items into the ocean may accelerate algal blooms due to high nutrient loading. If the nutrient concentration around the shipwreck remains high, toxic algal blooms may appear, thus unbalancing the ecosystem and feeding into the bioaccumulation of toxic chemicals.

The spill of powdered substances such as the aforementioned bright yellow sulphur along with urea, cement and quicklime affected the total suspended solids (TSS) encircling the shipwreck and thus increased the turbidity. A variety of other chemicals including an assortment of undefined organic and inorganic chemicals, HNS in perfumery products and paints were also released to the water when the ship sank. These heavy metals and other spilled chemicals can cause long-term harm to both marine life and human health through the bioaccumulation and bioconcentration that occurs along the food chain [3].

3.4.2 Plastic Nurdles – The Major Visible Sign of a Mass-scale Environmental Disaster

Another major cargo consignment that directly affected the marine life and beaches across the west coast of the island are the plastic nurdles consisting of linear low-density polyethylene (LLDPE), low-density polyethylene (LDPE) and high-density polyethylene (HDPE). The plastics that burned during the fire released complex organic pollutants such as BTEX (benzene, toluene, ethylbenzene, p-xylene), phthalates and polyaromatic hydrocarbons (PAHs) into the seawater. Those that did not fully burn within the fire were dislodged and fell into the ocean or sank with the ship. The sheer quantity of plastic released during this incident can veritably classify this as one of the worst ecological disasters faced by Sri Lanka. The plastic nurdles became somewhat of a salient identifier for the X-Press Pearl disaster as it was one of the first visible signs of an ecological catastrophe upon Sri Lanka's territorial waters. Sarakkuwa Beach was the most significantly affected by the washing ashore of burnt and unburnt plastic nurdles. A 30 cm thick covering of nurdles was seen after the first explosion and even to date these nurdles wash ashore and settle deep into the strata of the beach. Nurdles were identified to have settled around 1.2 m deep into the stratigraphy of the beach.

The monsoon season and prevailing currents transported the plastic spill over a larger area along the coastline. With this the western coastal belt of Sri Lanka and, to a certain extent, the southern and northern regions witnessed an unmanageable load of plastic polluting its shores and coastal ecosystems. This in turn affected the operating salterns in this region as microplastics as well as heavy metals mixed into the water which was sourced as a raw material for salt production. Another concern was the transport of containers with sea waves, allowing the nurdles to travel a further distance than expected. A container of LDPE nurdles was discovered in Galle, some 100 km or so away from the location of the wreck.

The pressing issue of microplastics has been of high concern in recent years and the X-Press Pearl disaster came to be during its height. The continued persistence of microplastics is one of the main issues being researched. Nurdles (sized < 5 mm), a primary microplastic, have the ability to further disintegrate into nanoplastics due to sunlight, wave action, chemical action, salinity and biofilm formation, thus making clean-up and removal nigh on impossible [4]. The burnt microplastics (comprised only of LDPE) released into the ocean were abundantly seen as a combination of molten and unmolten nurdles fused together to form large and small lumps. These burnt plastics are brittle and more prone to degradation, disintegrating to nanoplastic with greater effect than the unburnt. The plastic particles also act as vectors for potentially toxic elements (PTEs), carrying them further from the source of pollution [5].

The plastic nurdles thus contaminating the ocean and the shores of the island will directly affect marine life, human health, tourism and the local fishing industry [6]. As the colour and size of the nurdles provoke marine life to ingest it, mistaking some of the coloured nurdles to be plankton, these plastics which are burnt and chemically contaminated will accumulate along the food chain. This was seen in the fish samples analyzed by Marine Biological Resource Division of NARA, immediately after the disaster. Depending on their feeding behaviour, various species have died due to obstruction of the digestive tract.

Other than fish species, sea turtle carcasses were found on the beaches with plastic nurdles and burned plastic debris ingested through mouth. Though most of the plastics tend to float due to their density and thus make it to shore, a certain quantity can, and have, settled in the sea floor, after mixing with the sand particles. With time they will degrade to nanoplastics, ultimately negatively affecting the demersal organisms.

Other than the cargo, the shipwreck itself acts as a pollutant, due to the degradation of its structure. Paints and surface treatment agents as well as the corroded and heat-damaged metals of the ship will add to the pollution of the surrounding waters. With this disaster, over 8,000 MT of chemicals, 11,000 MT of plastics, 3,000 MT of metals and ~5,000 MT of various other cargo have been consumed in fire, thrown overboard and washed up onto the shore, dissolved in the ocean or lie far beneath on the seabed. Recovering this cargo that has fallen into the ocean is an arduous task, as the ocean currents and continuous movements may pose a hurdle in correctly identifying the exact locations of each cargo consignment. As it is difficult to exactly estimate the quantity of cargo burnt and the quantity thrown overboard it will be even more difficult to ensure complete removal of waste generated due to this incident. It is also of utmost importance that the waste be carefully handled and managed to avoid any further environmental and health impacts.

REMARKS

The MV X-Press Pearl incident involved fire, explosions and the destruction of the ship and its cargo, which included hazardous substances that contributed to the fire and explosions and had negative impacts on the marine ecosystem and the livelihoods of the local fishing community. In this chapter authors have critically examined the hazardous cargo on the X-Press Pearl and the complex chemical effects it had on the marine environment and local community.

REFERENCES

[1] Datta, A. "A review of fire hazards and fire protection concerns of commercial fuel cycle facilities in the United States." Nuclear engineering and design 125.3 (1991): 315–323.

[2] Aliko, V., et al. "Get rid of marine pollution: bioremediation an innovative, attractive, and successful cleaning strategy." Sustainability 14.18 (2022): 11784.

[3] Hawkes, L. A., et al. "Investigating the potential impacts of climate change on a marine turtle population." Global Change Biology 13.5 (2007): 923–932.

[4] Koelmans, A.A., Besseling, E., Shim, W.J. Nanoplastics in the Aquatic Environment. Critical Review. In: Bergmann, M., Gutow, L., Klages, M. (eds) *Marine Anthropogenic Litter*. Springer (2015), pp 325–340.
[5] Rochman, C. M., et al. "Ingested plastic transfers hazardous chemicals to fish and induces hepatic stress." Scientific Reports 3.1 (2013): 1–7.
[6] Andrady, A. L. "Microplastics in the marine environment." Marine Pollution Bulletin 62.8 (2011): 1596–1605.

4 The Fate of the Pearl

Deshai Botheju

4.1 INTRODUCTION

X-Press Pearl, a container ship carrying approximately 1500 containers, arrived at Colombo in the late hours of the 19th of May 2021. Awaiting entry to the port, it then anchored outside the harbor about 9.5 NM (about 18 km) away from the shoreline. An onboard fire incident was reported during late hours the following day, to which the Colombo port's fire service vessels promptly responded. Despite swift firefighting and early success, the fire was not able to be completely extinguished. In spite of strenuous firefighting efforts, mostly using fire cannons on Ports Authority tugboats and the SL Navy ships, the fire began to escalate after several days, partly supported by monsoon winds and dry weather conditions. On the morning of the 25th of May, that is, 5 days after fire initiation, a violent explosion occurred onboard the vessel. At that point, the vessel crew was evacuated with two crew members suffering minor injuries in the process. Eight containers were initially reported to be thrown overboard as a result of the explosion. However, later evidence (based on debris washed ashore) pointed towards many more cargo being dislodged from the ship and ended up in the sea, as a result of this initial explosion or due to the ongoing fire. Within 24 hours after this first explosion, debris started to be found ashore along the coastline from Colombo to Negombo. The initial debris washed ashore included a whole shipping container deformed clearly due to an explosion, many sacks of low-density polyethylene beads (nurdles/pellets), some rolls of yarn, food items, etc. (Botheju, 2021). Soon the shoreline was covered with polymer beads and a gummy textured dark substance which was suspected to be self-polymerized resins.

By the 26th of May, despite continued firefighting efforts, the fire intensified and escalated to engulf the whole ship from the forward section to the aft. The firefighting vessels had to retreat due to the intense heat. By the 27th of May, the fire had largely consumed much of the flammable substances aboard the ship and the intensity of the fire subsided. The firefighting efforts were resumed with fire cannons, as well as with air drops of dry chemical packs. By this time, progressively more plastic beads were being found on the beaches, and they were found even further south from Colombo, in areas like Hikkaduwa and Galle, along the western coastline of Sri Lanka. Also, many carcasses of dead fish and other sea creatures were found on the beaches, including a number of sea turtles. By the 27th of May, the fire had been reduced to a pool fire

The Fate of the Pearl

FIGURE 4.1 Remaining pool fire on the ship, reaching the end of the fire episode. Photo credit: SL Airforce Media.

deep down in the ship's interior (possibly in the fuel tanks; see Figure 4.1). By the 28th of May, the fire had been mostly extinguished and only smoke plumes were visible. On the same day, at around 2300 hours' time, a second explosion was reported on the ship, but no additional intensification of fire was noticed. By the 1st of June 2021, the fire had been completely extinguished, and the ship was sufficiently cool, allowing salvors to go aboard the ship and to inspect it. On the 2nd of June, attempts were made to tow the ship towards deep sea. Already at the beginning of that attempt, a large section of the hull was disintegrated, and the engine room area was observed to be flooded. Shortly after that the ship started to sink. By the end of the day, the aft section of ship had been settled on the sea floor at a depth of about 21 m, while the forward section was still slowly sinking. The middle section had been completely submerged in water, while a pile of container rubble was still visible above the water level. The living quarter structure located at the aft continued to be visible above the sea surface due to its height. It was evident that the ship was now a wreck in the SL waters that needed to be further dealt with.

4.2 INITIATION AND PROPAGATION OF FIRE

On the 11th of May 2021, about 8 days prior to the arrival at Colombo, an acid leak had been observed on a container carrying 25 tonnes of nitric acid, when the ship was at Hamad port in Qatar. Attempts of discharging this leaking container at Hamad Port were rejected by the port authorities. A similar rejection happened when the ship was at Hazira, India, 4 days later (i.e. on the 15th of May). The continued leaking of acid is considered to be the precursor to the subsequent fire that occurred 5 days later at Colombo.

There could be several exothermic chemical reactions responsible for the initiation of this primary fire. The nitric acid itself does not burn, but as a strong oxidizer it can violently react with many organic substances as well as with some metals (see Reaction 4.1).

$$C + HNO_3 \dashrightarrow CO_2 + NO_2 + NO + H_2O \quad \text{(Reaction 4.1; reaction is not balanced)}$$

This is a highly exothermic reaction, and the resulting heat further accelerates the reaction rate. If the heat is not removed (e.g. by firefighting using a lot of water) the reaction can become a runaway reaction or, in other words, a "thermal explosion". Mere use of a small amount of water can also worsen the situation due to the exothermic water dissociation of the concentrated acid (see Reaction 4.2).

$$HNO_3 + H_2O \dashrightarrow H_3O(+) + NO_3(-) \quad \text{(Reaction 4.2)}$$

Generation of nitrogen dioxide (NO_2) was evident during the first day of the fire initiation (i.e. the 20th /21st of May), as reddish-brown smoke was emanating from the ship. NO_2 is a toxic gas that causes severe air pollution, although it is not flammable. Meanwhile, nitric oxide (NO) is a colorless, flammable gas.

The fire itself can also thermally decompose HNO_3 to generate NO_2 (see Reaction 4.3). The releasing oxygen (O_2) may provide oxygen for further combustion, even if the location is not well ventilated (e.g. deep down in a cargo hold, or inside a container).

$$HNO_3 \dashrightarrow NO_2 + O_2 + H_2O \quad \text{(Reaction 4.3)}$$

As shown in Figure 4.2, there was only one nitric acid container stored in cargo hold no. 2. This is likely to be the place where the acid leak started. Some of the adjacent containers were filled with epoxy resin and high-density polyethylene (HDPE). Also, the nearby containers had vinyl acetate, methanol and lithium-ion batteries; all are flammable and potentially explosive substances. In the cargo hold below the deck level, more methanol had been stored.

The epoxy resin itself is flammable, and if mixed with nitric acid, it will decompose exothermically, giving off more heat and starting a fire.

The reaction between nitric acid and methanol produces methyl nitrate (see Reaction 4.4), which is a strong explosive and highly volatile too.

$$CH_3OH + HNO_3 \dashrightarrow CH_3NO_3 + H_2O \quad \text{(Reaction 4.4)}$$

Lithium in lithium-ion batteries may react with nitric acid to produce lithium nitrate and NO. The reaction is exothermic (see Reaction 4.5).

$$Li + HNO_3 \dashrightarrow LiNO_3 + NO + H_2O \quad \text{(Reaction 4.5)}$$

Further, Li-ion batteries can rapidly catch fire, and once started it is very difficult to extinguish such fires. Even if it is managed to extinguish initially, the fire could re-ignite once the firefighting/cooling is terminated.

The ship's cargo had several other flammable and combustible materials stored at different places. Some of them are mentioned below:

Vinyl acetate: Flammable substance

The Fate of the Pearl

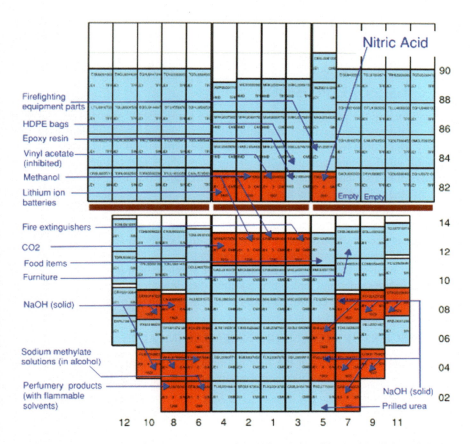

FIGURE 4.2 Overview of the contents stored near the nitric acid container onboard XPP.

Aluminum smelting byproducts: These metal processing byproducts can react with seawater to generate highly flammable and toxic gases such as hydrogen, acetylene, methane, ammonia and phosphine. Hwang et al. (2006) experimentally observed vigorous H_2 release when some types of aluminum smelting by-products are mixed with water. These wastes are quite complicated and diverse in nature. Therefore, the exact chemical reactions occurring when mixed with water can depend on the specific type of aluminum waste. This means in the case of XPP, even the fire water used in firefighting could have led to further escalation of the fire, had such water reached the locations where these aluminum by-product wastes were stored.

Sodium methylate: Reacts with water-generating methanol (which is flammable/potentially explosive itself).

HDPE and LDPE bags: These are combustible polymer beads.

The ship's cargo contained many more nonspecified organic liquids and various other combustible materials such as furniture, food items and automobile parts including tires.

Further, ship had approx. 620 MT of heavy fuel oil and less than 50 MT of marine diesel fuel for use in the ship's engines. A significant quantity of lube oils and also some quantities of fuel sludge were present.

4.3 CAUSES OF EXPLOSIONS

As mentioned in the previous section, there were many agents onboard the ship that could have created explosions. In addition to nitric acid's highly violent oxidation property (that could lead to thermal runaway–type explosions), other agents such as methanol, vinyl acetate, aluminum remelting by-products and sodium methylate (in alcohol) could also have contributed to the occurrence of the first explosion. Among them, methanol and aluminum remelting by-products can be identified as direct contenders. Note also that due to the very high congestion on the ship (which was packed with containers), a gas cloud explosion could have occurred when a flammable gas is accumulated within a partially or fully confined space, such as below deck cargo holds, or even within a single container.

The second explosion reported on the 28th of May can be due to the extinguished pool fires in the ship's fuel tanks. It is a known fact that an extinguished or partially extinguished ventilation-controlled pool fire can lead to gas explosions.

4.4 PRECURSORS TO ENVIRONMENTAL IMPACTS

The environmental impacts of the XPP accident are discussed in detail in the other chapters of this book. The following description provides only a background overview of this subject.

As per the cargo load list that appeared on open media, XPP contained a large amount of different substances including a number of environmentally hazardous organic substances, such as acetyl cedrane used in perfumery products that is categorized as a substance posing acute and chronic toxicity for marine life. However, it is likely that at least a part of such organic substances is disintegrated and burnt in the fire event, thereby reducing the likelihood of releasing them into the marine environment. According to visual evidence, the fire temperature should have reached quite high values, as evident from the sagged crane boom on the ship (see Figure 4.3; note, however, that the strength of steel drastically reduces at temperatures well below its melting temperature; thereby sagging could occur due to the own weight of the boom (Franssen & Real, 2012). Under high-temperature conditions, many organic substances completely break down into relatively harmless basic ingredients. However, any cargo stored close to the ship bottom could have been shielded from the fire heat due to submergence in accumulated firewater (that filled in the ship bottom).

The ship was reported to have close to 700 MT of fuels and lube oils. It can reasonably be expected that a part of this would have also been burnt in the fire. However, some unburnt fuel amounts might still be trapped in the fuel tanks of the ship (which are either not ruptured open during the fire, or the pool fires at those deep-down locations in the ship were ventilation controlled and also possibly extinguished by firefighting). In that case, these oils (if remaining) can be leaking into the sea at a later time (e.g. during any salvage operation).

The Fate of the Pearl

FIGURE 4.3 Sagged steel crane of X-Press Pearl. Photo credit: Ishara Kodikara/AFP.

While the ship was still afloat during the fire, the bottom sections were evidently filled with a large amount of water (i.e. the water used in firefighting, and the water leaked through the damaged hull) contaminated with various toxic and environmentally harmful chemical ingredients, potentially including many persistent and bioaccumulative organic substances. As the ship is sunk now, this cocktail of liquid chemicals (mixed with water) must have already been released into the sea. The observations made on the 2nd of June 2021 indicated a yellowish substance floating near the ship, suspected to be originating from a sulfur stock transported on the ship. This is indicative of the release of any liquid contents or other unburnt substances collected at the bottom sections of the ship (including the release of any acidic or alkaline chemicals such as nitric acid and sodium hydroxide which were present onboard the ship). Nevertheless, the actual extent of the impact zone of that chemical cocktail release was never accurately estimated, as the wave action and the sea currents should lead to dispersion and dilution of harmful chemical concentrations after a certain distance limit.

The ship's cargo load list also indicates a large amount of scrap metals and lead ingots. These heavy metals can cause a considerable marine pollution and cause long-term harm to human health (Mitra et al., 2022), as many of these metals can be slowly released (due to chemical and biochemical reactions) to the seawater and then bioaccumulate and bioconcentrate along the food chains, eventually entering the human body via seafood consumption.

The fire has caused a substantial air pollution near the coastline of Sri Lanka. A large part of the fire could have propagated at relatively lower temperatures as evident from the prolonged black smoke emissions (even though there were some 1–2 days of high-intensity fire duration). Such incomplete burning of plastics (specially the ones containing chlorine, such as PVC) can release various harmful fumes including polychlorinated dibenzo-p-dioxins and polychlorinated dibenzofurans when subjected to uncontrolled open fires (Zhang et al., 2015). These dioxins and furans are among some of the very harmful and persistent pollutants ever created due to manmade activities. The smoke plume generated from the fire carried directly towards the shoreline, due to the prevailing weather conditions. Therefore, the pollutants are basically scattered and settled within the island of Sri Lanka. The smoke plume should have had many other toxic gases and components such as NO_2, NO, PH_3, CO, NH_3, VOCs, etc. as well as particulate matter.

Large-scale marine and beach pollution has happened due to the scattering of microplastics in the form of polymer beads. It was reported that some dead fish were found with these polymer beads in their intestines. Also, a large amount of partially polymerized masses had been found in the sea as well as on the beach. However, apart from some dead fish and several sea turtle carcasses, no mass-scale fish death has been reported. The observed few deaths could have been mostly caused by the reduced dissolved oxygen levels in the water. Increased temperatures in the immediate locality of the ship would cause reduction in oxygen levels. However, large temperature increases cannot happen much farther away from the burning ship as the mixing action of waves and currents quickly reduce the sea water temperature to ambient condition. In addition, chemical toxicity effects and slightly changed pH levels could also have contributed to some destruction of sea life near the location of the shipwreck. The large consignment of solid caustic (sodium hydroxide – NaOH) on XPP might have had a role to play in explaining some fish and marine animal deaths reported in later days of the accident (after sinking). In this regard, the slow dissolving of large amounts of solid caustic is a more prominent cause for concern compared to the single container of nitric acid.

4.5 IMPORTANT ASPECTS REGARDING THE SHIPWRECK REMOVAL

The X-Press Pearl shipwreck continues to be a source of marine pollution as it degrades further at its current location and as it is still likely to contain a considerable amount of hazardous chemical substances and other potential persistent pollutants. In addition, it is plausible that some oils are still trapped in the wreck. Therefore, it is highly recommended that the shipwreck be removed from its current location as soon as possible. This will nevertheless be a complex and expensive engineering undertaking. The process needs to be executed by a well-experienced and qualified salvage vendor having access to heavy machinery needed for the work. In addition, a separate group of experts comprising chemical hazard management, marine pollution mitigation, oceanography and marine biology must observe and advise during the whole process from planning to execution. The duration may be about 3–6 months (if not longer).

The shipwreck may be removed using one of the two approaches:

(A) It might be possible to lighten up the ship and have the hull repaired, refloated and towed away to a salvage yard or dry dock for dismantling (refloating strategy)

(B) The ship may be removed piece by piece at the location (piecemeal strategy)

If it is found to be technically feasible after a detailed engineering evaluation, the approach (A) above is recommended as the preferred cause of action as it will lead to the minimum escalation of the pollution situation. The piecemeal approach can lead to additional pollution due to ship breakage and added disturbance. However, if detailed engineering assessments would prove that approach (A) is infeasible, then approach (B) may be the only viable option (considering leaving of shipwreck at its current location is not an acceptable option).

The whole process requires meticulous planning, and the execution must be precise. The following key steps need particular attention.

4.5.1 Underwater Survey

The shipwreck and the surrounding area must be subjected to an extensive underwater survey to identify the location and the nature of the underwater debris field. It is understood that there are many items of debris (including damaged containers) spread over a certain perimeter around the current location of the shipwreck. The locations of all these items must be identified and accurately mapped so that each and every item can be removed later. However, these underwater objects can move along the sea floor due to the action of sea currents and waves, especially under rough sea conditions. Therefore, underwater debris maps will not be held accurate over an extended period of time. That means the debris removal process must be completed within a reasonable duration after the survey. The actual rate of debris movement is hard to predict without hydrodynamic data and the actual weight and locations of the debris.

Preliminary survey and mapping can be carried out using an underwater survey ship equipped with sonar. However, at later stages more accurate visual observations must be made either using underwater ROVs or using human divers. Such visual observation is a necessity prior to conducting the actual debris removal. The identification of the damages to the X-Press Pearl's hull will also require precise underwater observations using human divers or ROVs.

4.5.2 Contingency Planning

As mentioned before, the shipwreck removal process will disrupt the settled status of the ship and can lead to an escalating secondary pollution event. A comprehensive contingency plan must be prepared to address this risk. Further details on this aspect are given in other subsections below and also in Chapter 17 (*Lessons from the Pearl*).

4.5.3 Debris Removal

Removal of individual containers lying on the seabed can be carried out using a vessel equipped with a light crane, or using a sheerleg barge. The sea depth at the area is known to be shallow (20–30 m), and therefore, this operation will be feasible. However, assistance from skilled divers may be needed to attach cables to the debris items/containers. If any lumps of free debris are discovered (i.e. debris not contained with the enclosures/containers), then special techniques will be needed to remove such items (and can be quite complicated). The debris-collecting vessel must have secured cargo holds to store hazardous substances. Any events of further spills or fire events associated with such debris must be expected and prepared for, and suitable safeguards must be available (such as suitable firefighting means). Certain hazardous materials can have pyrophoric characteristics, which means they could undergo spontaneous combustion when taken out of the seabed and become dry.

4.5.4 Removal of Remaining Cargo

Removal of remaining cargo sitting on the XPP shipwreck will be an extensive task in this salvage operation. Many of the remaining containers may have to be removed one by one. This task will require divers' assistance (or robotic ROVs) to attach lifting cables. Some of the debris may be directly removed using *clamshell grapple* type dredging grabs. It is expected that at least one or two *sheerleg barges* will be needed on-site to carry out required lifting operations effectively and efficiently. Once again, the lifted debris must be secured within a cargo hold of a vessel (instead of a flatbed barge).

4.5.5 Oil Removal

Even though it is quite likely that large parts of the oils onboard XPP (i.e. heavy fuel oil, marine gas oil, lube oils, and oil sludge) must have been burnt or evaporated due to the intense heat of the fire, there is a possibility of some trapped oils still contained within some of the tanks in the ship bottom. This must be thoroughly studied using underwater observations, and if such oils are detected, then they must be removed using pumping-out techniques before any debris removal operation begins. Reducing the weights on the ship will destabilize it again and can escalate into an oil spill, if such oils are not removed first.

4.5.6 Hull Repair

If the intention is to refloat the ship (after lightening it up by the removal of debris load), the damaged hull needs to be repaired temporarily. If the hull damage is not that extensive, underwater hull repairing is feasible. This requires skilled divers with hull repair kits. Metal panels lined with rubber sheeting can be fixed to the existing hull using special piercing techniques.

4.5.7 Pumping out of Water

After removing debris load, and after completing the required hull repair job, the water filled in the ship needs to be pumped out using large pumps while trying to refloat the ship.

4.5.8 Refloating

The ship might naturally refloat after debris removal, potential hull repair, and pumping out of water. If that is realizable, it can be the most favorable way to refloat the ship. The partially sunk, uncapsized state of the ship is an advantage in this case.

In contrast, mechanically lifting a ship of this size will be a formidable challenge due to the limited capacities of the sheerleg barges (cranes). Possibility of any underwater jack-up technique may be evaluated as an alternative.

If the damage to the ship hull is extensive (i.e. unrepairable in situ) and it is not possible for the ship to be refloated, then the other option will be to break it into pieces and lift it piece by piece (piecemeal approach). This operation can lead to more in situ pollution. Therefore, such piecemeal approach must only be used if all other avenues are proved to be infeasible.

4.5.9 Tow Away

If the ship is refloated, then it must be inspected properly, secured any remaining cargo or hazardous materials, and carried out any additional repairs, before being carefully towed towards a suitable salvage yard where it can be broken down and recycled. Due to the condition of the ship, she may not be able to face a long voyage again. Therefore, the closest possible salvage yard or dry dock must be used to process the ship.

4.6 CONCLUDING REMARKS

The immediate cause for the XPP disaster is linked to its hazardous consignment containing a number of flammable and explosive substances; some of them may have undergone exothermic reactions triggered by a leaking nitric acid container, releasing heat and causing a recalcitrant fire. Potential procedural, management, and incident control failures have apparently contributed to the eventual catastrophe. Chapter 17 of this book (*Lessons from the Pearl*) further discusses these aspects.

The XPP fire and explosion disaster had evidently caused a large environmental, human health, and socioeconomic damage to Sri Lanka as well as to the nearby countries in the region and to the marine ecosystems in international waters. The consequences will be spread over a long period of time. Strenuous recovery efforts will be needed in order to even partially restore the situation.

The shipwreck must be safely removed from its current location as soon as possible. Continued presence of it will lead to long-term marine pollution and potential health hazards to Sri Lankan as well as regional populations. A properly managed engineering operation executed by a skilled salvage vendor with access to suitable heavy machinery and expert support can safely get the shipwreck removed with minimum escalation risk.

ACKNOWLEDGMENT

Some of the contents in this chapter reflect the viewpoints presented in Botheju, D. (2021), courtesy of "ft.lk".

REFERENCES

Botheju, D. (2021). X-Press Pearl: What went wrong? The scientific basis. Available online: www.ft.lk/opinion/X-Press-Pearl-What-went-wrong-The-scientific-basis/14-719287

Franssen, J. and Real, P. (2012). Fire design of steel structures (Eurocode 1). European Convention for Construction Steelwork. Wiley online library: https://onlinelibrary.wiley.com/doi/book/10.1002/9783433601570

Hwang, J. Y., Huang, X., and Xu, Z. (2006). Recovery of metals from aluminum dross and saltcake. Journal of Minerals & Materials Characterization & Engineering, Vol. 5. No. 1, pp. 47–62.

Mitra, S., Chakraborty, A. J., Tareq, A., M., Emran, T. B., Nainu, F., Khusro, A., Idris, A. M., Khandaker, M. U., Osman, H., Alhumaydhi, F. A., and Simal-Gandara, J. (2022). Impact of heavy metals on the environment and human health: Novel therapeutic insights to counter the toxicity. Journal of King Saud University – Science, Vol. 34, No. 3.

Zhang, M., Buekens, A., Jiang, X., and Li, X. (2015). Dioxins and polyvinylchloride in combustion and fires. Waste Management & Research. Vol. 33, No. 7, pp. 630–643.

5 Onshore Responses during the X-Press Pearl Disaster

K.P.G.K.P. Guruge, T.S. Ranasinghe, I.P. Amaranayake, K.R. Gamage, T. Maes, and P.B.T.P. Kumara

ABBREVIATIONS AND ACRONYMS

NM	Nautical Miles
MEPA	Marine Environment Protection Authority
NGOs	Nongovernment Organizations
NARA	National Aquatic Resources Research and Development Agency
SLN	Sri Lanka Navy
SLCG	Sri Lanka Coast Guard
SLPA	Sri Lanka Ports Authority
NOSCOPS	National Oil Spill Contingency Plan
DMC	Disaster Management Centre
CPC	Ceylon Petroleum Cooperation
CPSTL	Ceylon Petroleum Storage Terminal Limited
CC&CRMD	Coast Conservation and Coastal Resources Management Department
DFAR	Department of Fisheries and Aquatic Resources
SL Army	Sri Lanka Army
SLAF	Sri Lanka Air Force
CEA	Central Environment Authority
ICP	Incident Command Post
OSD	Oil Spill Dispersants
CCC	Central Coordination Centre
MoD	Ministry of Defence
SAS	Shoreline Assessment Survey
SPIP	Shoreline Planning Information Process
ITOPF	International Tanker Owners Pollution Federation Limited
DOR	Daily Operational Report

DOI: 10.1201/9781003314301-5

OSRL	Oil Spill Response Limited
SLCDF	Sri Lanka Civil Defence Force
NBRO	National Building Research Organization
PID	Photoionization Detectors
CBRN	Chemical Biological Radio Nuclear
CEFAS	Centre for Environment Fisheries and Aquaculture Science
SoPs	Schemes of Procedures

5.1 INTRODUCTION

The MV X-Press Pearl, a Singaporean-flagged container cargo vessel sailing from Port Hazira, India, to the Port of Colombo, was confronted with an onboard fume emission at 9.5 nautical miles (NM) from the Colombo Outer Harbour area on 20 May 2021 (Partow, Lacroix, Le Floch and Alcaro, 2021; Pattiarachchi and Wijeratne, 2021), signaling the onset of the disaster. There were 1486 containers on board at the time of the incident; 81 containers were filled with hazardous materials (i.e., 25 tonnes of nitric acid, caustic soda, methanol) and 87 containers with different plastic pellets (i.e., high-density and low-density polyethylene, etc.). In addition, the ship contained 348 tonnes of bunker fuel oil (Partow, Lacroix, Le Floch and Alcaro, 2021; Withanage, 2021).

The ship reported a fire on deck on 21 May 2021. Environmental experts made predictions about the anticipated air and marine water pollution due to the burning of chemicals and hazardous materials. A more in-depth assessment revealed a potential threat posed by the dangerous cargo and bunker fuel onboard to cause irreparable loss and damage to the marine and coastal environment, including impacts to the health and safety of the people nearby. In addition, the disaster would negatively impact socioeconomic activities if proper remedial actions were not taken urgently. Therefore, firefighting responses were put in place to control the fire, and the situation was managed until 24 May 2021 (Perera, 2021; Partow, Lacroix, Le Floch and Alcaro, 2021). The fire spread across the entire ship, leading to explosions on 25 May 2021 (Partow Lacroix, Le Floch and Alcaro, 2021; Pattiarachchi and Wijeratne, 2021).

Hazardous materials and bunker oil that were on board spilled out into the marine water during the ship's explosion on 25 May 2021, creating the worst disaster in Sri Lanka's maritime history, mainly caused by the 87 containers of plastic pellets on board, of which some fell out of the ship during the explosion, causing a massive deposition of plastic pellets on the beaches in the North-Western, Western, and Southern provinces of Sri Lanka (Partow, Lacroix, Le Floch and Alcaro, 2021; Saplakoglu, 2021). The pellets were made from low-density polyethylene plastic nurdles (Withanage, 2021). The plastic nurdle spill that occurred due to the MV-X-Press Pearl maritime disaster was estimated to be the single-largest release of plastic nurdles and chemical pollution into the ocean (UNEP, 2022). The burnt/bleached carapace and the disintegrated shells and skin of some dead turtles indicated that they had been in contact with the nitric acid and/or caustic soda, indicating that a chemical spill from the ship occurred (Partow Lacroix, Le Floch and Alcaro, 2021). Therefore, immediate disaster responses were applied to restore the affected regions.

The disaster response is defined as "Efforts to reduce the risks posed by an emergency by protecting people, the environment, and property, as well as efforts to return to pre-emergency conditions" (HELCOM, 2002). Responses can be categorized as predisaster and postdisaster (Srinivas, 2021). Predisaster responses aim for prevention, mitigation, and preparedness activities to reduce the loss of life and property (Harrison and Johnson, 2016). Response, recovery, and reconstruction are postdisaster activities to achieve early recovery and rehabilitation of the affected resources and regions (Sriniwas, 2021; He and Zhuang, 2019).

We defined the loud explosions and fire engulfing the whole ship on 25 May 2021, releasing large amounts of hazardous materials into Sri Lanka's marine water, as the "Disaster". The applied responses from 20 May 2021 to the time of the ship explosion (around noon) on 25 May 2021 (Partow, Lacroix, Le Floch and Alcaro, 2021) were defined as the "pre-disaster responses". The applied responses from the time of the explosion up to date were referred to as the "post-disaster responses".

The Marine Environment Protection Authority (MEPA) is the responsible government organization in charge of preventing, controlling, and managing marine pollution within the Sri Lankan Marine Environment (MEPA, 2021). Therefore, MEPA played a leading role in the disaster responses to prevent, control, and manage marine pollution from the MV X-Press Pearl disaster in collaboration with different government and nongovernment organizations (NGOs). Together they applied various pre- and postdisaster responses to control foreseen and observed marine pollution from the aforesaid incident within the Sri Lankan marine waters.

This chapter describes the onshore responses to the worst maritime catastrophe in Sri Lanka, the MV X-Press Pearl ship accident. It lists the most important proactive actions to be taken in response to such maritime disasters, determining the fate of sensitive ecosystems in the area.

5.2 PREDISASTER RESPONSES

5.2.1 JOINT VISUAL INSPECTIONS AROUND THE VESSEL

The main focus in the predisaster phase is the identification of potential hazards. Therefore, information and data collection at the incident site is essential to predict potential risks. The direct observation of the site is the most precise technique in this regard (Skavdal, 2003). The first joint visual observation was done on 21 May 2021 by the officials of MEPA, National Aquatic Resources Research and Development Agency (NARA), Sri Lanka Navy (SLN), Sri Lanka Coast Guard (SLCG), and Sri Lanka Ports Authority (SLPA). Despite the rough sea condition, a site inspection was done around the vessel, and photographs were taken to study the present situation of the ship. As per the team's observation, they noticed a fire in two containers. At that time, there was no high risk of an oil spill. The team observed that the other containers were not at risk of catching fire at that time, and continuous firefighting was carried out by the SLPA tug (Figure 5.1). SLN and SLCG ships remained in the vicinity to assist in further observations.

FIGURE 5.1 Firefighting responses to control the fire inside the ship.

MEPA and Sri Lankan Authorities conducted several joint visual inspections around the vessel to ascertain the situation during and afterward.

5.2.2 Implementation of NOSCOP

Small explosions were recorded in the containers causing a fire by 22 May 2021 (Partow, Lacroix, Le Floch and Alcaro, 2021), signaling the likelihood of a potential oil spill from the 348 tonnes of stored bunker fuel oil. As the state agency responsible for protecting the marine environment on the island, MEPA took immediate action to implement the National Oil Spill Contingency Plan (NOSCOP) by 22 May 2021, considering the expected oil spill as the worst-case scenario of the MV X-Press Pearl ship accident.

Three tiers are classified in the NOSCOP for oil spill risks and the required responses, taking into account the extent of the oil spill and the proximity to a response center (NOSCOP, 2005).

i) Tier 1 = A relatively small oil spill (up to 50 tonnes) is categorized as Tier 1. It requires local responses; therefore, the local oil pollution emergency plan will be applied using local resources and capabilities.
ii) Tier 2 = A medium spill of between 50 and 100 tonnes of oil, which requires the implementation of NOSCOP. It operates with the aid of resources from different sources (i.e., government and industry resources).
iii) Tier 3 = A large oil spill with above 100 tonnes of oil. These spills require international assistance, and NOSCOP has to be implemented with the support of respective authorities/international organizations.

MEPA made its disaster preparedness plan for the Tier 2 oil spill responses following the given guidelines in the NOSCOP. Planning is a key factor in preparing for and responding

Onshore Responses during the X-Press Pearl Disaster

to a disaster (Brown, 2018). MEPA appointed an Incident Management Team (IMT) following the guidelines in the NOSCOP to make comprehensive plans for mitigation, preparedness, response, and recovery for the foreseen oil spill from the MV X-Press Pearl ship burning. The IMT consisted of qualified members from the following institutes:

i) Disaster Management Centre (DMC)
ii) Ceylon Petroleum Cooperation (CPC)
iii) Ceylon Petroleum Storage Terminal Limited (CPSTL)
iv) National Aquatic Resources Research and Development Agency (NARA)
v) Sri Lanka Ports Authority (SLPA)
vi) Coast Conservation and Coastal Resources Management Department (CC&CRMD)
vii) Department of Fisheries and Aquatic Resources (DFAR)
viii) Sri Lanka Coast Guard (SLCG)
ix) Sri Lanka Army (SL Army)
x) Sri Lanka Navy (SLN)
xi) Sri Lanka Air Force (SLAF)
xii) Central Environment Authority (CEA)
xiii) Sri Lanka Police
xiv) Department of Wildlife and Conservation (DWLC)
xv) Department of Meteorology
xvi) Merchant Shipping Secretariat

The key responsibilities of IMT for managing the foreseen oil spill are:

i) to identify sensitive ecosystems nearby and prioritize the expected adverse effects on the environment and humans,
ii) to make models for predicting the distribution of oil spill,
iii) to implement the NOSCOP in the Incident Action Plan (IAP),
iv) to supply required tools and equipment for the oil spill responses in the affected sites, and
v) to mobilize the workforce or technical specialists in affected locations considering the evolution of the oil spill with time.

Several IMT meetings were conducted at the MEPA head office to prepare (IAP) as their major task during the disaster preparedness phase.

5.2.2.1 Preparation of IAP

IMT prepared a list of available equipment in their institutes and equipment was kept on standby mode to manage the expected oil spill. Another list of currently unavailable but required equipment for a wider disaster was prepared. They made several assumptions to make IAP based on the South-West monsoon wind and current patterns, availability of resources, and the oil slick movement models:

i) oil will flow in the South-West direction from the location of the burning ship;

ii) limited sea responses could be applied due to rough sea conditions; therefore, it is essential to prepare for onshore responses;
iii) oil will reach the coast after the emulsification, and it will take 4–5 hours to reach the coast from the time of the spill; and
iv) oil will reach the Western coast of Sri Lanka, and it will spread on the beaches from Negombo to Dickovita during the initial phase of the spill.

In the next stage, they set objectives and strategies to prepare the IAP for the oil spill responses, considering available resources and the above assumptions.

The objectives of the IAP were:

i) to monitor the movement of oil slick,
ii) to disperse as much oil offshore as possible using an oil dispersant to create oil patches, minimizing the oil impact on the sensitive ecosystems in coastal waters,
iii) to contain and recover oil slicks using oil boom and skimmers if weather and sea conditions permit, and
iv) to get ready for the onshore responses (i.e., beach cleaning) as limited offshore responses could be applied (i.e., booms, skimmers) due to rough sea conditions.

IMT identified tasks for the responsible authorities/institutes to get ready to manage the predicted massive oil spill. They used six strategies: monitoring, dispersing oil slicks, containment and recovery, beach cleaning, waste management, and community awareness.

5.2.2.1.1 *Monitoring*

To make the best decisions on oil spill responses, decision-makers and respondents need precise information about the oil spill (i.e., the extent of the oil spill, the distribution pattern of the oil spill, effects of the oil spill, effectiveness of the applied responses, identification of the most vulnerable marine ecosystems, etc.) (Australian Maritime Safety Authority, 2020). Air, marine, and coastal surveillance is an essential requirement in oil spill responses to monitor the oil spill's present status and selecting offshore or onshore responses. SLAF was identified to do the air surveillance during the foreseen disaster period. SLN and SLCG took the responsibility to conduct coastal and marine surveillance, respectively. Moreover, since the ocean current patterns and wind affect the oil slick movement on the seawater surface, predictions about the oil slick movement are one of the most crucial factors for deciding the most appropriate responses (Shen, Perrie and Wu, 2019; Kim, Yang, Oh and Ouchi, 2014). The responsibility of the Meteorological Department of Sri Lanka and NARA in the NOSCOP was to provide oceanographic data during the expected oil spill (i.e., ocean current speed and direction) for predicting the oil slick movement (Table 5.1).

5.2.2.1.2 *Dispersing Oil Slick*

Dispersants are chemicals that can quickly remove large amounts of oil from the surface as they help the breakdown of the oil slick into smaller oil droplets. These

TABLE 5.1
The Initially Prepared IAP during the Predisaster Period to Combat the Foreseen Oil Spill

Strategy	Assigned Institute or Organization	Assigned Task
Monitoring	SLN and SLCG for **marine Surveillance**	• to use their boats in marine surveillance and provide information on the oil slick's size, thickness, and direction.
	SLAF for **air surveillance**	• to use their helicopters in air surveillance from Dickovita to Negombo Lagoon • to provide information on the size, thickness, color, and direction of the oil slick.
	SLN and SLCG for **coastal surveillance**	• to identify any oil patches in coastal waters from Dickovita to Negombo Lagoon
	SLCG, SLN, SLCG, SLAF, SL Army for Shoreline surveillance (Foot Patrol)	• to identify any oil patches in shoreline from Dickovita to Negombo Lagoon.
	Meteorological Department and NARA	• to provide oceanographic data to predict the oil slick movement.
Disperse oil slick	SLCG and SLPA	• to use vessel (in SLCG) and tugboat (in SLPA) to spray OSD • OSD will be provided by the MEPA.
Containment and recovery	SLPA, CPC, CPSTL SLN, SLCG SLN and MEPA	• to contain and recovery at deep sea using tug boat, booms, and skimmers • to contain and recover at coastal area using booms and skimmers.
Beach cleaning	SLN, SL Army, SLCG, SLAF and CC&CCRMD	• to conduct beach clean-up programs using available equipment (PPE has to be worn during the clean-up programs due to prevailing COVID-19 pandemic condition and hazardous chemicals/materials).
Waste management	CEA	• to coordinate with relevant local Authority and arrange temporary oily waste storage sites • to prepare waste management plan.
Community awareness	DFAR	• to inform fishing communities about the oil spill incident
	DMC and SL Police	• to create community awareness in the affected area and to avoid unnecessary gatherings • to disseminate correct information about the incident.

smaller oil droplets can easily mix with water, decreasing the impact of the oil slick (Response Techniques, 2022). It was decided to apply oil dispersants offshore, aligning with the second IAP objective of minimizing the impact of oil spills on sensitive coastal ecosystems. In case the MV X-Press Pearl ship created an oil spill, MEPA had to get ready to provide the oil spill dispersants (OSDs) for the SLN and SLCG to spray them from their vessels and boats (Table 5.1).

5.2.2.1.3 Containment and Recovery

Containment and recovery are the two major responses to control oil spills (Oil Program Center, 1999). These techniques aim for physical oil removal from the marine environment using booms and skimmers (ITOPF, 2022). Booms are floating physical barriers to stop the spreading of the oil spill. The boom is deployed from a rigid arm extended from the ship, making "U"- or "J"-shaped pockets, and oil can be collected into these pockets (Oil Program Center, 1999). Skimmers are used to remove floating oil, and they can be self-propelled. Sometimes they can be used from shore or operated from boats. There is a pump in a skimmer to transfer recovered oil and water for storage (ITOPF, 2022; Cumo, Gugliermetti and Guidi, 2007). SLPA, CPC, CPSTL, SLN, and SLCG took responsibility for containment and recovery operations in the open ocean using tugs, boats, booms, and skimmers. SLN and MEPA were identified for the containment and recovery operations in the coastal region (Table 5.1).

5.2.2.1.4 Beach Cleaning

If containment and recovery are not executed flawlessly in the open ocean, some of the spilled oil (i.e., tar balls) may enter the coastal area where ecologically sensitive ecosystems are located (i.e., lagoons, estuaries). Beach clean-ups following the oil spill have to occur to preserve the area from the oil spill damage, protecting the species' habitats and ensuring the area's aesthetic value. Therefore, the beach clean-up is considered a significant component of the oil spill response (Oil Program Center, 1999). The responsibility for beach clean-up programs was given to the SLN, SL Army, SLAF, SLCG, and CC&CRMD.

5.2.2.1.5 Waste Management

The proper removal of oil and oily debris to a temporary storage, treatment facility, or final disposal is essential in the oil spill responses. It is referred to as waste management (POSOW, 2016). The CEA was the responsible government organization for waste management. They were asked to coordinate with relevant local authorities to prepare a waste management plan and arrange a temporary oily waste storage site to manage the generated oily waste from the expected oil spill (Table 5.1).

5.2.2.1.6 Community Awareness

Community awareness is one of the innovative tools in disaster management, and it improves the community's resilience to the impacts of the disaster (Joyce, Kunguma, and Andries, 2013). If the burning MV X-Press Pearl ship would create an oil spill, DFAR was responsible for informing the fishing community in the affected area about the disaster. Moreover, DMC and SL Police were identified to disseminate the correct information about the incident and control unnecessary gatherings in the affected regions (Table 5.1).

Moreover, IMT classified the vulnerable areas to the foreseen oil spill according to high, medium, and low risk, considering the direction of the oil spill and the probability of the severity (Table 5.2, Figure 5.2). It was decided to take priority actions based on the risk level.

Map A shows the whole of Sri Lanka, with the Gampaha and Colombo districts highlighted in red. The red star offshore in map A and B indicate the location of the MV X-Press Pearl ship. Map B illustrates the enlarged section of the Gampaha and Colombo districts, showing the high-, medium-, and low-risk regions.

The purpose of oil spill responses is to remove oil at sea before oil enters the coastal regions, protecting more sensitive coastal marine ecosystems (i.e., seagrass beds and mangroves) (National Academies of Sciences, Engineering, and Medicine, 2013). Therefore, it was planned to deploy booms and skimmers offshore to prevent oil slicks from entering the high-risk areas.

The IMT team determined different responses for the identified vulnerable zones to the foreseen oil spill based on the risk level and the availability of resources (Table 5.3).

5.2.3 The Organization Chart and the Mobility Plan

As the responsible government organization in the NOSCOP implementation, MEPA played a leading role in the points discussed under 2.2, preparing and coordinating the external resources to combat the foreseen massive oil spill. Moreover, they were

TABLE 5.2
The Risk Level Classification and the Vulnerable Regions to the Foreseen Oil Spill

Risk Classification	Vulnerable Regions to the Foreseen Oil Spill
High	Dungalpitiya Beach
	Kepungoda Beach
	Pamunugama Beach
	Uswatakeyyawa Beach
	Thaldiyawatta Beach
Medium	Negombo Lagoon mouth
	Morawella Beach
	Pitipana Beach
	Dikowita fishery harbor
	Prithipura Beach
	Maha Oya mouth
Low	Kelani River mouth
	Craw Island Beach
	Mutwal Beach
	Colombo harbor

76 Maritime Accidents and Environmental Pollution

FIGURE 5.2 The locations of predicted high-, medium-, and low-risk areas to foreseen oil spill during predisaster phase.

internally ready for this mission by developing the organization chart and the mobility plan for the Incident Command Post (ICP). MEPA prepared an organizational chart and mobility plan following the guidelines in the NOSCOP in case the stored bunker oil in the MV X-Press Pearl ship caused an oil spill (Figure 5.3).

5.2.3.1 Oil Spill Management Decision Making Process (OSMDMP)

Director/NOSCOP: Director/NOSCOP is the chief officer in the prepared organizational chart (Figure 5.3). Some of the main responsibilities and duties of Director/NOSCOP are:

i) preparing the NOSCOP based on the National Disaster Management and Emergency Operational Plan

TABLE 5.3
The Identified Oil Spill Responses for the High-, Medium-, and Low-Risk Zones

Location	Recommended Response	Responsible Agency
High-Risk Zone		
Dungalpitiya and Pitipana Beach	• boom and skimmer deployment offshore direct to Dungalpitiya Pitipana, coastal line • beach clean-up programs (if offshore responses do not work perfectly)	• SLN • SLPA • CPC • SLAF
Kepungoda and Pamunugama	• boom and skimmer deployment offshore direct to Kepungoda and Pamunugama coastal line • beach clean-up programs (if offshore responses do not work perfectly)	• SLN • MEPA • SLPA • SLN
Thaldiyawatta and Uswetakeyawa Beach	• boom and skimmer deployment offshore direct to Thaldiyawatta, Uswetakeyawa coastal line • beach clean-up programs (if offshore responses do not work perfectly)	• SLN • CPSTL • SLPA • SLCG
MEDIUM-RISK ZONE		
Dikovita harbor	• fence boom deployment to close Dikovita fishery harbor mouth	• SLCG
Negombo Lagoon mouth	• close the Negombo Lagoon mouth	• SLCG • SLPA • MEPA
Morawella, Pitipana, and Prithipura beach	• depletion of oil slick to low-sensitivity area	• SLN • SLPA
Maha Oya	• beach clean-up programs	• SLCG • SL Army
LOW-RISK ZONE		
Colombo harbor	• the necessary actions should be taken following the Colombo harbor oil spill contingency plan	• SLPA
Kelani River mouth, Crow Island, and Mutwal Beach	• beach clean-up programs	• SLN

ii) coordinating with relevant authorities/institutes to plan the NOSCOP implementation
iii) appointing the incident commander
iv) overall coordination of the response with the Incident Commander and Advisory team

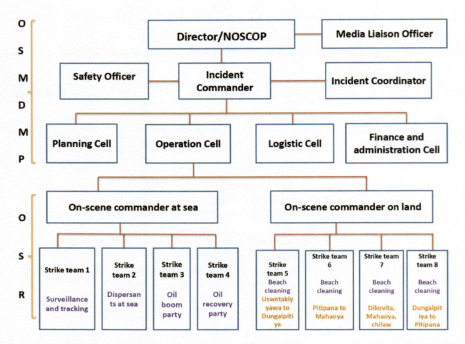

FIGURE 5.3 The prepared organization chart and mobility plan (OSMDMP = Oil Spill Management Decision Making Process and OSR = Oil Spill Responses).

v) deciding the spilled oil clean-up methods and waste management processes
v) directing the polluter for the restoration of the affected sites to their previous state (if it is practicable)
vi) requesting international support through the relevant ministry (considering the status and extent of the spill) (NOSCOP, 2005).

There was a unit that could directly contact the Director/NOSCOP, referred to as the **Media Liaison Cell** (Figure 5.3). The main responsibility of this unit was preparing and validating the press reports about the ongoing oil spill responses following the Director/NOSCOP's advice (NOSCOP, 2005).

Incident Commander (IC): IC worked under the instructions of the Director/NOSCOP (Figure 5.3). Some of the key responsibilities of the IC are:

i) managing oil spill responses on land and sea
ii) participating in advisory meetings with Director/NOSCOP
iii) coordinating the cell activities (planning, operation, logistics and finance and administration) and approving the prepared IAP
iv) instructing the on-scene commanders about the prepared IAP and the updated response strategies based on the evolution of the oil spill with time
v) overall coordination in the containment, recovery, clean-up, storage, and disposal processes (NOSCOP, 2005).

Onshore Responses during the X-Press Pearl Disaster 79

The planning, operations, logistics, and finance and administration cells were asked to join with the IC for the oil spill management decision-making process (Figure 5.3).

The planning cell should analyze the oil spill situation considering all the relevant information (i.e., source and size of the spill, etc.). They have to define the oil spill response strategy afterward with the aid of available information (i.e., surveillance reports) to predict the oil spill distribution and vulnerable areas to the oil spill (NOSCOP, 2005).

The logistics cell is responsible for identifying and procuring all the resources (i.e., equipment, instruments, etc.) for the oil spill responses (NOSCOP, 2005).

The finance and administration cell is responsible for:

i) appointing relevant personnel for the appropriate oil spill responses with a list of their duties and a work plan
ii) accessing finance during the oil spill response period
iii) establishing expenses approval procedures and the preparation of claim documents for the compensation (NOSCOP, 2005).

Operation cell: After the Director/NOSCOP has approved the IAP, the operation cell has to implement the IAP. Some of the most important responsibilities of this cell are to:

i) mobilize required resources to implement IAP
ii) coordinate the applied oil spill response on land and at sea and inform the Director/NOSCOP about the progress and effectiveness of applied strategies to manage oil spill (NOSCOP, 2005).

5.2.3.2 Oil Spill Responses (OSR)

On-scene commanders have the most prominent role in the oil spill responses. They are responsible for directing response actions and coordinating all required efforts at the oil spill scene. On-scene commanders directly connected with the operation cell to execute the predecided responses in the affected regions. They lead their teams in applying oil spill responses at the affected locations (NOSCOP, 2005) in case the stored bunker oil in the MV X-Press Pearl ship created a massive oil spill in Sri Lankan waters.

Two on-scene commanders were identified in the prepared IAP as **on-scene commander at sea** and **on-scene commander on land**. Altogether, eight groups were assigned to two on-scene commanders, providing four groups for each. All four groups were allocated to beach clean-up programs for the on-scene commander on land. In contrast, surveillance and tracking, dispersants at sea, oil booms, and recovery groups were given to the on-scene commander at sea (Figure 5.3).

5.2.4 BANNING FISHING ACTIVITIES

A large number of dead fish were found on the Western coast of Sri Lanka during the predisaster phase, possibly due to the debris and toxic chemicals from the MV

X-Press Pearl ship. Therefore, the DFAR banned coastal fishing from Panadura Ganga (Bolgoda Lake) in Kalutara district to Maha Oya in Gampaha district through Colombo harbor and Negombo Lagoon (Figure 5.4) by 22 May 2021 (Rubesinghe et al., 2022; Waravita, 2021). The expert committee, including MEPA and NARA, were appointed to look into the causes of fish mortality and decide on the time frame

FIGURE 5.4 The "No Fishing Zone" by 22 May 2021. Map A indicates the whole of Sri Lanka. The red star offshore in maps A and B indicates the location of the MV X-Press Pearl Ship. Map B illustrates the enlarged section of the "No Fishing Zone" by 22 May 2021.

for banning fishing activities (Mallawaarachchi, 2021). In the meantime, DFAR conducted awareness-raising programs within the fishing communities about the impacts of the MV X-Press Pearl maritime disaster.

5.2.5 Assessing Air Pollution Impact

The smoke plume created by the fire was the most obvious impact during the predisaster period, indicating considerable air pollution (Figure 5.5). As a consequence of the nitric burning, the emitted yellow smoke indicated the production of acidic nitrogen dioxide gas. The smoke plume was dispersed to the inland regions depending on the wind conditions. Since there were dangerous goods on the ship, there was a possibility of making a pollutant cocktail in the smoke plume (i.e., nitrogen oxides, sulfur dioxide, carbon monoxide, dioxins, furans, etc.), which could cause human respiratory diseases when entering the coastal and inland areas (Partow, Lacroix, Le Floch and Alcaro, 2021). It emphasizes the importance of conducting air quality surveys around the burning ship to determine the pollutant concentrations. The National Building Research Organisation (NBRO) conducted air monitoring near the MV X-Press Pearl (about 50 m distance from the burning vessel) with the aid of photoionization detectors (PIDs) (National Building Research Organization, Ministry of Defence, 2021; Partow, Lacroix, Le Floch and Alcaro, 2021), and the results indicated high pollution levels near the fire (National Building Research Organization, Ministry of Defence, 2021).

The incident analysis model identified the critical area; 120 km^2 of the area was defined as a vulnerable region to short-term exposure to the pollutants (Partow,

FIGURE 5.5 The release of black smoke from a fire onboard the MV X-Press Pearl cargo ship. Photo credit: SLAF Media Unit.

Lacroix, Le Floch and Alcaro, 2021; Rubesinghe et al., 2022). Further analysis revealed that a significant amount of pollutants were not recorded at ground level and were within the standard concentrations (National Building Research Organization, Ministry of Defence, 2021; Partow, Lacroix, Le Floch and Alcaro, 2021).

5.2.6 Legal Initiatives

Since the vessel rested in Sri Lankan waters and in view of the high possibility of potential chemical and bunker oil spill/leakage at any time, as per Section 24 of Marine Pollution Prevention No. 35 of 2008, the following directives were issued to the local agent of the ship, the captain of the ship, and the shipping company.

i) Take urgent action to control/extinguish fire on board to avert toxic air emission.
ii) Take all necessary actions in order to avert any potential chemical and bunker oil spill/leakage in Sri Lankan waters.
iii) Obtain the assistance of foreign experts, if it is evident that the capacity of the local stakeholders is not sufficient to manage the situation.
iv) Make necessary contingency arrangements to respond to any possible chemical or oil spill.
v) It is the advice of this Authority to tow away the vessel from Sri Lanka waters, if the situation goes beyond control, which might pose severe threat to the environment, port operations, human health, and socioeconomic activities of the country at large, after having taken all the appropriate measures as mentioned above.

5.3 POSTDISASTER RESPONSES

5.3.1 Waste Spill

5.3.1.1 The Influence of the Oceanographic Properties on Waste Distribution

During the disaster period, the ocean currents and wind patterns around Sri Lanka played a main role in distributing the waste on beaches. This maritime disaster happened during the beginning of the South-West monsoon; thus, the onshore wind speed was >10 ms^{-1} while the wave's height was ~2 m. However, a decrease in wind speed to <5 ms^{-1} was recorded after 2 June 2021 (Pattiarachchi and Wijerathna, 2021).

The West Indian Coastal (WIC) current is a southward-flowing Eastern boundary current, and it flows across the Gulf of Mannar during the South-West monsoon. Furthermore, it flows southward and eastward on the West and South coasts of Sri Lanka, respectively (Shankar, N Vinayachandran and S Unnikrishnan, 2002; Pattiarachchi and Wijerathna, 2021). It drove the distribution of waste generated from the MV X-Press Pearl disaster, southward from the burning ship's location on the Western coast and eastward on the Southern coast. Once past Sri Lanka, the currents form an anticlockwise eddy, creating a recirculation of water. It is defined as the Sri Lanka Dome (SD, Figure 5.6). SD connects the West and East coast water

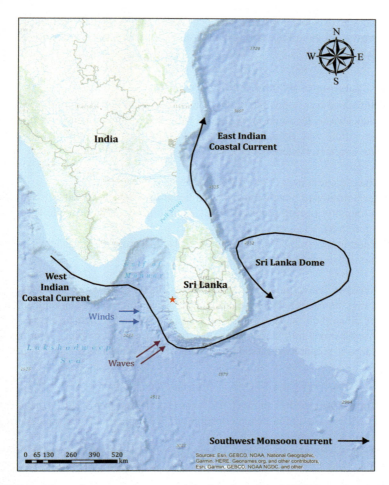

FIGURE 5.6 The main current patterns and the location of the MV X-Press Pearl ship in Sri Lankan waters.

in Sri Lanka, suggesting the distribution of waste from the West coast, where the burning ship was located, to the East coast (Vinayachandran and Yamagata, 1998; Pattiarachchi and Wijerathna, 2021).

Moreover, the Yaas tropical cyclone formed in the Bay of Bengal, and it started to flow northward by 22 May 2021(Figure 5.6). It influenced the changes in mean water level at Trincomalee (increase in mean water level) and Colombo (decrease in mean water level), reversing currents to flow in the East and North directions on the South and West coast of Sri Lanka, respectively, after 30 May 2021 (Pattiarachchi and Wijerathna, 2021). Therefore, the Yaas tropical cyclone's influence on the ocean current pattern reflected the distribution of waste (plastic nurdles) up to the Mannar region on the West coast (northward to the location of the burning ship) by the first week of June 2021.

The red star offshore indicates the location of MV X-Press Pearl ship. Blue and purple arrows indicate the existing winds and wave directions, respectively, during the disaster.

5.3.1.2 Planning of Beach Clean-up Programs

The spilled waste was deposited on the coastal region in North Colombo, specifically from Sarakkuwa to Dunugalpitiya, by 25 May 2021. The epicenter of this massive waste deposition was Sarakkuwa Beach (Partow, Lacroix, Le Floch and Alcaro, 2021). The waste consisted of:

- damaged containers (Figure 5.7),
- huge amounts of plastic nurdles with linear low-density polyethylene (LLDPE), low-density polyethylene (LDPE), and high-density polyethylene (HDPE) (Figure 5.8),
- burned lumps (melted pollutants with plastic nurdles, Figure 5.9), and
- other debris and cargo, etc.

This massive deposition of waste, including plastic nurdles, triggered an alarm for implementing intense beach clean-up programs. Given that all the planning was made to control the oil spill during the predisaster phase, MEPA had to adapt quickly to manage the unexpected waste spill in the marine environment. In the first phase, a primary operational plan was prepared immediately to recover the accumulated waste from beaches.

5.3.1.2.1 Shoreline Clean-up Operational Process

The Central Coordination Centre (CCC) was established in the MEPA head office in Colombo. MEPA appointed beach coordinators from their staff for the clean-up sites (Table 5.4).

MEPA identified the Sri Lanka Coast Guard and Sri Lanka armed forces (SL Army, SLN, and SLAF) as the leading human resource to engage in the clean-up programs, bearing in mind the prevailing mobility restrictions due to the COVID-19 pandemic condition in the country and the proximity of their military camps to the affected locations. They made an official request to the Ministry of Defence in Sri Lanka (MoD) through the Disaster Management Centre (DMC), requesting for the help of the required workforce.

5.3.1.2.2 Shoreline Assessment Survey (SAS)

The nurdle distribution pattern on the beaches had to be studied; thus, MEPA formed a "MEPA Shoreline Assessment Team". The objective of the shoreline survey was to assess the extent of every part of the potentially affected beaches to apply clean-up mechanisms. SAS team conducted the comprehensive shoreline survey collaborating with volunteering scientists, graduates, and university undergraduates. SAS team engaged daily with the survey, covering different parts of the affected beaches. It covered Puttalam, Mannar, Gampaha, Colombo, Galle, Hambanthota, Ampara, Batticaloa Trincomalee, Mullaitivu, Kilinochchi, and Jaffna districts. Furthermore,

FIGURE 5.7 The washed-up broken containers from the burning MV X-Press Pearl cargo ship.

Source: http://www.adaderana.lk/news.php?nid=4012&mode=head

daily site inspections at a regular distance (500 m) were done along all parts of the accessible beaches in the impacted areas to study the evolution of waste distribution over time. During the initial phase of the survey, the extent of the plastic nurdle accumulation on beaches was categorized into high, medium, low, and no contamination with the aid of visual observations.

86 Maritime Accidents and Environmental Pollution

FIGURE 5.8 The deposited plastic nurdles on the beach.

FIGURE 5.9 The deposited burnt lumps on the beach.

Source: http://www.adaderana.lk/news.php?nid=4012&mode=head

TABLE 5.4
The Tasks of the Central Coordination Center and Beach Coordinators in the Beach Clean-up Programs

Category	Task
Central Coordination Center	• Making decisions about the clean-up program (i.e., required instruments, distribution of workforce) based on the distribution of wastes around the Sri Lankan coastal zone • Facilitating the required materials and instruments for the clean-up sites • Overall coordination of the beach clean-up programs.
Beach Coordinators from MEPA	• Coordinating logistic arrangements in the site (i.e., providing meals and personal care equipment to the people who engaged in clean-up programs) • Recording data on the site.

5.3.1.2.3 Shoreline Planning Information Process (SPIP)

The SAS team submitted data to the Shoreline Planning Information Process (SPIP). MEPA beach coordinators were also asked to submit their site-specific requirements to the SPIP. SPIP was asked to develop a report with the aid of all the gathered information which was submitted to the CCC for decision making. The decisions taken by the CCC were transferred to the workforce at the site through beach coordinators (Figure 5.10).

5.3.1.2.4 Initiation of Beach Clean-up Programs

The coastal belt in the Gampaha district on the Western coast was severely impacted by mass shoreline strandings of waste due to its proximity to the burning location of the MV X-Press Pearl cargo ship (Partow, Lacroix, Le Floch and Alcaro, 2021). MEPA initiated the beach clean-up program from Sarakkuwa to Dunugalpitiya on 26 May 2021 in collaboration with the Sri Lanka Armed Forces (SLN, SLAF, and SL Army) and SLCG. In the meantime, MEPA established the Sarakkuwa operational center to record the field data and facilitate the logistic arrangements (i.e., provide food and personal care equipment to the people engaged in clean-up programs) at the clean-up sites. Even though people wanted to voluntarily join the clean-up programs, it was impossible at the initial stage due to the prevailing COVID-19 lockdown condition in the country. Additionally, fishing was banned in the affected area, as discussed in 2.4; therefore, the fishing community joined at the later stages of the clean-up programs. The Sri Lanka Red Cross supported the beach clean-up efforts, providing workforce in the Gampaha, Puttalam, Colombo, and Matara districts (Rubesinghe et al., 2022). The Sri Lanka Civil Defence Force (CDF) also joined at the later stage of the clean-up programs.

Waste, specially plastics, does not follow geographic boundaries as it can be distributed to any place with wind, waves, and currents. The waste distribution was extended from the West coast to the North-West and South coastal belt by 29 May 2021 (Figure 5.10) due to the oceanographic properties mentioned in Section 5.3.1.1. MEPA regional officers engaged with the three forces and SLCG to continue the beach

FIGURE 5.10 The initial beach clean-up operational plan.

clean-up programs from the North-West to the Southern coastal region. Gampaha and Puttalam districts showed a higher accumulation rate for plastic nurdles (Table 5.5) and proved to be a long-term plastic nurdles deposition as well. MEPA is still organizing frequent beach clean-up programs in Puttalam and Gampaha districts to remove plastic nurdles and other washed-off waste from the MV X-Press Pearl cargo ship. Since the Mannar, Kirinda, and Rekawa coastal regions showed the lowest concentration of plastic pellets (few pellets), clean-up programs were not conducted there.

5.3.1.3 The Contribution of International Tanker Owners Pollution Federation Limited (ITOPF) to Improve the Initial Beach Clean-up Plan

In the later phase, ITOPF collaborated with MEPA to enhance the efficiency and quality of the beach clean-up programs. ITOPF made a few modifications to the initial plan. They joined with CCC to make decisions on the ongoing beach clean-up programs, and ITOPF members participated in the beach clean-ups, providing technical assistance. Moreover, ITOPF introduced two apps to classify the plastic nurdles accumulation state on beaches and to produce Daily Operational Reports (DOR).

- Analysis of the plastic nurdle accumulation state was done by the "survey 123 app".
- The mechanism of the "survey 123" app: A person has to pick plastic nurdles within 2 minutes into a given cup from each affected area. The weight of the collected nurdles was calculated, and the weight had to be inserted into the app. The plastic nurdles accumulation state of the site was automatically generated afterward (Table 5.5, Figure 5.11).

TABLE 5.5
The Classification of Plastic Nurdles Accumulation Levels in the Different Districts and Regions

Classification	District	Regions
High Contaminated	Puttalam	Thoduwawa, Chilaw, Ambakadawila, Marawila, Kurusapalliya, Katuneriya
	Gampaha	Pamunugama, Sarakkukanda, Kepungoda, Bimpadura, Uswetakeiyawa, Elangoda, Dungalpitiya, Thalahena, Seththapaduwa, Negombo Lagoon, Negombo Beach Park, Negombo fish market, Poruthota
Medium Contaminated	Matara	Matara, Walliwala, Midigama
	Colombo	Galle Face, Wellawatta, Moratuwella, Mt. Lavinia, Rathmalana, Dehiwala, Egodauyana, Moratuwa
Light Contaminated	Galle	Paraliya, Thelwatta, Kaikawala, Warahena
	Hambanthota	Kirinda, Rakawa
	Mannar	Mannar Arippu, Wankalei, Silawathhrai, Kokkupudayan
Noncontaminated	Ampara, Batticaloa Trincomalee, Mulatiuv Kilinochchi, Jaffna	

Map A shows the whole of Sri Lanka, and the Puttalam, Gampaha, Colombo, Kaluthara, Galle, Matara, and Hambantota districts are highlighted in red. The red star at sea in maps A and B indicates the location of the MV X-Press Pearl Ship. Map B illustrates the enlarged section of the Puttalam, Gampaha, Colombo, Kaluthara, Galle, Matara, and Hambantota, showing the high-, medium-, and low-contaminated beaches for plastic nurdles pollution.

The second app was introduced to make the DOR. The DOR was produced with the aid of the information submitted by the SAS team about the extent of the plastic nurdles distribution and the site-specific report submitted by the regional and local beach coordinators.

The CCC made decisions following the DOR about the ongoing beach clean-up programs, and decisions were transferred to the workforce at the site through regional and local beach coordinators (Figure 5.12).

Many beach clean-up techniques were used to restore the affected beaches. Labor-intensive, improvised methods were quickly identified to remove washed-off waste from the burning ship during the initial phase of the clean-up programs. Nurdles and solid waste collection methods improved over time, considering the availability of resources and site-related practical issues.

90 Maritime Accidents and Environmental Pollution

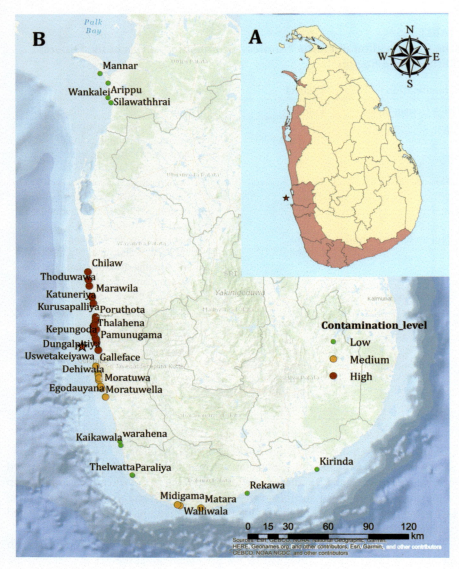

FIGURE 5.11 Plastic nurdles distribution on the beaches in North-Western, Western, and Southern provinces of Sri Lanka.

FIGURE 5.12 The updated beach clean-up operational chart with support of ITOPF and Oil Spill Response Limited (OSRL).

5.3.1.4 Application of Simple Techniques to Remove Deposited Waste on the Beaches

At the initial stage of the clean-up program, simple techniques were used due to the unavailability of advanced instruments.

5.3.1.4.1 Waste Removal by Hand (Figure 5.13), Shovels (Figure 5.14), Garden Brooms, and Dust Pans (Figure 5.15)

5.3.1.4.2 Construction Machinery

Construction machinery was used to remove the bulk of plastic nurdles and other waste accumulated from the MV X-Press Pearl ship accident (Figure 5.16).

However, it was difficult to remove only waste from the sands using the aforementioned techniques. Different techniques had to be applied to separate the waste, specially plastic nurdles, from the sand.

5.3.1.5 The Separation of Waste from the Sand

Manual sieves, mechanical sieves, trommel, and baskets were used for the separation of collected waste from the sand.

Manual and mechanical sieve: waste remained on the sieve because sand was filtered through the sieve (Figures 5.17 and 5.18).

Baskets: The baskets were used to separate waste from seawater. The basket was dipped in the seawater (Figure 5.19). Seawater was drained out through the tiny spaces in basket; therefore, the waste (plastic nurdles) remained inside the baskets (Figure 5.20).

Trommel: Plastic nurdles remained at the last stage of the trommel and it was collected into the water-filled bucket afterward. Plastic nurdles float on the water-filled

FIGURE 5.13 Plastic nurdles removal by hand.
Source: www.facebook.com/photo?fbid=3790057704436789&set=pcb.3790070247768868

FIGURE 5.14 The waste removal by shovels.

Onshore Responses during the X-Press Pearl Disaster

FIGURE 5.15 Plastic nurdles removal by garden brooms.

FIGURE 5.16 The removal of bulk amount of collected waste sacks.

94   Maritime Accidents and Environmental Pollution

FIGURE 5.17   The use of manual sieving to separate waste.
Source: www.facebook.com/photo?fbid=211932414428647&set=pcb.211932627761959

FIGURE 5.18   The mechanical sieve.

bucket due to density separation; thus, floating plastic nurdles were removed from the bucket using a hand filter (Figures 5.21 and 5.22).

    Water bath: The collected waste with sand was put into the water basket. Plastic nurdles like low-density waste floated on the water's surface, separating them from the sand (Figure 5.23).

Onshore Responses during the X-Press Pearl Disaster

FIGURE 5.19 The dipped baskets in the water to trap floating wastes. Photo Credit: MEPA.

FIGURE 5.20 The separated waste from the seawater is trapped inside the basket.

5.3.1.6 Application of Advanced Techniques to Remove Accumulated Waste on the Beaches

More advanced techniques were applied at some locations to separate and collect the waste from the sand; they were more efficient to remove plastic nurdles and other waste than the methods we explained in 5.3.1.5.

FIGURE 5.21 The use of trommel to separate plastic nurdles.
Source: www.dailymirror.lk/caption_story/In-search-of-debris/110-216036

5.3.1.6.1 BeachTech 2000
A trailing machine with a sieving belt connects with a tractor in BeachTech 2000. The driver has to drive the tractor across the polluted area. In the meantime, sand with waste (including plastic nurdles) enters the sieve belt. Waste remains on the belt and is directed to the waste storage component in the instrument. The sand is filtered from the sand belt and deposited on the beach. Therefore, BeachTech 2000 was used to remove waste more efficiently, and the clean-up process could be done with minimum human interaction with the hazardous materials.

5.3.1.6.2 Beach Sweepers
The operation mechanism is almost the same as the BeachTech 2000, but the machine is not connected to a tractor; the operator has to walk behind the machine (Figure 5.24).

However, BeachTech 2000 and beach sweepers were used on dry sand in large flat areas.

5.3.1.6.3 Blue Machine
Erosion and the accretion of sands on beaches are natural processes. Sand is removed from one location during the erosion period and is deposited on other beaches, balancing the coast's energy impacts (Rathnayake et al., 2018). Therefore, beaches are exposed to the erosion and deposition phases from time to time. The collected core samples from Sarakkuwa Beach showed the layers of waste inside

FIGURE 5.22 The separation of the plastic nurdles by hand filters at the last stage of trommel.
Source: www.dailymirror.lk/caption_story/In-search-of-debris/110-216036

the sand, reflecting that Sarakkuwa Beach was in the accretion phase during this maritime disaster.

A local investor produced the blue machine to separate waste in sand layers, and the machine used gravity separation to differentiate six contaminant types from the sand. The blue machine used sand excavated from the beach (Figure 5.25a) and freshwater for the density separation (Figure 5.25b). The extent of the separation varied from large plastic pieces to fine foam (Rubesinghe et al., 2022).

5.3.1.7 Plastic Nurdles Removal Mechanism in Sensitive Ecosystems

IMT identified sensitive regions (i.e., lagoons, estuaries) in the affected areas using a preprepared map about the "Sensitive ecosystems in Sri Lanka". They applied different techniques to remove plastic nurdles from sensitive areas. Plastic nurdles were deposited in some places in the Negombo Lagoon (i.e., mangrove cover in "Kadol kele"); vacuum pumping was used to suck out the plastic nurdles from the site, causing minimum disturbance to the ecosystem (Figure 5.26).

FIGURE 5.23 The use of water bath to separate plastic nurdles and other debris.

Source: www.facebook.com/photo?fbid=3953227101453181&set=pcb.3953240118118546

FIGURE 5.24 The beach sweepers.

Source: www.facebook.com/photo?fbid=3962467697195788&set=pcb.3962469497195608

Onshore Responses during the X-Press Pearl Disaster

(a)

(b)

FIGURE 5.25 Sand excavation (a) and the particle separation in the blue machine (b).

Sources: (a) www.facebook.com/photo?fbid=3953227364786488&set=pcb.3953240118118546 and (b) www.dailymirror.lk/caption_story/In-search-of-debris/110-216036

FIGURE 5.26 Application of vacuum pumping to remove plastic nurdles in mangrove region.

5.3.1.8 Chemical Waste Removal

The onshore chemical spill response was a critical task due to the unavailability of trained people in chemical spill management. There were washed-off mixed hazardous chemicals (i.e., gas cylinders and intact or damaged chemical packs, etc.) on the shoreline. The workers engaged in the clean-up programs did not know how to properly remove them from the shoreline without exposing them to the chemicals and causing gas emissions or blasting. There is a chemical response team in the SL Army and SLAF, referred to as the **C**hemical **B**iological **R**adio Nuclear (CBRN) team. CBRN team supported the proper removal of washed-off chemical waste from the shoreline.

5.3.1.9 Waste Management and Sampling of Waste for the Chemical Analysis

People who had engaged in the clean-up processes collected waste in sacks (Figure 5.27), and these sacks were initially stored on the beaches (Figure 5.28) and in the container yard at Wattala, Sri Lanka. MEPA used 42 containers to store the collected waste. In a later stage, a warehouse in Pamunugama, Sri Lanka, was tasked to hold the collected waste (Figure 5.29). MEPA transferred stored waste sacks from the container yard to the warehouse. Moreover, packed waste sacks were repacked in jumbo bags to provide additional protection for the collected waste.

Since hazardous materials formed part of the collected waste, the CEA monitored and regulated the whole waste management process. In total, 1600 metric tonnes of washed-off waste from the MV X-Press Pearl maritime disaster had been collected during the beach clean-up programs by 14 July 2021 (Gunawardana, 2021).

Onshore Responses during the X-Press Pearl Disaster

FIGURE 5.27 The waste collection in sacks.

FIGURE 5.28 The storage of collected waste sacks on the beach.

FIGURE 5.29 The proper storage of waste at the warehouse at Pamunugama, Sri Lanka.

The samples from the collected waste in the containers had to be taken for further chemical analysis to evaluate the environmental damage as a result of the MV X-Press Pearl maritime disaster. Random sampling was done on the collected waste (i.e., plastic nurdles, hazardous chemicals, etc.), following the Centre for Environment Fisheries and Aquaculture Science (CEFAS) guidelines. Opening the containers was challenging due to the hazardous nature of the stored chemicals; therefore, the chemical response team in the SLPA took part in the container opening process to support the MEPA officers during this random sampling.

Salvage operations had been planned to remove the shipwreck and the settled debris on the ocean floor. The Resolve Marine Company did the debris removal under the monitoring of the MEPA officials, and it was recorded approximately 78% of debris had been removed before the beginning of the South-West monsoon period in 2022. The debris from the sunken ship was collected in containers (Figure 5.30) and brought to the land; subsequently, containers were stored in the Container Yard at Wattala (MEPA, 2021; Farzan, 2022). The debris storage process was also done under the supervision of MEPA officials to ensure safe storage (MEPA, 2021).

However, shipwreck removal operations have been delayed due to the South-West monsoon period in 2022. It is planned to recommence the shipwreck removal operations during the calm sea state (after the South-West monsoon in 2022), and Shanghai Salvage Company will engage with the process (Farzan, 2022).

5.3.2 Oil Spill Responses

Despite the prepared IAP for the expected oil spill in the predisaster phase, another two plans were prepared for the oil spill responses, considering the predicted oil spills after the explosion and sinking of the ship on 25 May and 2 June 2021, respectively.

FIGURE 5.30 Collection of debris from sunken ship in containers at sea.

Source: https://mepa.gov.lk/removal-of-debris-and-removal-of-wrecked-mv-x-press-pearl/

The same procedures discussed in 2.2 were followed to prepare action plans during the postdisaster phase.

5.3.2.1 The Prepared Oil Spill Action Plan after the Explosion and Fire Engulfing the Whole Ship

The ship explosion signaled the start of a hazardous chemical and a large oil spill from the stored 348 tonnes of bunker oil; thus additional action plans were prepared by 25 May 2021, considering the large oil and chemical spills as the worst-case scenario.

The IMT team identified the appropriate oil spill responses and responsible institutes/organizations to execute actions in the affected sites in case the explosion and fire engulfed the whole ship and created an oil spill (Table 5.6). The same assumptions, objectives, and strategies in the IAP that had been prepared during the predisaster phase (in Section 5.2.2.1) were used for this plan. The high-, medium-, and low-risk regions were identified following the oil simulation models (Table 5.7 and Figure 5.31).

Since two chemical containers had fallen into the sea by 25 May 2021, the IMT team asked the CBRN team to engage in managing hazardous chemical spill responses (Table 5.6).

Map A indicates the whole of Sri Lanka, with the Gampaha and Colombo districts highlighted in red. The red star at sea in map A indicates the location of the MV

TABLE 5.6
The Prepared Emergency Action Plan for the Foreseen Oil and Chemical Spill after the Ship Explosion and Fire Engulfing the Whole Ship

Responsible Agency	Recommended Action
DMC	• Arrange community awareness about chemical and oil spill hazard and unnecessary gathering at Dikovita to Negombo Lagoon • Arrange health assistance from Dikovita to Negombo
CPC/CPSTL	• Arrange suitable tugs to disperse oil slick at open ocean • Arrange required dispersant
NARA	• Arrange damage assessment survey from Dikowita to Negombo
CC&CRMD	• Arrange beach clean-ups from Dikovita to Negombo
SLPA	• Arrange boom and tug to deplete oil slick to avoid entering Negombo lagoon • Arrange for any chemical response from Dikovita to Negombo
DFAR	• Make arrangement to raise awareness with fisheries communities on oil and chemical spills • Make suitable arrangements to obtain information on oil and chemical spill evidence from fisheries communities
SLCG	• Arrange troops to foot patrol from Dikovita to SLN Kelani • Arrange coastal booms to close Negombo harbor mouth • Arrange coastal clean-up teams to Dikovita to SLN Kelani
SL Army	• Arrange CBRN Unit to respond to any chemical hazard if chemical container arrives at Pitipana to Negombo Lagoon mouth • Arrange troops to foot patrol from Pitipana to Negombo Lagoon mouth • Arrange beach clean-up teams to clean beaches from Pitipana to Negombo Lagoon
SLN	• Arrange marine surveillance to monitor oil slick movement at open ocean • Arrange coastal surveillance to monitor oil slick movement at coastal waters • Arrange foot patrols from SLN Kelani to Dungalpitiya • Arrange chemical response teams to respond from Dikovita to Dungalpitiya
SLAF	• Undertake air surveillance from Dikovita to Negombo • Arrange foot patrols from Dungalpitiya to Pitipana • Arrange beach clean-up teams from Dungalpitiya to Pitipana • Arrange chemical response teams for any chemical hazard from Dungalpitiya to Pitipana
CEA	• Arrange temporary oily waste storage facilities at selected places from Dikovita to Negombo • Arrange waste transportation facility • Coordinate INSEE to dispose of oily waste at final destination
MEPA	• Arrange interagency coordination • Technical support • Logistic support

TABLE 5.7
The Risk Level Classification and the Vulnerable Coastal Regions to the Foreseen Oil Spill during the Ship Explosion and Burning Phase

Risk Classification	Vulnerable Regions to the Foreseen Oil Spill
High	Negombo Lagoon mouth
	Morawella Beach
	Pitipana Beach
	Dungalpitiya Beach
	Kepungoda Beach
	Pamunugama Beach
	Uswatakeyyawa Beach
	Thaldiyawatta Beach
Medium	Dikowita fishery harbor
	Prithipura Beach
	Maha Oya mouth
Low	Kelani River mouth
	Craw Island Beach
	Mutwal Beach
	Colombo harbor

X-Press Pearl Ship. Map B illustrates the enlarged section of the Gampaha and Colombo districts, showing the high-, medium-, and low-risk regions.

5.3.2.2 The Prepared Oil Spill Action Plan after the Sinking of the Ship

Several assumptions were made to prepare the action plan:

i) oil will flow in the South-West direction from the location of the burning ship
ii) oil will reach the Western coast of Sri Lanka and spread on the beaches from Negombo to Dickovita during the initial phase of the spill
iii) oil will reach the coast after the emulsification, and it will take 4–5 hours to get to the coast from the time of the spill
iv) since the present sea state is between 2 and 3, suitable containment booms and recovery equipment can be deployed
v) in case the oil slick starts to spread toward ecologically and economically valuable locations such as the Negombo Lagoon mouth and the Colombo harbor, OSD must be sprayed to disperse the oil slick.

The objectives of the plan were set for the expected massive oil spill as:

i) to monitor the slick movement
ii) to contain oil leaking by encircling the vessel with offshore oil booms (if the ship is grounded)

FIGURE 5.31 The locations of predicted high-, medium-, and low-risk areas to predict oil spill after the explosion and fire engulfed the whole ship.

iii) to disperse as much oil as possible into small droplets using dispersants in the deep sea; if the boom does not work perfectly to trap the spilled oil in the deep sea
iv) to protect other ecologically and economically valuable areas in the affected coastal regions using oil boom and skimmers
v) to conduct beach cleaning activities; if containment and recovery do not work optimal in the open ocean. The clean-up program priority will be given to the most sensitive regions on the coastline.

IMT planned to apply almost the same strategies discussed in Section 5.2.2.1.1. They inserted one additional strategy to assess the impact of the foreseen oil and chemical spill on marine life. The DWLC was responsible to identify impacted species and conduct further investigations to find the oil spill's impact on them.

The high-, medium-, and low-risk locations were identified based on the oil slick movement prediction model, taking into account ocean and wind current patterns after the end of May 2021 in Section 5.3.1.1 (Table 5.8, Figure 5.32). The priority was given to applying oil spill responses to the high-risk regions. However, oil spill responses were implemented at the identified medium- and low-risk regions as well (Table 5.9).

Map A shows the whole of Sri Lanka with the Gampaha and Colombo districts highlighted in red. The red star at sea in map A indicates the location of the MV X-Press Pearl Ship. Map B illustrates the enlarged section of the Gampaha and Colombo districts, showing the high-, medium-, and low-risk regions.

5.3.2.3 Application of Onshore Oil Spill Responses

Negombo Lagoon is one of the most productive lagoons in Sri Lanka, covered by mangroves; thus, the lagoon has a strong connection with fishing activities, providing significant ecological and economic values to the society (Wimalasiri, Perera and Jayathilaka, 2018). MEPA took immediate actions to make barriers to stop entering oil into the Negombo Lagoon, such as the deployment of boomers in

TABLE 5.8
The Risk Level Classification and the Vulnerable Coastal Regions to the Foreseen Oil Spill after the Sinking of the Ship

Risk Classification	Vulnerable Regions to the Foreseen Oil Spill
High	Maha Oya mouth
	Poruthota Beach
	Negombo Lagoon mouth
	Dungalpitiya Beach
	Kepungoda Beach
	Pamunugama Beach
	Uswatakeyyawa Beach
	Thaldiyawatta Beach
Medium	Morawella Beach
	Pitipana Beach
	Dikowita fishery harbor
	Prithipura Beach
	Coastal region from Galle Face to Moratuwa
Low	Kelani River mouth
	Craw Island Beach
	Mutwal Beach
	Colombo harbor
	Coastal region from Galle Face to Wellawatta

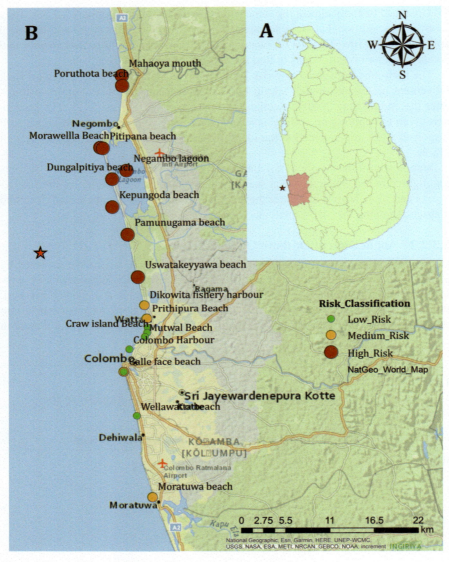

FIGURE 5.32 The locations of predicted high-, medium-, and low-risk areas to foreseen oil spill during ship sinking phase.

the Negombo Lagoon area with the support of the "Iron Man" voluntary team and SLCG (Figures 5.33 and 5.34). The booms were deployed across the lagoon mouth by 27 May 2021 to protect the Negombo Lagoon and associated ecosystems from the predicted oil spill after the explosion and fire engulfing the whole ship. Booms were deployed in several places to make barriers for the oil slick's entrance to the sensitive marine and coastal ecosystems.

TABLE 5.9
The Identified Oil Spill Responses for the High-, Medium-, and Low-Risk Zones after the Sinking of the Ship

Location	Recommended Response	Responsible Agency
High-Risk Zone		
Negombo Lagoon	• Close lagoon mouth	• SLCG • SLPA • MEPA
Negombo harbor to Maha Oya mouth	• Beach cleaning	• SL Army
Thaldiyawatta and Uswetakeyawa Beach	• Offshore boom and skimmer deployment perpendicular to Thaldiyawatta, Uswetakeyawa coastal line • Disperse oil slick at deep sea • beach clean-up programs (if offshore responses are not perfect)	• SLN • CPSTL • SLPA • SLN • MEPA • SLCG
Dungalpitiya and Pitipana Beach	• offshore boom and skimmer deployment perpendicular to Dungalpitiya-Pitipana, coastal line • Disperse oil slick at deep sea • beach clean-up programs (if offshore responses do not work optimally)	• SLN • SLPA • CPC • SLCG • SLAF
Kepungoda and Pamunugama	• offshore boom and skimmer deployment perpendicular to Kepungoda and Pamunugama coastal line • beach clean-up programs (if offshore responses do not work optimally) • Disperse oil slick at open ocean	• SLN • MEPA • SLPA • SLN • SLPA
At incident Site	• circle around ship accident	• SLPA • CPC • SLN
Medium-Risk Zone		
Dikovita Harbor	• fence boom deployment to close off Dikovita fishery harbor mouth	• SLCG • CPSTL • SLN
Coastal Region From Galle Face to Moratuwa	• Beach cleaning	• SLCG • SL Army
Ma oya	• Beach clean-up programs • Deploy improvised boom	• SLCG • SL Army • SLCG
Low-Risk Zone		
Colombo Harbor	• The necessary actions should be taken following the Colombo harbor and Colombo dockyard oil spill contingency plan	• SLPA

FIGURE 5.33   Boom deployment in the Negombo lagoon.
Source: www.facebook.com/photo?fbid=3784813771627849&set=pcb.3784814264961133

FIGURE 5.34   Deployed boom in the lagoon to prevent oil entrance to the inner lagoon.
Source: www.facebook.com/photo?fbid=3784813968294496&set=pcb.3784814264961133

However, oil sheen was observed soon after the major fire broke out. The RADARSAT-2 satellite detected an area of 0.51 km^2 diameter and a 3.23-km-long oil slick near the ship on 8 June 2021. Moreover, the Sentinel 1 satellite imagery indicated the oil sheen extended lengthwise to approximately 4.3 km on 14 June 2021. The aerial survey confirmed the spreading of the oil sheen on 20 June 2021 (Partow, Lacroix, Le Floch and Alcaro, 2021; Pattiarachchi and Wijeratne, 2021). In the meantime, MEPA conducted the offshore oil spill responses with SLCG (i.e., application of dispersants) and got ready for the onshore responses as described in Table 5.9. Fortunately, a significant amount of oil did not enter the coastal regions and only a small amount of tar balls were deposited on some areas of the West coast. The deposited tar balls were manually removed by the beach clean-up programs.

5.3.3 Uplift of Fishing Ban

The coastal regions subjected to the fishing ban mentioned in 2.4 were gradually reduced (Partow, Lacroix, Le Floch and Alcaro, 2021). DFAR introduced a revised no-fishing zone by 1 December 2021 (Figure 5.35), excluding the Kalutara district and Negombo Lagoon (Rubesinghe et al., 2022).

Map A indicates the whole of Sri Lanka. The red star at sea in maps A and B indicate the location of the MV X-Press Pearl Ship. Map B illustrates the enlarged section of the "No Fishing Zone" by 1 December 2021.

5.4 COMMUNITY AWARENESS PROGRAMS

The MV X-Press Pearl cargo ship emitted black smoke due to the burning of chemicals and hazardous materials on the ship. Since the ship contained a huge amount of acids (i.e., 25 tonnes of nitric acid), burning indicated the production of acidic gases (i.e., nitrogen dioxide). As a consequence of the massive release of acidic gases into the atmosphere, there was a possibility to create acid rains. Moreover, exposure to toxic chemicals and hazardous materials can produce chronic and acute human diseases. Therefore, awareness programs were conducted through media (i.e., television, newspapers) and social (i.e., Facebook) media, informing the community:

i) to wear a wet mask to avoid the inhalation of this toxic plume;
ii) to stay away from the contaminated beaches due to the washed-off hazardous materials and toxic chemical spill; people were advised to stay at home and stop unnecessary gatherings at the affected sites to minimize their exposure to hazardous chemicals, materials, and acid rains;
iii) to stop fishing in the affected regions; and
iv) not to collect hazardous materials washed off from the burning ship.

5.5 DATA RECORDING ABOUT THE IMPACTED MARINE SPECIES

Dead marine animals (Figures 5.36 and 5.37) were recorded on the shore during and after the disaster. The beach survey assessment team recorded the number of

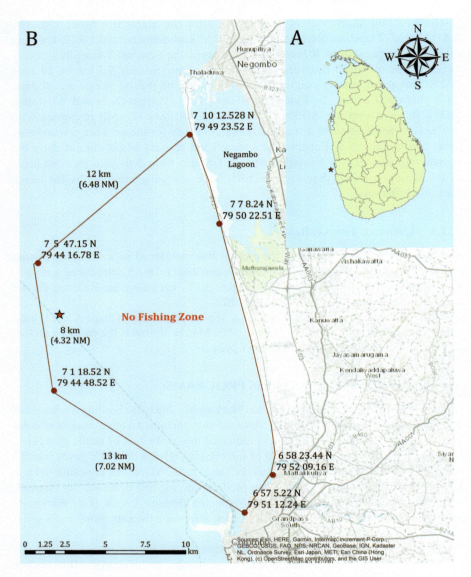

FIGURE 5.35 The revised no fishing zone by the DFAR by 1 December 2021.

incidents of marine animals found on the beach. They submitted the collected data to the DWLC and NARA for further investigations. In summary, 436 marine species were recorded with adverse effects by 12 December 2021, including 359 turtles (i.e., Olive redly, Hawks bill, Green, Loggerhead, Hawksbill turtles) and 66 marine mammals (i.e., dolphins, whales, dugongs).

The cause of the marine animals' death was investigated to determine if it happened due to the impact of the MV X-Press Pearl maritime disaster or any

Onshore Responses during the X-Press Pearl Disaster

FIGURE 5.36 The dead turtles.

Source: www.adaderana.lk/news/75777/x-press-pearl-disaster-court-briefed-on-death-toll-of-sea-animals

FIGURE 5.37 The dead body of the dolphin.

Source: www.dailymirror.lk/top_story/Over-10-turtles-dolphin-seabirds-several-fish-found-dead-after-MV-X-Press-Pearl-disaster/155-213540

other reason. A committee was appointed, including experts from universities (i.e., University of Ruhuna, University of Moratuwa, and University of Colombo, etc.), research institutes (i.e., NARA, NBRO), and government organizations (i.e., MEPA) to conduct the scientific analysis, investigating causes for the sudden death of marine animals. After the postmortem investigation of the animals, the committee completed the report and submitted it to the Attorney General's Department to proceed with the legal activities associated with the created environmental disaster from the MV X-Press Pearl ship accident.

5.6 CHALLENGES ASSOCIATED WITH IMPLEMENTATION OF ONSHORE RESPONSES

Implementation of the onshore responses for the consequences of the MV X-Press Pearl maritime multidisaster was challenging due to several factors such as:

i) Finding a workforce to engage in the beach clean-up program, to remove washed-off waste, was a major challenge due to the existing national lockdown to control the spread of the COVID-19 virus. At the initial stage of the beach clean-up programs, SL armed forces and SLCG provided the workforce. Their prompt, timely, and effective contribution to the intense cleanup programs in removing washed-off waste, including hazardous materials, was highly appreciated. However, in the later phase, SL armed forces had to engage with the COVID-19 community vaccination process as they had to prioritize the vaccination process over the clean-up programs to control the spreading of COVID-19 among Sri Lankans, creating a vacuum in the beach clean-up workforce. SLCDF joined with the SLCG for beach cleanup programs afterward to fill that vacuum in the workforce.

ii) The lack of equipment, technology, previous knowledge, and experience in managing plastic nurdle spills. This event was the first recorded plastic nurdle spill in Sri Lankan marine waters. Furthermore, no trained people were available to manage the hazardous chemical spill. The sudden, relatively high number of dead turtles and dolphins indicated the acute toxic impacts of the MV X-Press Pearl maritime disaster. Since then, legal actions have been taken against the shipping company; the dead marine life provides good evidence of the created environmental disaster by the ship accident. However, there were no proper freezing and storage facilities to preserve all the dead marine animals available on the shoreline after the ship's explosion. The lack of training in sampling animals and postmortem analysis, especially for the turtles and dolphins, was another challenging issue. Therefore, a team of Italian scientists visited Sri Lanka and conducted the training program at the University of Peradeniya, Sri Lanka, for the postmortem analysis of turtles and dolphins.

iii) The prevailing COVID-19 pandemic conditions in the country. The country was in national lockdown due to the relatively high infection rate of the COVID-19 virus. COVID-19 pandemic condition severely affected the mobilization of the workforce for the intense clean-up processes. Some of the affected beaches were closed without applying clean-up processes due to the spread of the COVID-19 virus among the people engaged in clean-up programs. The supply of required materials and instruments was disrupted from the local and international markets due to delays in procurement and transport processes.

iv) Finding a bulk amount of PPE for the people engaged in clean-up processes. Hazardous chemicals were released into the marine water with the ship's explosion. Plastic nurdles and other washed-off waste were contaminated with these hazardous chemicals; therefore, people were asked to wear PPE during the beach clean-up process. There was a shortage of PPE in the country by that time due to its high usage in the medical sector during the COVID-19 pandemic.

v) Controlling the unnecessary gatherings and theft incidents in the marine litter–accumulated areas. Intact plastic pellet sacks piled up on beaches, and villagers collected these packs of hazardous materials and washed-off parts of the burning ship. In some places, collected marine litter sacks were stolen from the temporarily stored sites; thus, a special mechanism was arranged to control the theft incidents.

vi) The prevailing rough sea conditions during the South-West monsoon period caused disruptions to the beach clean-up programs and oil spill responses. The monsoonal winds and ocean currents pushed debris and cargo from the wreck toward the shoreline and vice versa. Rough sea conditions significantly interrupted the boom deployment on the shoreline for oil spill responses.

vii) No specific vessels in relevant civil authorities to deploy booms and skimmers offshore were present; thus, containment and recovery operations did not work optimally in offshore waters.

viii) The process of meeting financial requirements was not an easy task. Although the oil spill responses had been laid down carefully, the real scenario was totally different. The immediate hiring and processing of new equipment and services for beach and coastal clean-ups was an extremely difficult and tedious process.

ix) Finding a proper place for waste storage was a challenge, as discussed in Section 5.3.1.8; waste was stored on the beaches during the initial phase of the beach clean-up programs. However, proper waste management system is crucial in postdisaster shoreline responses; the warehouse in the Pamunugama was used for waste storage following the cabinet approval during the later phase of the postdisaster responses.

5.7 LESSONS LEARNED AND RECOMMENDATIONS

Sri Lanka did not have previous experience managing this type of multidisaster in marine water. Therefore, as a nation, we learned many lessons from shoreline responses in the MV X-Press Pearl maritime multidisaster.

i) We have a NOSCOP to apply responses to the oil spill, but we did not have national contingency plans to adapt to the chemical or waste spill at sea. We did not know how to safely collect and store hazardous materials. Moreover, we did not have knowledgeable people and high-tech equipment to manage a dangerous chemical spill. Therefore, ad-hoc decisions had to be taken to manage postdisaster responses. It highlighted the importance of:
    - conducting more training programs to practice "how to manage the hazardous chemical spill or waste spill/capacity building"
    - introducing guidelines for plastic and hazardous chemical spill responses.

ii) It would be better if we could establish wildlife rescue teams along the beaches around Sri Lanka. Dead and adversely impacted marine animals were recorded all around the country and animals were transferred to the DWLC offices for further investigations and treatments. The time to move animals from the beach to the DWLC offices proved to be a crucial factor for the survival rate of the animals. If there was a rescue team at the sites, we could have treated them in advance at the beach.

iii) We collected impacted marine life from this maritime disaster once they showed the signs of toxic effects or after their death. If we could do the sampling for the sentinel species (i.e., filter feeder bivalves) in the affected regions during the early days of the disaster, we could have taken an idea about the acute impacts of hazardous materials and plastic nurdles on them. It would be beneficial for predicting the effects of created environmental disaster on the marine life at higher trophic levels before they show lethal symptoms to get precautionary practices to save their lives.

iv) During NOSCOP planning all procedural approvals and routings had been clearly established, and NOSCOP is updated annually; however, during the real disaster incident, the respondents and disaster management teams had to go through the same approval process in several cases. This preplanning and Schemes of Procedures (SoPs) have to be updated and practiced routinely, and identified lapses of the SoPs should be corrected immediately.

v) A well-trained crew has to be on board for the ship's casualty management (i.e., major structural failure, loss of life, collision, etc.), and we suggest introducing a casualty management plan for the cargo ships. We could have minimized the created ecological and economic disaster of the MV X-Press Pearl disaster if the ship had a proper casualty management plan and a well-trained crew for casualty management.

vi) There were overlaps among government institutes for the given provisions in the government acts to manage the onshore responses. Since it reduces the efficiency of the process, revising the conditions in government acts to assign specific tasks (minimizing the overlaps) for the relevant authorities in the onshore responses is recommended.

vii) Officials who engaged in the shoreline responses had to engage not only in the operation planning and executing the shoreline responses but also in some other works (i.e., procurement processes) due to a lack of human resources. It distracted the efficiency of the shoreline responses due to the burden created by the high workload. Establishing a permanent dedicated group under one umbrella ("one-stop-shop") for 24 hours × 7 days of services is strongly recommended during the disaster management period, and they should work only in their assigned positions to manage the disaster. IMT, the legal cell, and the damage assessment teams are suggested to be included in this group. We do not have an operational room to handle marine disasters; thus, this influences the marine disaster management process, decreasing the efficiency and productivity of the whole process. Establishing a common marine disaster operational center under DMC is highly recommended for the better management of marine disasters in the future.

ACKNOWLEDGMENTS

The authors acknowledge the SLAF Media Unit, MEPA, Ada Derana, and Daily Mirror webpages for granting permission to use their pictures for this publication, and Ms. Sandamali Galwaduge from Uva Wellassa University for her assistance in making maps for the publication.

REFERENCES

Australian Maritime Safety Authority. 2020. *Oil spill monitoring.* [online] Available at: www.amsa.gov.au/marine-environment/pollution-response/oil-spill-monitoring (Accessed 5 June 2022).

Brown, H., 2018. *Managing disaster preparedness and response for hybrid collections in Australian national and state libraries.* [online] Taylor & Francis. Available at: https://doi.org/10.1080/24750158.2018.1539903 (Accessed 25 April 2022).

Cumo, F., Gugliermetti, F. and Guidi, G., 2007. Best available techniques for oil spill containment and clean-up in the Mediterranean Sea. *Water Resources Management*, [online] 103. Available at: www.witpress.com/Secure/elibrary/papers/WRM07/WRM07049FU1.pdf (Accessed 5 June 2022).

Farzan, Z., 2022. *X-Press Pearl: Debris removal commences.* [online] Sri Lanka News – Newsfirst. Available at: www.newsfirst.lk/2022/01/03/x-press-pearl-debris-removal-commences/ (Accessed 23 July 2022).

Gunawardana, M., 2021. *Tackling Sri Lanka's Nurdle Problem.* [online] Roar.media. Available at: https://roar.media/english/life/environment-wildlife/tackling-sri-lanka-nurdle-problem (Accessed 4 June 2022).

Harrison, S. and Johnson, P., 2016. Crowdsourcing the Disaster Management Cycle. *International Journal of Information Systems for Crisis Response and Management*, 8(4), pp.17–40.

He, F. and Zhuang, J., 2016. Balancing pre-disaster preparedness and post-disaster relief. *European Journal of Operational Research*, 252(1), pp.246–256.

HELCOM, 2002. *Response to accidents at sea involving spills of hazardous substances and loss of packaged dangerous goods.* [online] https://helcom.fi/media/publications/HELCOM-Manual-on-Co-operation-in-Response-to-Marine-Pollution-Volume-2.pdf. Available at: https://helcom.fi/media/publications/HELCOM-Manual-on-Co-operation-in-Response-to-Marine-Pollution-Volume-2.pdf (Accessed 2 May 2022).

ITOPF. 2022. *Containment & Recovery*. [online] Available at: www.itopf.org/knowledge-resources/documents-guides/response-techniques/containment-recovery/ (Accessed 5 June 2022).

Joyce, C., Kunguma., and Andries, J., 2013. Public Awareness Campaigns, a Disaster Risk Reduction Strategy for Fire and Flood Hazards in the Western Cape, South Africa. *World Conference on Disaster Management*. 10.13140/2.1.3880.7687

Kim, T., Yang, C., Oh, J. and Ouchi, K., 2014. Analysis of the Contribution of Wind Drift Factor to Oil Slick Movement under Strong Tidal Condition: Hebei Spirit Oil Spill Case. *PLoS ONE*, 9(1), p.e87393.

Mallawaarachchi, A., 2021. *Fishing prohibited in sea area around X-Press Pearl*. [online] Available at: www.dailynews.lk/2021/05/29/local/250369/fishing-prohibited-sea-area-around-x-press-pearl (Accessed 12 May 2022).

MEPA, 2021. *Marine Environment Protection Authority-Sri Lanka*. [online] Available at: https://mepa.gov.lk/ (Accessed 25 April 2022).

National Academies of Sciences, Engineering, and Medicine. 2013. *An Ecosystem Services Approach to Assessing the Impacts of the Deepwater Horizon Oil Spill in the Gulf of Mexico*. Washington, DC: The National Academies Press. https://doi.org/10.17226/18387

National Building Research Organization, Ministry of Defence, 2021. *Air Pollution Impact Due to the Firing of MV X-press Pearl Ship Nearby Colombo Port*. [online] Nbro.gov.lk. Available at: www.nbro.gov.lk/index.php?option=com_content&view=article&id=336:air-pollution-impact-due-to-the-firing-of-mv-x-press-pearl-ship-nearby-colombo-port&catid=8&Itemid=190&lang=en (Accessed 24 June 2022).

NOSCOP, 2005. *The Sri Lanka National Oil Spill Contingency Plan*. [ebook] Marine Environment Protection Authority. Available at: www.coastguard.gov.lk/assets/images/regulations/oil_spill_prevention_plan/National_Oil_Spill_Contingency_Plan.pdf (Accessed 18 May 2022).

Oil Program Center, 1999. *Understanding Oil Spills And Oil Spill Response*. EPA Office of Emergency and Remedial Response.

Partow, H., Lacroix, C., Le Floch, S. and Alcaro, L., 2021. *X-Press Pearl Maritime Disaster Sri Lanka*. [online] UN Environmental Advisory Mission. Available at: www.unep.org/resources/report/x-press-pearl-maritime-disaster-sri-lanka-report-un-environmental-advisory-mission (Accessed 23 May 2022).

Pattiarachchi, C. and Wijeratne, S., 2021. *X-Press Pearl Disaster: An Oceanographic Perspective*. [online] Groundviews. Available at: https://groundviews.org/2021/06/08/x-press-pearl-disaster-an-oceanographic-perspective/ (Accessed 30 May 2022).

Perera, K., 2021. *The X-press Pearl Debacle – Event Flow*. [online] Environment Foundation (Guarantee) Limited. Available at: https://efl.lk/x-press-event-flow/ (Accessed 30 May 2022).

POSOW, 2016. *Oil Spill Waste Management Manual*. Malta.

Ratnayake, N., Ratnayake, A., Azoor, R., Weththasinghe, S., Seneviratne, I., Senarathne, N., Premasiri, R. and Dushyantha, N., 2018. Erosion processes driven by monsoon events after a beach nourishment and breakwater construction at Uswetakeiyawa beach, Sri Lanka. *SN Applied Sciences*, 1(1).

Response Techniques, 2022. *Dispersants*. [online] ITOPF. Available at: www.itopf.org/knowledge-resources/documents-guides/response-techniques/dispersants/ (Accessed 5 June 2022).

Rubesinghe, C., Brosché, S., Withanage, H., Pathragoda, D. and Karlsson, T., 2022. *X-Press Pearl, a "new kind of oil spill" consisting of a toxic mix of plastics and invisible chemicals*. [ebook] International Pollutants Elimination Network (IPEN). Available at: https://ipen.org/sites/default/files/documents/ipen-sri-lanka-ship-fire-v1_2aw-en.pdf (Accessed 20 June 2022).

Saplakoglu, Y., 2021. *Tons of toxic pellets blanket Sri Lanka beaches, causing environmental disaster.* [online] livescience.com. Available at: www.livescience.com/toxic-plastic-pellets-sri-lanka-beaches.html (Accessed 30 May 2022).

Shankar, D., Vinayachandran, P. and Unnikrishnan, A., 2002. The monsoon currents in the north Indian Ocean. *Progress in Oceanography*, 52(1), pp.63–120.

Shen, H., Perrie, W. and Wu, Y., 2019. Wind drag in oil spilled ocean surface and its impact on wind-driven circulation. *Anthropocene Coasts*, 2(1), pp.244–260.

Skavdal, T., 2003. *Introduction to Disaster Assessment and Assessment Methodologies.* [online] International Training Program on Total Disaster Risk Management. Available at: www.adrc.asia/publications/TDRM2003June/16.pdf (Accessed 5 April 2022).

Srinivas, H, Pre- and Post-Disaster Management: Environmental Management Tools to Reduce Disaster Risks. GDRC Research Output – Management Tools Series E-118. Kobe, Japan: *Global Development Research Center*. Available at: www.gdrc.org/uem/disasters/disenvi/tools/pre-post.html

UNEP, 2022. *Oil, acid, plastic: Inside the shipping disaster gripping Sri Lanka.* [online] UNEP. Available at: www.unep.org/news-and-stories/story/oil-acid-plastic-inside-shipping-disaster-gripping-sri-lanka (Accessed 19 April 2022).

Vinayachandran, P. and Yamagata, T., 1998. Monsoon Response of the Sea around Sri Lanka: Generation of Thermal Domes and Anticyclonic Vortices. *Journal of Physical Oceanography*, 28(10), pp.1946–1960.

Waravita, P., 2021. *"No fishing zone" to remain – The Morning – Sri Lanka News.* [online] The Morning – Sri Lanka News. Available at: www.themorning.lk/no-fishing-zone-to-remain/ (Accessed 5 July 2022).

Wimalasiri, H., Perera, H. and Jayathilaka, R., 2018. The current status of Negombo Lagoon fishery. National Aquatic Resources Research and Development Agency (NARA) International Scientific Sessions *2018*,

Withanage, H., 2021. *The X-Press Pearl Fire – A Disaster of Unimaginable Proportions.* [online] Groundviews. Available at: https://groundviews.org/2021/06/03/the-x-press-pearl-fire-a-disaster-of-unimaginable-proportions/ (Accessed 30 May 2022).

Figure References

Daily Mirror, 2021. In search of debris. [online] Dailymirror.lk. Available at: www.dailymirror.lk/caption_story/In-search-of-debris/110-216036 (Accessed 30 April 2022).

In search of debris ... – caption story: Daily Mirror (2021) *Caption Story | Daily Mirror.* Available at: www.dailymirror.lk/caption_story/In-search-of-debris/110-216036 (Accessed: April 30, 2022).

Marine Environment Protection Authority-Sri Lanka, 2022.

MEPA, 2022. MEPA Facebook page. [online] Facebook. Available at: www.facebook.com/photo?fbid=3790057704436789&set=pcb.3790070247768868 (Accessed 15 June 2022).

MEPA, 2022. MEPA Facebook page. [online] Facebook. Available at: www.facebook.com/photo?fbid=211932414428647&set=pcb.211932627761959 (Accessed 15 June 2022)

MEPA, 2022. MEPA Facebook page. [online] Facebook. Available at: www.facebook.com/photo?fbid=3953227101453181&set=pcb.3953240118118546 (Accessed 15 June 2022)

MEPA, 2022. MEPA Facebook page. [online] Facebook. Available at: www.facebook.com/photo?fbid=3962467697195788&set=pcb.3962469497195608 (Accessed 15 June 2022)

MEPA, 2022. MEPA Facebook page. [online] Facebook. Available at: www.facebook.com/photo?fbid=3953227364786488&set=pcb.3953240118118546 (Accessed 15 June 2022)

MEPA, 2022. MEPA Facebook page. [online] Facebook. Available at: www.facebook.com/photo?fbid=3784813771627849&set=pcb.3784814264961133 (Accessed 15 June 2022).

Over 10 turtles, dolphin, seabirds, several fish found dead after MV X-Press pearl disaster – breaking news: Daily Mirror (2021) Breaking News | Daily Mirror. Available at: www.dailymirror.lk/top_story/Over-10-turtles-dolphin-seabirds-several-fish-found-dead-after-MV-X-Press-Pearl-disaster/155-213540 (Accessed: June 30, 2022).

Removal of debris and removal of wrecked MV X press pearl. Marine Environment Protection Authority-Sri Lanka (2022). Available at: https://mepa.gov.lk/removal-of-debris-and-removal-of-wrecked-mv-x-press-pearl/. (Accessed: 29 June 2022)

Sri Lanka cleans up after massive container ship fire | pictures (2021) *Reuters*. Thomson Reuters. Available at: www.reuters.com/news/picture/sri-lanka-cleans-up-after-massive-contai-idUSRTXCQEUG (Accessed: June 20, 2022).

"X-press pearl" debris wash up on Sri Lanka Shores (2022) *Sri Lanka News*. Available at: www.adaderana.lk/news.php?nid=4012&mode=head (Accessed: June 20, 2022).

6 Societal and Media Response

Hemantha Withanage and Chalani Rubesinghe

6.1 INTRODUCTION

The X-Press Pearl ship accident is already nominated as the most damaging chemical spill in history. Undoubtedly, considering the burning and spillage of 1486 containers loaded with a variety of goods including 348 tons of bunker oil, 9700 tons of plastic pellets,[1] and 81 containers carrying dangerous goods, including 25 tons of nitric acid, 1040 tons of caustic soda, and 210 tons of methanol and various other heavy metals and persistent organic compounds that have many adverse impacts on the environment as well as humans.[2]

Sri Lanka was inexperienced in handling similar spills. The decision-making was slow as they were depending on mere predictions. Sri Lanka invested the best it could and obtained the support of every accessible source to put down the fire and ultimately decided to pull the burnt ship to the deep sea. But unfortunately, it sank amid the coral bed during the effort. Sri Lankan serene beach was soon covered with burnt plastic lumps, millions of plastic nurdles, and items washed ashore from toppled containers and sunken ship while all the chemicals on board mixed up with the water and sand.

6.2 READINESS TO MARITIME DISASTERS

Maritime disasters include both natural and manmade disasters. Natural disasters include tsunamis, flooding, coastal erosion, earthquakes, storms, variations in the climate, and coastal resource degradation. In contrast, manmade disasters include chemical, biological, radiological, and/or nuclear (CBRN) emergencies that involve the intentional or accidental release of toxic industrial chemicals, biological agents such as bacteria, viruses, fungi, and parasites or their products or parts of their products or chemical products that has the ability to release ionizing radiation.[3]

Amid these, concerns arise about the fact of readiness and resilience of both coastal communities and the regulatory authorities in facing these disasters. The most vulnerable group exposed to direct consequences of maritime disasters is the coastal community. This includes around 23% of the world's population of around 1.2 billion people living within 100 m of sea level and 100 km of a shoreline [3]. By no means does this exclude the community outside, as manmade accidents leave exigent remarks on the ocean life and physical factors that the whole population depends on. The best examples are the fish harvest affected by spills or the change in wave direction resulting from building or removing physical structures that determine the direction of the wind and thereby the waves. Studies reveal the factors behind this vulnerability to be high population density, poor and elderly portion of the community and degraded environmental conditions in hazard-prone areas, lack of awareness of maritime disasters, unequal and insufficient accessibility to basic services, and insufficiently built infrastructure as well as insufficient capacity to respond to maritime emergencies [3].

A video produced by the Community Conservation research network hosted at Saint Mary's University in Halifax, Canada, reveals that around 205 million people are affected annually by weather-related disasters.[4] It further talks about improving resilience (the capacity to recover quickly from hazards) and improving adaptability [the change by humans, societies (or other species) to become better suited to the state of the environment] [4].

Yet, around the world, while the planning and research on community resilience and adaptation are ongoing, their implementation and practical application are still a question. In order to ensure this resilience, the local authorities need to understand and implement the most appropriate mechanisms for the local context depending on the available resources, while the community stays aware of the possible hazards and measures to be taken in an event of a maritime disaster.

Sri Lanka has already demonstrated this lack of readiness during major maritime disasters it faced in recent history. Back in 2004 when Sri Lanka faced tsunami, people were unaware of tsunamis. Instead of quacking digging the information and disseminating warnings to the coastal community, media kept reporting what was going on. It was seen that people were trying to mark the portions of the shoreline when the seawater line was retrieved and carried televisions instead of essentials when the first wave came. During the New Diamond ship fire, the authorities did not issue any warning to the general public; rather, they were preparing a contingency plan to face the oil spill that would require a total of 2500 m of booms*[5] instead of the 1500 m of booms currently in hand[6] [5]. There was no mechanism in practice to raise awareness for the coastal community, especially fishermen, on potential threats or measures to be taken if an oil spill was to occur.

6.3 SOCIETAL AND MEDIA RESPONSES TO MARITIME ACCIDENTS AROUND THE WORLD

A review presented by Joseph Scanlon on "Research about the Mass Media and Disaster: Never (Well Hardly Ever) The Twain Shall Meet" explains the behavior and context of media in reporting disasters, pointing out the importance of media in communicating a disastrous situation before, during, and after the incident emphasizing the effective communication of warning. He also draws a lot of examples to argue that media is a tool to handle with care since it has the ability to distort information, resulting in miscommunication and unnecessary panic [6].[7]

The engagement of mainstream media and social media is vital in communicating the warning effectively. They are the major tools that would communicate about a disaster encouraging protective behavior among public in preparation for disasters, give warnings, trigger particular behaviors, and disseminate information about a hazard, allowing to minimize its impact as well as to make decisions [7].

Media can be the channel to build resilience among people. In disaster-prone areas or to vulnerable communities, the media can help disseminate information on actions to be taken in a time of disaster or protective behaviors frequently so that people are able to react immediately to a particular situation. Information on the types of disasters, warning signs, and consequences that triggers certain protective behavior can also be disseminated through media. Both electronic and social media can reach the public to disseminate early warnings.

During the X-Press Pearl ship disaster, second to the observation of fishermen, mainstream media was the channel that first reached the coastal community with the message of a burning ship. Had the community been informed about the potential threats of a chemical spill, the exposure to chemicals could have been reduced. But this information was not available to media. The authorities themselves were unaware of the potential hazards of the fire and the potential consequences of issuing necessary warnings to people through the media. The involvement of science in the analysis and deriving of information in this particular situation seemed to be very slow.

Media alone cannot address disaster risk reduction. The media is the bridge between information and the general public for sure. But it is essential that the media is given correct information before they generate information that could mislead to suspicion, resulting in outrage.[8]

If the relevant Authority/ies have a risk communication plan to disseminate this information before the media derive stories out of context and out of public or personal opinion, the outrage could be managed. Applying the context to the X-Press Pearl incident, the inappropriate behavior in people, like collecting items potentially carrying toxic contaminants or fishing in the affected zone, could have been avoided if the authorities went to media with: expression of empathy, clarification of facts/call for action (who will be doing what, when, where, how and why), plans of the state authorities to find out the missing information, a statement of commitment, and clear referrals for information to media as well as the affected community [8].

During the major maritime disasters Sri Lanka previously faced, it was quite obvious that the coastal community as well as the authorities were unaware of the measures to be taken in such instances. In the case of tsunami, media was not effectively used to communicate the warnings to people. Rather they kept reporting what was happening, while the authorities could have used mainstream media to evacuate people from the area, if proper notices were issued and properly guided. Symonds Peter in his compilation on "The Asian tsunami; why there were no warnings" had analyzed the responses to tsunami in four countries. He highlights that none of the four countries (Indonesia, Sri Lanka, Thailand, and India) issued an official warning, making people completely helpless in the face of approaching waves[9].

In relation to incidents of pellet spills, it seems that around the world, this has become one of the common maritime accidents; in many cases, governments and nongovernmental organizations and the community get together in clean-up efforts. But the one special feature of the accident that occurred in Sri Lanka was that it was associated with the burning of other hazardous and some persistent chemical pollutions, and these particles too were washed up ashore. Therefore, this media reporting was supposed to have some scientific explanations taken from toxicologists. In fact, most of the reports from other countries do explain the impact of plastic pollution when they report nurdle spills, but they hardly talk about the other potentially toxic contaminants washing ashore along with these tiny plastic bodies.[10,11,12,13] This gap in knowledge transfer clearly hinders the information flow to the community that receives information through social or mainstream media. In Sri Lanka, it is also observed that language is another barrier to the flow of information as different languages report incidents on different scales.

6.4 THE AFFECTED COMMUNITY AND THE AWARENESS OF THE HAZARDS OF THE SHIP ACCIDENT

The coastal community was little aware of the incident and its possible consequences. Although some of them observed the fumes coming from the ship[14] (Kamilas Perera, Negombo), they were unaware of the real incident, until they heard the story from media. It must be noted that although there were communication systems that made the fishermen aware of certain weather conditions, this particular incident was not informed through those communication systems.

It was evident that people immediately started collecting items washed ashore. Because this was a period when people were suffering due to the COVID-19 pandemic as well as "Yaas" cyclone. Thus, even though they were unaware of a particular use of the bags of plastic pellets and other items that were carried by ship, they still collected them until the authorities deployed NAVY and police to banned the area restricting access. However, the correct information on why they should not be touched or collected was not communicated. Therefore, while some people returned the collected goods in fear of getting fined, some kept collecting pellets from sand with the intention of selling them as seen in Figure 6.1.

Although there was a fear of eating fish first, eventually fishermen convinced the public to eat fish while relevant authorities who conducted analysis on biological samples kept their lips sealed.

During interviews with the community, almost 100% admitted that they did not know of the harmful effects of the chemical spill that occurred.

Due to lack of awareness of harmful chemicals carried ashore with these items, people were exposed to them through touching (skin contact), inhaling, and eventually ingestion. People were not even afraid of having a bath in the sea despite the shoreline carrying considerable amounts of pellets washed ashore as seen in Figure 6.2.

Since the people were desperate for money in the absence of a proper income from fishing, many women willingly engaged in the "money for work" program that deployed women for cleaning pellets as well as other plastic pieces from the beach. These women were often observed handling sand without protective gear, although they were provided to them as seen in Figure 6.3.

FIGURE 6.1 A woman collecting sand to separate out the plastic pellets. Photo Courtesy: Mr. Hemantha Withanage.

6.5 THE CONSEQUENCE OF THE SHIP ACCIDENT ON LIVELIHOODS

The impact made on the livelihood of people in the area by this accident was different from that of the COVID-19 pandemic. Although people were restricted from the individual sale of fish due to restricted movement and gatherings, fishermen were still able to go fishing and to sale within limitations. But, after the accident, people were restricted from fishing due to the chemical contamination of the massive spill of plastic nurdles and burnt particles washed ashore making the beach inaccessible.

Fishing equipment such as fishing nets was damaged. Fishermen who went fishing during the accident showed the damaged fishing nets. This had a huge impact on their industry. Some whitish substances similar to melted plastics or burnt pieces were entangled in the nets. It was not easy to remove them and fishermen had to abandon these nets. Some are which cost around LKR40,000 per two kits of fishing nets (M.R. Fernando, Negombo). Replacing these nets was a huge loss in itself. Fishermen complained that until they themselves experienced it, authorities did not make them aware of such consequences of going fishing (Mr. K.A. Karunaratne, Kalutara).[15]

The fish catch was affected. Dead fish with pellets in their bodies were found on the beach as well as in water. Carcasses of dolphins, turtles, and many other species including massive amounts of dead mussels were found on the seashore. But

FIGURE 6.2 Despite the contamination, sunken ship, and the pellets washing ashore, people were enjoying bath in the sea. Photo courtesy Harshani Abayawardhana.

authorities did not reveal to people a reason or at least issue any guidelines on fish consumption. Leaving people freely exposed to contaminated seafood, sand, and seawater.

The compensation for the ship accident does not seem to have considered all strata among the coastal inhabitants. The most affected among the coastal inhabitants were those that directly and indirectly depended on fishing.

Direct dependents:

- Offshore multiday boat operators. This group was affected for a shorter period, during the imposed fishing ban that extended up to 16 km from the shoreline

Societal and Media Response

FIGURE 6.3 Unaware of contaminants contained or contaminated in the sand and pellets, some women engage in beach cleaning for the sake of payment. Sadly, these "money for work" programs expose employees to potentially harmful chemicals despite the absence of health scan programs to reveal the health impact. Picture was taken by Mr. Janaka Kumara.

toward the sea as seen in Figure 6.4. But eventually, it was reduced up to 8 km as shown in Figure 6.5.
- Nearshore boat operators. Since the ship sunk on a coral reef, this directly impacted the catch. On the other hand, all fishing operations conducted within 8 km of the shoreline were stopped due to the fishing ban (Figure 6.5). In addition, massive amounts of pellets, burned lumps, and plastic chips made the shoreline inaccessible for fishermen for some time. Although the undetermined chemical contamination was not a concern, this fact indirectly affected the catch.
- Helpers of the boat. When the boats weren't sailing, fishing assistants who would join the crew on payment lost their income.
- Fish sorters. Once the catch is brought ashore, there's a set of people to sort fish from the fishing net and get paid. Amid those who lost their income due to inability to go for fishing were these people too.
- Fish vendors. Sellers in wholesale, retail, and small-scale fish selling lost their lifeline due to the ship accident. Even amid the COVID-19 pandemic, they were able to make sales and make some income to sustain themselves because boats were able to go fishing. But after the accident, on one hand, the fishermen were prevented from fishing and on the other hand, people avoided eating fish (for all the correct reasons). Among the worst impacted sellers are women who were in some households the only breadwinner of the family. But compensation didn't reach this level.
- Traditional *Theppam* (type of traditional raft) operators are another group of fishermen who lost their income for restrictions on nearshore fishing.

128 Maritime Accidents and Environmental Pollution

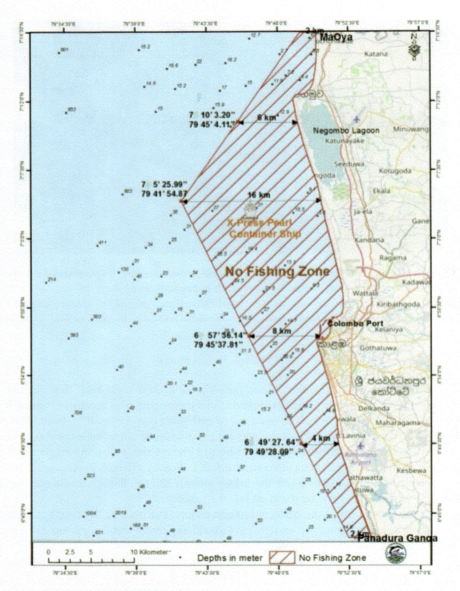

FIGURE 6.4 First fishing ban imposed.

- Hindered the grounds used for some livelihood operations. There were a group of people who have been using the coastal areas for various livelihoods. Making dry fish was one of them in which the beach was used to dry the fish. But as the beach was all covered with plastic nurdles (visually) and other burned pieces and tiny pieces of plastic and other material (below 5 mm in

Societal and Media Response

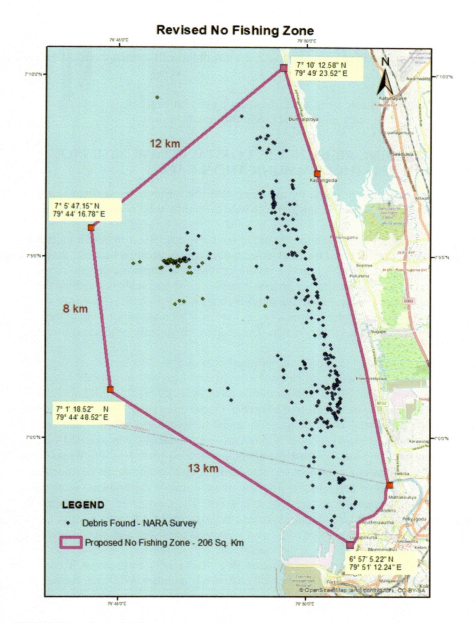

FIGURE 6.5 Revised fishing ban imposed.

size) this group lost their place and thereby were unable to make dry fish. Some of them cleaned the beach themselves and gathered the content (getting exposed to chemicals in them) in gunny bags but were unaware of a way for their disposal.

Indirect dependents:

- People who have been providing services to these fishermen, such as food, lost their income.
- Tourists.
- People who sold items to tourists coming to visit beaches.
- Divers that catch ornamental fish in the nearshore area.

6.6 BUILDING RESILIENCE IN COASTAL COMMUNITIES TO ADDRESS THE COMMUNICATION GAPS

Even in the idealistic scenario, assuming that the information is disseminated to media and through them to the public, would the public react correctly to a maritime disaster?

Several factors influence the situation, including the extent to which individuals are aware of the impact of the disaster or hazard, communication barriers, challenges faced by individuals with hearing, visual, or cognitive impairments who might not receive appropriate warnings or process complex information about potential threats. Additionally, limited access to warning facilities like mobile devices or televisions, along with personal beliefs in misconceptions or false information, can hinder individuals from taking necessary action. These factors frequently heighten vulnerability to disasters.

Whether people use electronic, printed, or social media channels for receiving information as well as which channel/s in particular greatly contributes to communication gaps. Distribution of information through social media would not benefit a community that has no access to the internet. On the other hand, less educated women or blind people who do not engage in social networking platforms read or write messages on their mobile phones will not receive messages communicated via short message service (SMS) or other social communication platforms like WhatsApp or Viber. Thus, it is important that communication pay attention to the composition of the target audience and employ different communication methods to communicate the message to as wider an audience as possible. The message in all languages used by the target/affected community; message in the forms of text messages, fact sheets, press releases, recorded video clips; and voice cuts to stream and broadcast through mainstream media are some of the tools.[16]

It is also important that information flow works both ways when it comes to responsible government authorities and coastal communities. While the coastal community can provide first-hand information to the authorities, the authorities can elaborate, examine, and analyze this information and make informed decisions as well as issue correct statements to the public and media.

A huge gap in communication was clearly visible throughout the series of events related to the ship accident.

First, the people were not timely made aware of the incident and were not prevented from fishing activities or instructed to be on standby for a possible hazard. They were only made aware of the ship fire through electronic media after the fire was almost out of control. Even then, they were not informed about engaging in fishing activities.

Once people started collecting items, authorities instructed them to refrain from collecting items washed ashore and in some areas, people were restricted from reaching the beach inundated with plastic nurdles, but they were not properly instructed on the reason. As a result, it could be seen that some people were collecting the plastic nurdles contaminated with hazardous chemicals. Some people were found to be spreading them over their home gardens despite the heavy metals BPA (bisphenols A, an endocrine disruptive chemical) and PAHs (polyaromatic hydrocarbons, a carcinogen and an endocrine disruptive chemical) carried, as revealed by the study conducted by the Centre for Environmental Justice [2]. Therefore, it is necessary that communications on similar maritime disasters where a chemical spill is associated be accompanied by warnings on possible short-term and long-term health effects of chemical contaminants.

In this particular incident, it was observed that even the media were restricted with the information given out to the public rather than revealing the science behind it. In similar maritime disasters, it is important to follow a procedure in assuring the safety of seafood consumption. But this detail was not properly communicated during this incident. Thus, we believe that not only people residing by the beach but also those who live across the island who consume seafood from the south to north, the coastline of the western part of the country was possibly exposed to chemical contamination through seafood to some extent. Therefore, this information on "recommendations to consume seafood and engage in fishing" should be incorporated into communication plans for building better resilience among the coastal community to better prepare for maritime disasters. This would be the key to ensuring safe consumption of seafood consumption; as well, fishermen are directed to fish where the catch is safe.

In summary, it could be said that, though this ship accident was immensely damaging to the environment as well as the coastal community in the area, it also gave an opportunity for Sri Lankan authorities, the coastal community, and the media to build resilience to face similar events in the future as being a country situated in one of the main shipping routes, Sri Lanka is more vulnerable to similar incidents.

NOTES

1. Partow, H.L., C., Le Floch, S., and Alcaro, L. 2021. X-Press Pearl maritime disaster Sri Lanka report of the UN Environmental Advisory Mission July 2021., UN Environmental Advisory Mission.
2. Rubesinghe, C., Brosché, S., Withanage, H., Pathragoda, D., and Karlsson, T. X-Press Pearl, a "new kind of oil spill" consisting of a toxic mix of plastics and invisible chemicals. International Pollutants Elimination Network (IPEN). (https://ipen.org/sites/default/files/documents/ipen-sri-lanka-ship-fire-v1_2aw-en.pdf)
3. U.S. Indian Ocean Tsunami Warning System Program. 2007. How resilient is your coastal community? A guide for evaluating coastal community resilience to tsunamis and other coastal hazards. U.S. Indian Ocean Tsunami Warning System Program supported by the United States Agency for International Development and Partners, Bangkok, Thailand. 144 p. (www.crc.uri.edu/download/CCRGuide_lowres.pdf)
4. Charles, A., Coastal communities at the ready, Nexus Media. Executive Producer: Anthony Charles. Funded by the Marine Environmental Observation,

Prediction and Response Network (MEOPAR), the Community Conservation Research Network (CCRN) and Saint Mary's University. (www.communityconservation.net/coastal-communities-at-the-ready/)

5 * A boom is a plastic contraption that is generally used to reduce the spread and consequent destruction of marine resources as a result of oil spills.

6 Jayawardhena, S. and Weerasinghe, T. 2020. Burning questions as fire onboard MT New Diamond simmers down. The Sunday Times. (www.sundaytimes.lk/200906/news/burning-questions-as-fire-onboard-mt-new-diamond-simmers-down-415525.html)

7 Scanlon, J. 2016. Research about the Mass Media and Disaster: Never (Well Hardly Ever) The Twain Shall Meet. (https://training.fema.gov/emiweb/downloads/scanlonjournalism.pdf)

8 There are several factors that contribute to public outrage. Such include; the nature of its origin (voluntary or coerced, natural or human origin, etc.), level of controllability, relevancy, and reaction from the authorized personnel/institutions. In a disaster situation, it is important that the authorities deploy methods to manage outrage.

9 Symonds, P. 2005. The Asian tsunami: Why there were no warnings. World Socialist Web Site. (www.wsws.org/en/articles/2005/01/warn-j03.html)

10 Reuters. 2012. Hong Kong government criticized over plastic spill on beaches. (www.reuters.com/article/us-hongkong-spill-idUSBRE87306J20120805)

11 The Great Nurdle Disaster: What to do if you find nurdles. 2018. Two Oceans Aquarium Cape Town, South Africa. (www.aquarium.co.za/blog/entry/the-great-nurdle-disaster-what-to-do-if-you-find-nurdles)

12 Nurdle hurdle: Clean-up of KZN beaches far from complete. 2018. news 24. (www.news24.com/News24/nurdle-hurdle-clean-up-of-kzn-beaches-far-from-complete-20180112)

13 Saliba, A., Frantzi, S. and van Beukering, P., 2022. Shipping spills and plastic pollution: A review of maritime governance in the North Sea. Marine Pollution Bulletin, Volume 181. (www.sciencedirect.com/science/article/pii/S0025326X2200621X)

14 "Our experiences on X-Press Pearl ship accident" – Part 1, Centre for Environmental Justice. (https://youtu.be/NADsIoVsBjI)

15 Our experiences on X-Press Pearl ship accident – Part 2. (https://youtu.be/MOWSgaO_Lm8)

16 Centre for Environmental Justice. 2021. Risk communication strategy for maritime disasters, United Nations Development Programme (UNDP) in Sri Lanka.

REFERENCES

1. Partow, H.L., C., Le Floch, S., and Alcaro, L. 2021. X-Press Pearl maritime disaster Sri Lanka report of the UN Environmental Advisory Mission July 2021., UN Environmental Advisory Mission.

2. Rubesinghe, C., Brosché, S., Withanage, H., Pathragoda, D., and Karlsson, T. X-Press Pearl, a "new kind of oil spill" consisting of a toxic mix of plastics and invisible chemicals. International Pollutants Elimination Network (IPEN). (https://ipen.org/sites/default/files/documents/ipen-sri-lanka-ship-fire-v1_2aw-en.pdf)

3. U.S. Indian Ocean Tsunami Warning System Program. 2007. How resilient is your coastal community? A guide for evaluating coastal community resilience to tsunamis and other coastal hazards. U.S. Indian Ocean Tsunami Warning System Program

supported by the United States Agency for International Development and Partners, Bangkok, Thailand. 144 p. (www.crc.uri.edu/download/CCRGuide_lowres.pdf)
4. Charles, A., "Coastal communities at the ready, Nexus Media. Executive Producer: Anthony Charles. Funded by the Marine Environmental Observation, Prediction and Response Network (MEOPAR), the Community Conservation Research Network (CCRN) and Saint Mary's University. (www.communityconservation.net/coastal-communities-at-the-ready/)
5. Jayawardhena, S. and Weerasinghe, T. 2020. Burning questions as fire onboard MT New Diamond simmers down. *The Sunday Times*. (www.sundaytimes.lk/200906/news/burning-questions-as-fire-onboard-mt-new-diamond-simmers-down-415525.html)
6. Scanlon, J. 2016. *Research about the Mass Media and Disaster: Never (Well Hardly Ever) The Twain Shall Meet*. (https://training.fema.gov/emiweb/downloads/scanlonjournalism.pdf)
7. Mao, Q., Li, N., Fang, D. 2020. Framework for modeling multi-sector business closure length in earthquake-struck regions. *International Journal of Disaster Risk Reduction*, Volume 51, 101916. (www.sciencedirect.com/science/article/pii/S2212420920314187)
8. Northwest Centre for Public Health Practice, School of Public Health, University of Washington. (n.d.). Emergency Risk Communication. (www.nwcphp.org/docs/riskcomm/toolkit/ercprint.pdf)
9. Symonds, P. 2005. The Asian tsunami: Why there were no warnings. *World Socialist Web* Site. (www.wsws.org/en/articles/2005/01/warn-j03.html)
10. Reuters. 2012. Hong Kong government criticized over plastic spill on beaches. (www.reuters.com/article/us-hongkong-spill-idUSBRE87306J20120805)
11. The Great Nurdle Disaster: What to do if you find nurdles. 2018. Two Oceans Aquarium Cape Town, South Africa. (www.aquarium.co.za/blog/entry/the-great-nurdle-disaster-what-to-do-if-you-find-nurdles)
12. Nurdle hurdle: Clean-up of KZN beaches far from complete. 2018. news 24. (www.news24.com/News24/nurdle-hurdle-clean-up-of-kzn-beaches-far-from-complete-20180112)
13. Saliba, A., Frantzi, S. and van Beukering, P. 2022. Shipping spills and plastic pollution: A review of maritime governance in the North Sea. *Marine Pollution Bulletin*, Volume 181. (www.sciencedirect.com/science/article/pii/S0025326X2200621X)
14. "Our experiences on X-Press Pearl ship accident" – Part 1. Centre for Environmental Justice. (https://youtu.be/NADsIoVsBjI)
15. Our experiences on X-Press Pearl ship accident – Part 2. (https://youtu.be/MOWSgaO_Lm8)
16. Centre for Environmental Justice. 2021. Risk communication strategy for maritime disasters. United Nations Development Programme (UNDP) in Sri Lanka.

7 Plastics, Nurdles, and Pyrogenic Microplastics in the Coastal Marine Environment
Implications of the X-Press Pearl *Maritime Disaster*

Madushika Sewwandi, Kalani Imalka Perera, Christopher M. Reddy, Bryan D. James, A.A.D. Amarathunga, Indika Hema Kumara Wijerathna, and Meththika Vithanage

7.1 PLASTIC SPILLS IN THE MARINE ENVIRONMENT

Plastic is ubiquitous in the ocean, which has become a significant repository for the material. One common form of plastic found in the ocean is the resin pellet or "nurdle". Nurdles are the raw material used at the production stage of plastic products and come in different sizes and colors (Fendall and Sewell, 2009). They are primary microplastics, being roughly 3 to 5 millimeters in diameter. Loss of nurdles during manufacturing, shipping, storage, and waste management are thought to be significant sources of them entering marine habitats (Karlsson et al., 2018). Millions of nurdles are discharged from production facilities into the nearby waterways every year (Plastic, 2022). Additionally, nurdles are released into the environment by frequent small-scale spills (nurdles leaking from shoddy-sealed bags that are simple to tear or puncture when handled manually or mechanically) and rarer large spills (from tons of pellets releasing into the environment at once) (Cabrera, 2022).

Over the years (from 20th century), roughly 1400 maritime accidents have occurred worldwide due to issues related to anchoring, crashes, fires, emissions, malfunctions, and machinery damage (Häkkien et al., 2013). Frequently, large plastic spills result from these accidents and often affect coastal areas nearby and far away (Annette and Gaëlle, 2020; Partow et al., 2021; Plastic, 2022). Volunteer labor is commonly used, in some capacity, during cleanup operations (Annette and Gaëlle, 2020; Partow et al., 2021).

In 1993, the bulk carrier SS *Hamada* capsized close to Abu Ghosun in Egypt's Wadi el Gemal National Park (Brümmer et al., 2022). The foundering caused the release of a portion of its plastic pellet cargo. In 2012, 150 tons of plastic pellets spilled into the ocean near Hong Kong when seven containers were lost from a container ship during Typhoon Vicente (Rochman, 2013). Six of the lost containers held polypropylene (PP) pellets (Releases, 2014; Rochman, 2013). Another spill happened in 2017 when containers were lost from the carrier MSC *Susanna* anchored in Durban, South Africa. An upper air cut-off low that caused the country's south coast to experience flash flooding and record rainfall caused the spill (Schumann et al., 2019). About 49 tons of nurdles littered beaches north and south of Durban, heavily polluting the northern KwaZulu-Natal coast. In 2019, approximately 700 locations along the Norwegian coast, as well as a few locations in Denmark and Sweden, were contaminated by plastic pellets after 270 containers fell off the MSC *Zoe* cargo ship (Annette and Gaëlle, 2020). A year later, a container was lost from the MV *Trans Carrier* in the North Sea around Norway, Denmark, and Sweden, spilling 13 tons of plastic pellets (Annette and Gaëlle, 2020). Following the incident, approximately 24 million nurdles washed onto the northern Dutch coastline, primarily harming Wadden Island's biodiversity (Annette and Gaëlle, 2020). Another spill in 2020 happened following the collapse of the container ship CSAV *Trancura*, with 30 tons of plastic pellets washed up on Cape Beach in South Africa (Annette and Gaëlle, 2020). That same year, two containers were lost from the MV CMA CGM *Bianca* in the Port of New Orleans, New Orleans, Louisiana, USA, spilling ~25 tons of polyethylene pellets (Dermansky, 2020; Dhanesha, 2022). The most recent container ship-related nurdle spill happened on May 20, 2021, when the cargo ship MV *X-Press Pearl* (*XPP*) caught fire while anchored 18 km northwest of the port of Colombo in the Sri Lankan Sea. The fire is believed to have been started following a nitric acid leak from one of the ship's containers.

7.2 PLASTIC AND NURDLES FROM THE *XPP* ACCIDENT

The *X-Press Pearl* shipwreck was the worst chemical and plastic maritime disaster ever from a single ship in Sri Lankan maritime history. According to the latest records for the cargo, more than 1750 tons of plastic pellets and 9700 tons of epoxy resin were onboard the *XPP* at the time of the event. The *XPP* cargo included 11939.2 metric tons of plastic: epoxy resins, synthetic resins, high-density polyethylene (HDPE), low-density PE (LDPE), linear low-density PE (LLDPE), polyvinyl chloride (PVC) films, "ALKYD" resins, formulations of PP, polymeric beads, plastic pellets, polycarbonate, unsaturated polymer resins, packaging materials, bare foam pig (drying or sweeping materials), and packages of polystyrene (PS) pellets, polybutadiene, vinyl copolymers, and vinyl acetate (Hassan et al., 2021).

During the ship fire, plastic was reported washing up along a ~300 km stretch of the western, northern, and southern coastline of Sri Lanka. Sarakkuwa Beach received the greatest amount of plastic, while the greatest concentration (plastics/m^2) was reported along the northern coastline of Colombo. Unlike previous plastic spills, a portion of the plastic was burned. Along with nurdles, washing onto the Sri Lankan coastline were pyrogenic plastic fragments or "pyroplastics" of varying size, shape, and color; other plastic debris; damaged sacks of nurdles; and foam of pyrogenic materials (Figure 7.1). Nurdles were the most abundant form of debris.

FIGURE 7.1 Various plastic-based contaminants spilled during the X-Press Pearl ship disaster.

7.2.1 Plastic Nurdles

Most of the plastic nurdles that were carried in containers on the ship were damaged and wrecked during the fire (Partow et al., 2021). Sometime during or after the event, one container of plastic pellets drifted over 100 km south and was stranded on an island (Sri Lanka) close to the coast. Thus, plastic nurdles spilled from the wreck and the drifted container. Though the nurdles washed ashore in greatest quantity during the fire, nurdles have continually been reported along the western and eastern coast. Nurdle deposition was not limited to the nearshore, but they have also been found in the backshore (about 1 m from the waterline). A few recent studies confirmed the nurdles were polyethylene, discussing whether they were HDPE, LDPE, or LLDPE (de Vos et al., 2021; Sewwandi et al., 2022).

7.2.2 Pyrogenic Microplastics

Brittle burned plastic fragments dark green or black in color also covered the beaches (Figure 7.1), though to a lesser extent than the nurdles. The pyrogenic microplastics washing ashore consisted of several polymer types, likely because of the wide variety of plastics that were onboard the ship at the time of the fire. A few studies identified the burned plastic as either LDPE or HDPE based on their density, IR spectra, and solvent-extractable materials (de Vos et al., 2021; Sewwandi et al., 2022). Notably, the pyrogenic plastic had ~3–5% extractable mass while the nurdles had 0.1–1% (de Vos et al., 2021). In comparison, nurdles spilled from the *Bianca* (a non-fire-related spill) had <0.1% extractable mass. Other studies found that some fragments had characteristic features of polyethylene terephthalate (PET) based on their IR spectra and thermogravimetric properties (Sewwandi et al., 2022).

7.2.3 LARGE, PARTIALLY PYROLYZED PLASTIC DEBRIS

Partially pyrolyzed large plastic debris also washed ashore. The textures of these large pieces differed, and their surfaces were likely critically damaged due to the fire (Figure 7.1). Some debris included melted nurdles agglomerated to partially melted plastic packaging. Chemical modification of the plastic's surfaces complicated the identification of the larger pyrolyzed pieces. Some of these pieces have been recognized as epoxy resins, PET, natural rubber, cyclic olefin copolymer, and aromatic polyamides. Because the pyrolyzed plastic debris is brittle, formation of microplastics is expected.

7.3 NURDLE POLLUTION AND DISTRIBUTION

7.3.1 NURDLE SPREADING ALONG SRI LANKA'S WESTERN COASTLINE

After the explosions on the *XPP* on 25 May 2021, containers started falling overboard. Plastic nurdles were spilled and spread over the sea due to the prevailing high current waves. On May 26, a massive amount of floating debris, including plastic pellets, had washed into the nearby coastal belt from the burning ship. Severe nurdle dispersion was reported two weeks after the nurdle spillage (Jayathilaka et al., 2022). Between 10 m and 400 m depth at the incident site's surface, the nurdles' predicted maximum mass flux was 3.5 kg/m^3 (Karthik et al., 2022). The rapid movement of waves can readily spread nurdles more than other water systems and the recovery is impossible (Lentz and Fewings, 2011). The nurdle dispersion suggested that the main factor influencing their transport was variations in ocean current patterns (Jayathilaka et al., 2022). At the time, the sea around Sri Lanka was rough, and the wind currents moved fast due to the prevailing southwest intermonsoon and the Yaas cyclone, which propagated toward the country through the northwest direction. Thus, the ocean circulation trends near Sri Lanka were influenced by the temporal and spatial wind fields and currents, which may also have impacted nurdle scattering (Jayathilaka et al., 2022). Storm surge incidents may have influenced the pattern of nurdle deposition, as well. High swell waves and surges formed by the strong wind changed tidal currents and likely caused the nurdles to scatter around the western coastline. According to analysis of nurdle trajectories, the nurdles spilled by the ship could have been transported a 1000 km seaward along the Sri Lankan coast in 60 days in the northwest and southeast direction by the predominance of strong monsoon winds (Karthik et al., 2022). Accordingly, rather than just the proximity of the source and its physical characteristics, extreme or storm events primarily regulated the accumulation and distribution of nurdles on to beaches. Swashing influenced the movement of nurdles on the beaches after their accumulation. Thus, beach elevation, the slope of the beach, and the direction where the high-energy waves face the beach affected the extent of nurdle deposition (Geographer Waves, 2022). Nurdle deposition is more likely to occur in beach areas with elevations greater than those calculated for the shoreline (Ferreira et al., 2021). Because the transport of nurdles and other floating debris in the ocean is altered by the wind and wave current changes that have occurred, such debris could not only negatively affect cleanup efforts but also challenge identification and removal of the debris from the coastal and marine environments.

On May 26, 2021, nurdles covered nearly one-third of the coastline from Galle to Kalpitiya. Reports described nurdles had accumulated on a few beaches along the northern coast above Colombo and the southeast coastline, as well (Partow et al., 2021; Perera et al., 2022). Compared to the other regions, the southern coast of Sri Lanka was identified as a hotspot with the highest nurdle mass flux of 3.5 g/m^3 (Karthik et al., 2022). Accordingly, during the *XPP* shipwreck period, the distribution patterns of the floating nurdles spread to the southern coast of Sri Lanka and offshore regions under the influence of wind fields and tides. During a beach survey we conducted just after the shipwreck incident (June to July 2021), nurdles were observed on Pesali Beach, Norochcholai Beach, Chilaw Beach Park, Marawila Beach, Negombo Beach, Dungalpitiya Beach, Sarakkuwa Beach, Wellawatta Beach, Moratuwa Beach, and Midigama Beach. Some beaches were entirely covered with nurdles (Figure 7.2). Sarakkuwa and Epamulla beaches were identified as the most polluted beaches among the 40 sites surveyed along a 17 km stretch of the western coast of Sri Lanka (beaches from Uswetakeiyawa to Basiyawatta) (Perera et al., 2022). Between Matara and Kalpitiya, about 40% of the coast was extremely polluted with plastic nurdles and pyrogenic plastic debris. Receipt of nurdles on the eastern coast of Sri Lanka implied transit through Sri Lanka Dome. Nurdles were not found to have accumulated along Sri Lanka's northern coastline, suggesting they had not traveled through the Polk Strait (Jayathilaka et al., 2022).

7.3.2 Nurdle Distribution to the Beaches Outside of Sri Lanka

The wind and seasonal changes can impact the transport of spilled plastic pellets from the *XPP* disaster site to elsewhere. Nurdles were able to transit in and around the Sri Lankan coast and into the open ocean rather than transporting to the southeastern coast of India due to the strong circulation features that were present in the southern beach locations along the Palk Bay and Gulf of Mannar at the time of the disaster. Nurdles that were dispersed beyond the Sri Lankan coast may have higher tendency to wash up on the shores of countries in the northern Indian Ocean, including Indonesia, India, the Maldives, Somalia, and others. But no evidence has been reported to date. However, months after the spills off the coast of South Africa, nurdles began washing up on the western coast of Australia (Pepper, 2019). Understanding nurdles' fate, distribution, and transboundary transport can be aided by modeling studies. One modeling study suggested that the nurdles moved south of the Sri Lankan coast and that there has been no significant accumulation of plastic from the spill on the Indian coast (Karthik et al., 2022).

Implications of the *X-Press Pearl* Maritime Disaster 139

A: Pesalai beach, B: Norochcholai Beach, C: Chilaw beach park, D: Marawila beach, E: Negambo beach
F: Dungalpitiya beach, G: Sarakkuwa beach, H: Wellawatta beach, I: Moratuwa beach, J: Midigama beach

FIGURE 7.2 Polluted beaches on Sri Lankan west coastline after the X-Press Pearl shipwreck incident (Photographs were taken from 1 to 15 June 2021).

7.3.3 NURDLES ON SARAKKUWA BEACH

7.3.3.1 During May 2021, September 2021, and October 2022

After the *XPP* shipwreck incident, nearby Sarakkuwa Beach was covered in mounds of nurdles (Figures 7.3a, 7.4b, 7.4c). Sarakkuwa was the closest beach to the shipwreck (Figure 7.4a). On 26 May, the entire beach was covered with a broad layer of nurdles and other pyrogenic plastic (Figure 7.3a). Nurdles white in color were most abundant; nurdles yellow, brown, and black in color were fewer. Possible causes of color change include burning, weathering, oxidation, and contaminant binding (James et al., 2022). The pellet pollution index (PPI) at the beach on 26 May was 228 nurdles per square meter (Sewwandi et al., 2022), indicating that Sarakkuwa Beach was highly polluted by nurdles (Fernandino et al., 2015). By September 2021, the average amount of nurdles on Sarakkuwa Beach had reduced to 32/m². Though the PPI decreased, the pollution status remained concerning; nurdles could still be found 6 months after the incident (Figure 7.3b). This could be due to the continuous deposition of nurdles by waves. Analyses of sand samples so far have found a considerable amount of nurdles, providing evidence for the long-term impact of the *XPP* maritime accident on beach pollution in Sri Lanka. All nurdles found at Sarakkuwa Beach were identified as PE from their IR spectra, which showed slight changes likely due

FIGURE 7.3 a) Presence of nurdles at Sarakkuwa Beach on 26 May 2021, b) presence of nurdles at Sarakkuwa Beach during October 2022, c) nurdles in different colors (found at Sarakkuwa Beach during May 2021), and d) i – spherical nurdles, ii – nurdles in square shape found on 26 May 2022, iii – nurdles found during September 2021, and iv – nurdles found during October 2022 at Sarakkuwa Beach after X-Press Pearl shipwreck.

to weathering. Apart from individual nurdles, agglomerated nurdles were also found, and their contamination by metals and petroleum products has been confirmed by several studies (de Vos et al., 2021; Sewwandi et al., 2022). Several metals have been found associated with the nurdles including Cr, Mo, Cu, Pb, Li, and Cd. Their source is thought to be from the metal cargo onboard the ship during the explosion. The source of the petroleum contamination could be any of the petroleum products stored onboard the ship as cargo at the time of the accident as well as from the ship's leaking fuel (de Vos et al., 2021).

7.3.3.2 Weathering Features of Nurdles

The shape of some nurdles collected at Sarakkuwa Beach during May 2021 (Figure 7.3d-ii) was different from those obtained from the packages that had washed ashore (Figure 7.3d-i). The observed deviations in shape could be due to any melting that occurred from high temperatures during the ship fire. Melted nurdles were found to agglomerate forming irregular shapes (de Vos et al., 2021). The presence of yellow, brown, and black colored nurdles is thought to be from the thermal degradation and burning of white nurdles (Figure 7.3c) (James et al., 2022; Sewwandi et al., 2022). Apart from fragmentation, PE nurdles can lose their white color and become yellowish with long-term exposure to high concentrations of UV irradiation and atmospheric oxygen (Corcoran, 2020). The yellowing index reveals the extent of the photooxidation of nurdles (Abaroa-Pérez et al., 2022). Accordingly, photooxidation might cause the formation of nurdles in different colors.

Environmental factors that marine microplastics are exposed to cause aging, weathering, surface cracking, yellowing, fragmentation, and degradation. Thus, weathering alters the plastic's chemical and physical characteristics. As illustrated in Figures 7.3d-i and iii, the sizes of PE nurdles found at the Sarakkuwa Beach during May 2021 were greater than those found four months after the wreck (during September 2021). The volume decreased by ~108 μm^3 over four months (Sewwandi et al., 2022). The nurdles found on the beach during October 2022 had reduced more in volume than the previously collected nurdles (Figure 7.3d-iv). PE degrades slowly in the environment.

The exhibited reduction in size could be from photooxidative weathering of nurdles due to exposure to sunlight and oxygen. Photooxidation can cause polymer chain scission, leading to nanoplastic and oligomer formation (Corcoran, 2020). Incorporation of oxygen into PE can be confirmed by the presence of adsorption bands for carbonyl or ketone functional groups (1500–1800 cm^{-1}) in IR spectrum of the polymer (Jiang et al., 2021). Carbonyl functional groups were identified at 1656 cm^{-1} (for –C=O stretching) and 1114–1040 cm^{-1} (for –C–O–C symmetric stretching) in the IR spectra of nurdles found at Sarakkuwa Beach during June and September 2021 (Sewwandi et al., 2022). Mechanical stress from wave and wind action in the marine environment could also weather the nurdles, forming surface cracks and grooves. The degree of surface erosion and the yellowness of the nurdles can be used to assess their weathering status (Jiang et al., 2021). The degree of erosion can be evaluated using the ratio between the eroded area and the whole area of a nurdle. Eroded and density reduced nurdles were within the collected nurdles at Pamunugama beach on May 25, 2021, after the *XPP* shipwreck (James et al., 2022). Besides, Nurdles found on the beach at Wadi el Gemal National Park, Egypt, twenty-nine years after spilling from the *SS Hamada* shipwreck in the Red Sea showed surface structural changes compared to those found inside the wrecked ship (Brümmer et al., 2022). The observation evidenced that saltwater causes a relatively lower weathering process than UV radiation and high beach temperatures. Highly weathered nurdles tend to interact with the other toxic pollutants in the surroundings, enhancing their environmental impact.

FIGURE 7.4 a) The locations of the X-Press Pearl ship (when it was brought into the deep sea), Sarakkuwa Beach, and Colombo Port; b) and c) Sarakkuwa Beach, covered by nurdles on May 2021; d) Sand sampling locations at Sarakkuwa Beach; e) The presence of buried nurdles at Sarakkuwa Beach; and f) Abundances of microplastics at different layers (at 30, 50, and 90 cm depth) of the Sarakkuwa Beach from the beach surface.

7.3.4 Buried Nurdles and Pyrogenic Microplastics at Sarakkuwa Beach

High tidal waves formed due to the monsoon and cyclone conditions caused changes in the morphodynamics of Sarakkuwa Beach. Most of Sri Lanka's beaches experience severe seasonal changes, especially due to variations in wave energy. Due to the rapid changes in beach composition, the nurdles and pyrogenic microplastics accumulated on the beach surface were mixed with beach sand and buried in the different layers of the beach (Figure 7.4e). From the study we performed in October 2021 by excavating sand samples from Sarakkuwa Beach (at eight sampling locations; Figure 7.4d), considerable amounts of nurdles and pyrogenic microplastics were found up to a 1 m depth below the beach surface. As illustrated in Figure 7.4f, nurdles or pyrogenic microplastics greater than 2 mm in size were the most abundant. The amount of buried nurdles and pyrogenic microplastics at a depth of 90 cm from the beach surface at E and F locations at Sarakkuwa Beach (Figure 7.4d) was higher than in other locations. Burial enhances the adverse effects of microplastics on beach crops and animals (Duncan et al., 2018; Menicagli et al., 2023). The nurdles buried in the uppermost layers of the beach may fragment more quickly because they are regularly mixed with beach sand.

7.3.5 Presence of Nurdles in the Negombo Lagoon

Nurdles can be distributed to inland waters due to tidal activities. Nurdle pollution can also enter mangrove forests, estuaries, and lagoons through channels and inlets directly connected with the ocean. After the *XPP* disaster in Sri Lanka, plastic pellets and pyrogenic microplastic fragments were found at the inlets to Negombo Lagoon (Figures 7.5a and b).

FIGURE 7.5 a) and b) Floating nurdles in Negombo Lagoon, c) Presence of nurdles in a water sample collected at the Kelani River, and d) Nurdles extracted from water samples collected from the Kelani River after X-Press Pearl shipwreck.

7.3.6 NURDLES FOUND IN THE KELANI RIVER

Nurdles could also be distributed to inland waters via tidal movements. More than eight months after the *XPP* shipwreck, during a study conducted in February and March 2022, several nurdles were found in the samples collected along the Kelani River in the western province from the river mouth (Modara) toward the inland waters. The similarity in features of the nurdles found in the water samples of the river and those spilled from the shipwreck suggested the possibility of nurdles scattering to the inland waters of Sri Lanka. Furthermore, nurdles were mainly PE (81.9%) and PP (18.1%). The nurdles from the *XPP* accident were identified as PE (de Vos et al., 2021). Because no plastic manufacturing industries are close to the Kelani River, the presence of nurdles with the same feature as the spilled nurdles from the shipwreck supports the notion that the nurdles from the spill scattered to inland waters (Figures 7.5c and d). Further studies should be conducted regarding the fate and transport of the nurdles via the river and their weathering, as some of them appear discolored, thermally altered, and reduced in size.

7.4 POSSIBLE ENVIRONMENTAL IMPACTS OF NURDLES AND PYROGENIC MICROPLASTICS

During the spill, nurdles were exposed to combustion, heat, and chemicals. Burned plastic resembled pyroplastics (Turner et al., 2019). Plastics are already considered to be diverse contaminants (Rochman et al., 2019) and are made more diverse by weathering (due to light, heat, and combustion). Thermal alteration and burning of plastic produces material with uncertain environmental impacts that have yet to be explored (James et al., 2022).

The presence of nurdles in the marine environment is a significant environmental problem. Nurdles persist and circulate in the ocean, continuing to wash ashore for decades. Nurdles also sorb hydrophobic organic contaminants and metals (Mato et al., 2001). Furthermore, removal of buried, floating, and deposited nurdles is complicated, requiring specialized machinery and extensive labor. Removing all spilled nurdles is impossible as wave energy and sand movement enhance the deposition of nurdles on the beaches and marine environment (Annette and Gaëlle, 2020; Hassan et al., 2021). Further, the removal process could disturb the naturally deposited sand profiles and organisms on the beach, causing damage to the ecosystem (Annette and Gaëlle, 2020).

According to the United Nations Environment Programme report, during the incident of *XPP*, the initial cleanup was done using intensive human resources. Once the visible plastics and other wreckages were cleaned, further cleanup efforts were harder (Bourzac, 2023; Hassan et al., 2021). Identifying the plastics became more challenging as they blended into the surroundings. The spilled plastic having various sizes, shapes, and colors makes remediation more difficult. Along with unburned nurdles, the debris also contained a variety of pyrogenic microplastics. The pyrogenic plastic is brittle and easily breaks into smaller pieces. Some of the spilled plastic resembled natural materials, making identifying the plastic more difficult (de Vos et al., 2021; James et al., 2022).

As mentioned before, during the *XPP* accident, billions of nurdles were spilled and spread hundreds of kilometers toward the western and southern coasts of Sri Lanka. Based on a recent study conducted by Sewwandi et al. (2022), it was evident that the nurdles from the incident could act as vectors for carrying Pb, Mo, Cu, and Cr toward the shoreline, which were some of the heavy metal elements released from the cargo ship.

7.4.1 CORALS

Coral reefs are among the planet's richest, but also most vulnerable, ecosystems. They shelter a variety of marine species. Sri Lanka has nearly 2% (or up to 32 km) of nearshore coral reefs that differ in quality. These reefs are predominantly fringing types located in nearshore waters or patch reefs on rocky surfaces at different distances from the shore on the continental shelf (Rajasuriya and White, 1995). Approximately 2% of the 1,585 km coastline is made up of fringing reefs. However, larger reef areas are found offshore in the Gulf of Mannar toward the northwest and also along the east coast (Rajasuriya et al., 2002). Sri Lanka's coral reefs support a total of 183 hard coral species and 6 species of spiny lobsters, along with various other invertebrates, sea turtles, and dolphins (Rajasuriya et al., 2002). Dugongs are found in inshore coral reef areas along the northwest coast of Sri Lanka in the Gulf of Mannar. The coral reef provides a house for the fishes that are of the highest economic importance in Sri Lanka (groupers, snappers, emperors, barracuda, jacks, sear and leatherskins, and fusiliers) and 35 species of butterfly fish, 6 species of large angelfish and pygmy angelfish, and many other reef fish that are important for the aquarium trade (Rajasuriya et al., 2002). The coral reef systems are at risk with the *XPP* disaster and associated chemical and nurdle spills.

Previous research has found that microplastics can have negative consequences on corals following exposure and ingestion (Chapron et al., 2018; Hall et al., 2015; Reichert et al., 2018). Further microplastic exposure can contribute to bleaching, necrosis, changes in photosynthetic efficiency, decreased growth and feeding rates, and increased mucous secretion in some coral species (Reichert et al., 2018). Microplastics threaten planktivorous organisms like corals because they are similar in size and form to zooplankton. Some organisms may experience reduced growth, fecundity, and feeding efficiency due to consuming microplastics (Besseling et al., 2013; Chapron et al., 2018; Sussarellu et al., 2016b). However, these consequences on corals are still not completely understood.

Furthermore, there is growing concern over the potential for large and small plastics to serve as vectors of contaminants and diseases (Goldstein et al., 2014; Lamb et al., 2018; Rotjan et al., 2019). When microplastics reach coral reefs, waves and currents constantly brush against the corals, damaging them (Bednarz et al., 2021). Further, corals may consume microplastics and have a false sense of "fullness", which prevents them from feeding on nourishing materials (Bednarz et al., 2021). Microplastics may obstruct the coral's digestive tract and harm its interior organs.

Additionally, microplastics can absorb contaminants and dangerous microbes from saltwater and pass them to corals.

Coral reefs are highly vulnerable to chemical contaminants, which can be lethal to reef formation (van Dam et al., 2011). In the western and southern seas of Sri Lanka, where there are more than 180 native reef species, hundreds of tons of chemicals were released along with the nurdles during the *XPP* cargo ship accident, potentially injuring reefs. The damage to the reefs is still being assessed.

7.4.2 SEAGRASS BEDS

Fifteen true seagrass species belonging to nine genera are distributed around the coastal water of Sri Lanka (Udagedara and Dahanayaka, 2020). Rare seagrass species, such as *Halophila beccarii,* are located in the Negombo estuary which is close to the *XPP* ship incident (Udagedara and Dahanayaka, 2020). Due to a microalgal proliferation caused by high nutrient loading, around 20% of the seagrass bed cover in the Negombo estuary has been lost (Leslie, 2011). Therefore, the chemical and nurdle spills of the *XPP* shipwreck would harm the seagrass population in Sri Lanka. Seagrass beds act as an effective trap of microplastics in the marine environment (de los Santos et al., 2021; Karthik et al., 2022; Zhao et al., 2022). Some studies identified that the form, color, and size of microplastics had no bearing on how seagrass trapped them (Huang et al., 2020). Entrapment by epibionts or adhesion via biofilms are examples of potential pathways for the buildup of microplastics in seagrass beds (Goss et al., 2018). As grazers consume seagrasses in the ecosystem, they could act as a pathway for microplastic pollution to enter marine food webs (Goss et al., 2018). The seagrass beds are also feeding and breeding grounds for many species, including fishes and shellfish (Karthik et al., 2022). Therefore, in the case of the *XPP* incident, there is a potential risk of the nurdles and the contaminants settling on the surface of seagrasses and macroalgae to be ingested by herbivores and then transfer up the food chain (Pantos, 2022).

7.4.3 MARINE ANIMALS

Nurdle spills directly impact the health and the existence of marine organisms in the marine environment (Partow et al., 2021). Numerous reports have stated that exposing to plastic nurdles and leaching chemicals from them have been shown to cause fatal abnormalities, including the development of the gut outside the body of different organisms including oysters and sea urchin larvae (McVeigh, 2022; Sussarellu et al., 2016a). Fish consume floating nurdles in various colors, assuming them to be their food. Tiny plastic particles are common in shallow benthic populations, and benthic invertebrates are more likely than pelagic and land-dwelling vertebrates to consume microplastics of all sizes (Graham and Thompson, 2009; Mladinich et al., 2022; Ward et al., 2019). Additionally, nurdles on beaches pose a threat to other organisms. Common gulls, puffins, and fulmars tend to consume nurdles as they resemble fish eggs (Van Franeker and Law, 2015).

Plastic nurdles have been found on beaches where turtles lay their eggs in Cyprus and gray seals breed in Norfolk, Britain (Duncan et al., 2019; Zhang et al., 2020). Since the 1990s, there have been reports of nearly 267 marine organisms suffering adverse effects from ingesting plastic, including sea turtles, fish, seabirds, whales, and pinnipeds (Graham and Thompson, 2009). Spills of virgin plastic pellets during marine incidents have the potential to be hazardous because they are more toxic than pellets left on beaches. For instance, leaching additives of virgin plastic pellets have posed toxic effects on the development of embryos of *Lytechinus variegatus* (Nobre et al., 2015b). Therefore, external exposure to plastic nurdles might create various health issues in marine life, affecting their ability to function biologically (Nobre et al., 2015a). Polystyrene microplastics showed negative effects on feeding, function, and fecundity in the marine copepod *Calanus helgolandicus* (Cole et al., 2015). Thus, exposure to secondary microplastics increased mortality, lengthened the interbreed period, and reduced reproduction of aquatic invertebrates, including zooplankton (Ogonowski et al., 2016). Microplastic particles have harmful effects on both phytoplankton and zooplankton, and these particles may have a negative impact on primary productivity in aquatic ecosystems (Troost et al., 2018). This implies that microplastics may impede the growth and survival of these organisms and ultimately reduce the amount of energy available for higher trophic levels.

Recovering fully from the effects, such as ensuring plastic nurdle-free seafood, could be a lengthy process. Accordingly, spilled plastic nurdles from the *XPP* accident may injure and kill marine organisms such as endangered turtles, dolphins, and several rare deep-sea species. After the shipwreck *XPP* incident, the carcasses of dead turtles have washed up on the beaches of Uswetakeiyawa, Panadura, Unawatuna, Wellawatte, Moratuwa, and Induruwa (Sivaramanan and Kotagama, 2022). Further, about 251 turtle carcasses and 33 marine animals, including whale and dolphin carcasses, were recovered on the coast near Sri Lanka. This could be due to dangerous chemical emissions from the cargo ship. Many turtle shells from the gathered bodies appear to have bleached, showing that the chemical harms marine life (Hassan et al., 2021). Experts from the Department of Wildlife Conservation, Sri Lanka, categorize turtle mortality into two primary stages. The first stage occurs during the initial week, where the turtle shells may appear bleached or sheared off, indicating possible chemical burns. The second stage involves a larger wave of mostly young turtles that show no signs of carapace damage and appear to be healthy based on investigations, with their digestive tracts full of food (Partow et al., 2021). According to the UN report by Partow et al. (2021), there are two important points to consider in relation to the mentioned deaths: firstly, they occurred during the monsoon season, which is known for a high occurrence of turtle stranding due to the strong western currents that wash carcasses ashore. Secondly, the reported increase in turtle deaths could be attributed to both a heightened awareness and increased monitoring by both governmental organizations and the general public. However, no published studies with scientific evidence support that those deaths were because of the *XPP* incident to date. Besides, nurdles were observed in fish's stomachs, gills, and mouths (Partow et al., 2021). For instance, plastic pellets were found in pharynx of Giant trevally (*Caranx ignobilis*) and in opercle of old silk sea bream (*Acanthopagrus berda*) (Partow et al., 2021).

Plastic pellets and burned and unburned material that washed into coastal areas after the *XPP* accident may have long-term environmental consequences. Plastic pellets can keep a high temperature, which may interfere with turtle nesting because the temperature is a significant component in determining turtle gender (Duncan et al., 2018). Furthermore, as contaminants are adsorbed and desorbed from the spilled plastic, the plastic may act as a conduit for potentially harmful metals to the terrestrial environment (Sewwandi et al., 2022).

7.4.4 SALT

While having an impact on marine organisms, nurdles could also have an impact on human health in different ways. Microplastic occurring via the degradation of plastic nurdles could lead to human ingestion through marine food, sea salt, and drinking water (Fadare et al., 2020; Guilhermino et al., 2021; Weber and Schunk, 2021). Salt production is mainly carried out in the coastal areas of Sri Lanka, particularly in the districts of Hambantota, Mannar, and Jaffna. In Sri Lanka, seawater is pumped or allowed to flow into evaporation ponds during the process of making sea salt. Hence, the spilled nurdles have higher tendency to contaminate the evaporation tanks. Recent studies conducted worldwide discovered microplastics in sea salt (Danopoulos et al., 2020; Sathish et al., 2020; Sewwandi et al., 2023). Moreover, the study by Kapukotuwa et al. (2022) revealed that salt is a crucial component of the human diet is that the highest concentration of microplastics were found in raw salt, followed by food-grade table salt and crystal salt. However, no evidence exists for the food contamination from plastic nurdles after the *XPP* incident to date. Further studies could be carried out to identify the dispersion and potential health impacts of plastic nurdles via human consumption (Asprey, 2022; Parker, 2018; Peixoto et al., 2019).

Studies are still being conducted on the *XPP* tragedy, and the environmental and health risks of the chemical and plastic spill are still uncertain. Further, the recovery may take many years and may never be completed (Bourzac, 2023). However, the research on the disaster can help to advance the scientific understanding of the health and environmental impacts of plastic nurdle pollution, being a case study for how to handle nurdle spills in the future.

7.5 SUMMARY AND FUTURE PERSPECTIVES

Hundreds of tons of white nurdles and pyrolyzed plastic debris washed onto Sri Lankan beaches after containers fell from the *XPP* when an explosion occurred on board. The plastic agglomerated, fragmented, charred, and was chemically altered due to exposure to combustion, heat, and chemicals, resulting in a complex spill of partially pyrolyzed nurdles, plastics, and pyrogenic microplastics. Nurdles deposited on the beaches were buried. Evidence suggests inland water may also have been contaminated with the nurdles. As of this writing, nurdles and pyrogenic microplastics continue to wash up on beaches, posing a significant obstacle to protecting the coastal zone. Beaches continue to be cleaned a year after the spill. The plastic is expected to be transported by ocean currents to beaches hundreds of kilometers away from the shipwreck and coastlines all over Sri Lanka. Accordingly, this requires continuous measurement of the flux of nurdles onto beaches. The long-term environmental impact could be mitigated through frequent beach cleanups. Without limiting the surface beach cleanups, the buried nurdles should be removed via suitable cleanup methods. Ongoing monitoring efforts for weathering and toxicity are necessary. A significant challenge to conducting any environmental assessment has been the lack of a baseline of the environmental conditions. Consequently, baseline conditions need to be implemented. There must be a specialized unit with resources in Sri Lanka to respond to future maritime incidents.

REFERENCES

Abaroa-Pérez, B., Ortiz-Montosa, S., Hernández-Brito, J.J. and Vega-Moreno, D. (2022). Yellowing, weathering and degradation of marine pellets and their influence on the adsorption of chemical pollutants. *Polymers* (Basel) 14(7). https://doi.org/10.3390/polym14071305

Annette, G. and Gaëlle, H. (2020) Plastic giants polluting through the backdoor: The case for a regulatory supply-chain approach to pellet pollution, *Surfrider Foundation Europe*, 22 Sep 2021, https://surfrider.eu/wp-content/uploads/2020/11/report-pellet-pollution-2020.pdf

Asprey, D. (2022) *New Study Finds Microplastics in Your Sea Salt*. Here's What You Need to Know, https://daveasprey.com/microplastics-sea-salt/

Bednarz, V., Leal, M., Béraud, E., Marques, J.F. and Ferrier, C. (2021) The invisible threat: How microplastics endanger corals. Frontiers, https://kids.frontiersin.org/articles/10.3389/frym.2021.574637

Besseling, E., Wegner, A., Foekema, E.M., van den Heuvel-Greve, M.J. and Koelmans, A.A. (2013). Effects of microplastic on fitness and PCB bioaccumulation by the Lugworm *Arenicola marina* (L.). *Environmental Science & Technology* 47(1), 593–600. https://doi.org/10.1021/es302763x

Bourzac, K. (2023). Grappling with the biggest marine plastic spill in history. C&en, https://cen.acs.org/environment/pollution/marine-plastic-spill-xpress-pearl-nurdle/101/i3?utm_source=LatestNews&utm_medium=LatestNews&utm_campaign=CENRSS&fbclid=IwAR32KqGHz507v_v77cm7OO7xI9uMhBiJt9GqERiLCFepKG3XlbHp-a3OQxM

Brümmer, F., Schnepf, U., Resch, J., Jemmali, R., Abdi, R., Kamel, H.M., Bonten, C. and Müller, R.-W. (2022). In situ laboratory for plastic degradation in the Red Sea. *Scientific Reports* 12(1), 11956. https://doi.org/10.1038/s41598-022-15310-7

Cabrera, J.S. (2022). Plastic pellet pollution can end through coordinated efforts, report shows. MONGABAY, https://news.mongabay.com/2022/12/plastic-pellet-pollution-can-end-through-coordinated-efforts-report-shows/

Chapron, L., Peru, E., Engler, A., Ghiglione, J.F., Meistertzheim, A.L., Pruski, A.M., Purser, A., Vétion, G., Galand, P.E. and Lartaud, F. (2018). Macro- and microplastics affect cold-water corals growth, feeding and behaviour. *Scientific Reports* 8(1), 15299. https://doi.org/10.1038/s41598-018-33683-6

Cole, M., Lindeque, P., Fileman, E., Halsband, C. and Galloway, T.S. 2015. The impact of polystyrene microplastics on feeding, function and fecundity in the marine copepod *Calanus helgolandicus*. *Environmental Science and Technology* 49(2), 1130–1137. https://doi.org/10.1021/es504525u

Corcoran, P.L. (2020). Handbook of Microplastics in the Environment. Rocha-Santos, T., Costa, M. and Mouneyrac, C. (eds), *Degradation of Microplastics in the Environment*, pp. 1–12, Springer International Publishing, Cham. https://doi.org/10.1007/978-3-030-10618-8_10-1

Danopoulos, E., Jenner, L., Twiddy, M. and Rotchell, J.M. 2020. Microplastic contamination of salt intended for human consumption: a systematic review and meta-analysis. *SN Applied Sciences* 2(12), 1950. https://doi.org/10.1007/s42452-020-03749-0

de los Santos, C.B., Krång, A.-S. and Infantes, E. (2021). Microplastic retention by marine vegetated canopies: Simulations with seagrass meadows in a hydraulic flume. *Environmental Pollution* 269, 116050. https://doi.org/10.1016/j.envpol.2020.116050

de Vos, A., Aluwihare, L., Youngs, S., DiBenedetto, M.H., Ward, C.P., Michel, A.P.M., Colson, B.C., Mazzotta, M.G., Walsh, A.N., Nelson, R.K., Reddy, C.M. and James, B.D. (2021). The M/V X-Press Pearl Nurdle Spill: Contamination of Burnt Plastic and Unburnt Nurdles along Sri Lanka's Beaches. ACS Environmental Au. https://doi.org/10.1021/acsenvironau.1c00031

Dermansky, J. (2020). Pollution Scientist Calls Plastic Pellet Spill in the Mississippi River 'a Nurdle Apocalypse'. DeSmog, www.desmog.com/2020/08/28/new-orleans-louisiana-plastic-spill-mississippi-river-nurdle-apocalypse/

Dhanesha, N. (2022) The massive, unregulated source of plastic pollution you've probably never heard of. Vox, www.vox.com/recode/23056251/nurdles-plastic-pollution-ocean-microplastics

Duncan, E.M., Arrowsmith, J., Bain, C., Broderick, A.C., Lee, J., Metcalfe, K., Pikesley, S.K., Snape, R.T.E., van Sebille, E. and Godley, B.J. (2018). The true depth of the Mediterranean plastic problem: Extreme microplastic pollution on marine turtle nesting beaches in Cyprus. *Marine Pollution Bulletin* 136, 334–340. https://doi.org/10.1016/j.marpolbul.2018.09.019

Duncan, E.M., Broderick, A.C., Fuller, W.J., Galloway, T.S., Godfrey, M.H., Hamann, M., Limpus, C.J., Lindeque, P.K., Mayes, A.G., Omeyer, L.C.M., Santillo, D., Snape, R.T.E. and Godley, B.J. (2019). Microplastic ingestion ubiquitous in marine turtles. *Global Change Biology* 25(2), 744–752. https://doi.org/10.1111/gcb.14519

Fadare, O., Wan, B., Zhao, L. and Guo, L.-H. (2020). Microplastics from consumer plastic food containers: Are we consuming it? *Chemosphere* 253, 126787. https://doi.org/10.1016/j.chemosphere.2020.126787

Fendall, L.S. and Sewell, M.A. (2009). Contributing to marine pollution by washing your face: Microplastics in facial cleansers. *Marine Pollution Bulletin* 58(8), 1225–1228. https://doi.org/10.1016/j.marpolbul.2009.04.025

Fernandino, G., Elliff, C.I., Silva, I.R. and Bittencourt, A.C. (2015). How many pellets are too many? The pellet pollution index as a tool to assess beach pollution by plastic resin pellets in Salvador, Bahia, *Brazil. Revista de Gestão Costeira Integrada-Journal of Integrated Coastal Zone Management* 15(3), 325–332.

Ferreira, A.T.d.S., Siegle, E., Ribeiro, M.C.H., Santos, M.S.T. and Grohmann, C.H. (2021). The dynamics of plastic pellets on sandy beaches: A new methodological approach. *Marine Environmental Research* 163, 105219. https://doi.org/10.1016/j.marenvres.2020.105219

Geographer Waves (02/08/2022). The geographer online, www.thegeographeronline.net/coasts.html

Goldstein, M.C., Carson, H.S. and Eriksen, M. (2014). Relationship of diversity and habitat area in North Pacific plastic-associated rafting communities. *Marine Biology* 161(6), 1441–1453. https://doi.org/10.1007/s00227-014-2432-8

Goss, H., Jaskiel, J. and Rotjan, R. (2018). *Thalassia testudinum* as a potential vector for incorporating microplastics into benthic marine food webs. *Marine Pollution Bulletin* 135, 1085–1089. https://doi.org/10.1016/j.marpolbul.2018.08.024

Graham, E.R. and Thompson, J.T. (2009). Deposit- and suspension-feeding sea cucumbers (Echinodermata) ingest plastic fragments. *Journal of Experimental Marine Biology and Ecology* 368(1), 22–29. https://doi.org/10.1016/j.jembe.2008.09.007

Guilhermino, L., Martins, A., Cunha, S. and Fernandes, J.O. (2021). Long-term adverse effects of microplastics on Daphnia magna reproduction and population growth rate at increased water temperature and light intensity: Combined effects of stressors and interactions. *Science of The Total Environment* 784, 147082. https://doi.org/10.1016/j.scitotenv.2021.147082

Häkkien, J., Posti, A., Weintrit, A. and Neumann, T. (2013). *Overview of Maritime Accidents Involving Chemicals Worldwide in the Baltic Sea.* CRS Press.

Hall, N.M., Berry, K.L.E., Rintoul, L. and Hoogenboom, M.O. 2015. Microplastic ingestion by scleractinian corals. *Marine Biology* 162(3), 725–732. https://doi.org/10.1007/s00227-015-2619-7

Hassan, P., Camille, L., Stephane, L.F. and Luigi, A. (2021) X-Press Pearl maritime disaster Sri Lanka-report of the UN environmental advisory mission JULY 2021, UN environmental advisory mission. UN. https://wedocs.unep.org/20.500.11822/36608

Huang, Y., Liu, Q., Jia, W., Yan, C. and Wang, J. (2020). Agricultural plastic mulching as a source of microplastics in the terrestrial environment. *Environmental Pollution* 260, 114096.

James, B.D., de Vos, A., Aluwihare, L.I., Youngs, S., Ward, C.P., Nelson, R.K., Michel, A.P.M., Hahn, M.E. and Reddy, C.M. (2022). Divergent forms of pyroplastic: Lessons learned from the M/V X-Press Pearl ship fire. *ACS Environmental Au* 2(5), 467–479. https://doi.org/10.1021/acsenvironau.2c00020

Jayathilaka, R.M.R.M., Weerakoon, W.R.W.M.A.P., Indika, K.W., Arulananthan, K. and Kithsiri, H.M.P. (2022). Spatio-temporal variation of plastic pellets dispersion in the coastline of Sri Lanka: An assessment of pellets originated from the X-Press Pearl incident during the Southwest monsoon in 2021. *Marine Pollution Bulletin* 184, 114145. https://doi.org/10.1016/j.marpolbul.2022.114145

Jiang, X., Lu, K., Tunnell, J.W. and Liu, Z. (2021). The impacts of weathering on concentration and bioaccessibility of organic pollutants associated with plastic pellets (nurdles) in coastal environments. *Marine Pollution Bulletin* 170, 112592. https://doi.org/10.1016/j.marpolbul.2021.112592

Kapukotuwa, R.W.M.G.K., Jayasena, N., Weerakoon, K.C., Abayasekara, C.L. and Rajakaruna, R.S. (2022). High levels of microplastics in commercial salt and industrial salterns in Sri Lanka. *Marine Pollution Bulletin* 174, 113239. https://doi.org/10.1016/j.marpolbul.2021.113239

Karlsson, T.M., Arneborg, L., Broström, G., Almroth, B.C., Gipperth, L. and Hassellöv, M. (2018). The unaccountability case of plastic pellet pollution. *Marine Pollution Bulletin* 129(1), 52–60. https://doi.org/10.1016/j.marpolbul.2018.01.041

Karthik, R., Robin, R.S., Purvaja, R., Karthikeyan, V., Subbareddy, B., Balachandar, K., Hariharan, G., Ganguly, D., Samuel, V.D., Jinoj, T.P.S. and Ramesh, R. (2022). Microplastic pollution in fragile coastal ecosystems with special reference to the X-Press Pearl maritime disaster, southeast coast of India. *Environmental Pollution* 305, 119297. https://doi.org/10.1016/j.envpol.2022.119297

Lamb, J.B., Willis, B.L., Fiorenza, E.A., Couch, C.S., Howard, R., Rader, D.N., True, J.D., Kelly, L.A., Ahmad, A., Jompa, J. and Harvell, C.D. (2018). Plastic waste associated with disease on coral reefs. *Science* 359(6374), 460–462. https://doi.org/10.1126/science.aar3320

Lentz, S.J. and Fewings, M.R. (2011). The wind- and wave-driven inner-shelf circulation. *Annual Review of Marine Science* 4(1), 317–343. https://doi.org/10.1146/annurev-marine-120709-142745

Leslie, J. (2011). *Fisheries and Environmental Profile of Chilaw Lagoon, Sri Lanka: A Literature Review*. Regional Fisheries Livelihoods Programme for South and Southeast Asia.

Mato, Y., Isobe, T., Takada, H., Kanehiro, H., Ohtake, C. and Kaminuma, T. (2001). Plastic resin pellets as a transport medium for toxic chemicals in the marine environment. *Environmental Science & Technology* 35(2), 318–324. https://doi.org/10.1021/es0010498

McVeigh, K. (2022) Plastic 'nurdles' stop sea urchins developing properly, *study finds. The Guardian,* www.theguardian.com/environment/2022/dec/15/plastic-nurdles-stop-sea-urchins-developing-properly-study-finds

Menicagli, V., Balestri, E., Fulignati, S., Raspolli Galletti, A.M. and Lardicci, C. (2023). Plastic litter in coastal sand dunes: Degradation behavior and impact on native and non-native invasive plants. *Environmental Pollution* 316, 120738. https://doi.org/10.1016/j.envpol.2022.120738

Mladinich, K., Holohan, B.A., Shumway, S.E., Brown, K. and Ward, J.E. (2022). Determining the properties that govern selective ingestion and egestion of microplastics by the blue mussel (*Mytilus edulis*) and eastern oyster (*Crassostrea virginica*). *Environmental Science & Technology* 56(22), 15770–15779. https://doi.org/10.1021/acs.est.2c06402

Nobre, C.R., Santana, M.F.M., Maluf, A., Cortez, F.S., Cesar, A., Pereira, C.D.S. and Turra, A. (2015a). Assessment of microplastic toxicity to embryonic development of the sea urchin *Lytechinus variegatus* (Echinodermata: Echinoidea). *Marine Pollution Bulletin* 92(1), 99–104. https://doi.org/10.1016/j.marpolbul.2014.12.050

Nobre, C.R., Santana, M.F.M., Maluf, A., Cortez, F.S., Cesar, A., Pereira, C.D.S. and Turra, A. (2015b). Assessment of microplastic toxicity to embryonic development of the sea urchin *Lytechinus variegatus* (Echinodermata: Echinoidea). *Marine Pollution Bulletin* 92(1–2), 99–104. https://doi.org/10.1016/j.marpolbul.2014.12.050

Ogonowski, M., Schür, C., Jarsén, Å. and Gorokhova, E. (2016). The effects of natural and anthropogenic microparticles on individual fitness in *Daphnia magna. PLOS ONE* 11(5), e0155063. https://doi.org/10.1371/journal.pone.0155063

Pantos, O. (2022). Microplastics: impacts on corals and other reef organisms. *Emerg Top Life Sci* 6(1), 81–93. https://doi.org/10.1042/etls20210236

Parker, L. (2018). Microplastics found in 90 percent of table salt, www.nationalgeographic.com/environment/article/microplastics-found-90-percent-table-salt-sea-salt

Partow, H., Lacroix, C., Le Floch, S. and Alcaro, L. (2021). X-Press Pearl maritime disaster Sri Lanka-report of the UN environmental advisory mission JULY 2021, UN environmental advisory mission. *UN.* https://wedocs.unep.org/20.500.11822/36608

Peixoto, D., Pinheiro, C., Amorim, J., Oliva-Teles, L., Guilhermino, L. and Vieira, M.N. (2019). Microplastic pollution in commercial salt for human consumption: A review. *Estuarine, Coastal and Shelf Science* 219, 161–168. https://doi.org/10.1016/j.ecss.2019.02.018

Pepper, F. (2019) Most plastic on our beaches could have come from anywhere. *But not the Durban nurdle. ABC News,* www.abc.net.au/news/science/2019-07-13/durban-nurdle-different-from-other-plastic-rubbish-beaches-wa/11268554

Perera, U.L.H.P., Subasinghe, H.C.S., Ratnayake, A.S., Weerasingha, W.A.D.B. and Wijewardhana, T.D.U. (2022). Maritime pollution in the Indian Ocean after the MV X-Press Pearl accident. *Marine Pollution Bulletin* 185, 114301. https://doi.org/10.1016/j.marpolbul.2022.114301

Plastic. (2022). *Plastic Pellet Pollution.* As You Sow, www.asyousow.org/our-work/waste/plastic-pellets

Rajasuriya, A. and White, A.T. (1995). Coral reefs of Sri Lanka: Review of their extent, condition, and management status. *Coastal Management* 23(1), 77–90. https://doi.org/10.1080/08920759509362257

Rajasuriya, A., Zahir, H., Muley, E., Subramanian, B., Venkataraman, K., Wafar, M., Khan, S. and Whittingham, E. (2002). *Status of coral reefs in South Asia: Bangladesh, India, Maldives, Sri Lanka.* Proceedings of the Ninth International Coral Reef Symposium, Bali, 23–27 October 2000, 2, 841–845.

Reichert, J., Schellenberg, J., Schubert, P. and Wilke, T. (2018). Responses of reef building corals to microplastic exposure. *Environmental Pollution* 237, 955–960. https://doi.org/10.1016/j.envpol.2017.11.006

Releases, P. (2014). Agreement on plastic pellet spill incident in Hong Kong waters. Hongkong Government.

Rochman, C.M. (2013). Plastics and priority pollutants: A multiple stressor in aquatic habitats. *Environmental Science & Technology* 47(6), 2439–2440. https://doi.org/10.1021/es400748b

Rochman, C.M., Brookson, C., Bikker, J., Djuric, N., Earn, A., Bucci, K., Athey, S., Huntington, A., McIlwraith, H., Munno, K., De Frond, H., Kolomijeca, A., Erdle, L., Grbic, J., Bayoumi, M., Borrelle, S.B., Wu, T., Santoro, S., Werbowski, L.M., Zhu, X., Giles, R.K., Hamilton, B.M., Thaysen, C., Kaura, A., Klasios, N., Ead, L., Kim, J., Sherlock, C., Ho, A. and Hung, C. (2019). Rethinking microplastics as a diverse contaminant suite. *Environmental Toxicology and Chemistry* 38(4), 703–711. https://doi.org/10.1002/etc.4371

Rotjan, R.D., Sharp, K.H., Gauthier, A.E., Yelton, R., Lopez, E.M.B., Carilli, J., Kagan, J.C. and Urban-Rich, J. (2019). Patterns, dynamics and consequences of microplastic ingestion by the temperate coral, *Astrangia poculata*. *Proc Biol Sci* 286(1905), 20190726. https://doi.org/10.1098/rspb.2019.0726

Sathish, M.N., Jeyasanta, I. and Patterson, J. (2020). Microplastics in Salt of Tuticorin, Southeast Coast of India. *Archives of Environmental Contamination and Toxicology* 79(1), 111–121. https://doi.org/10.1007/s00244-020-00731-0

Schumann, E.H., MacKay, C.F. and Strydom, N.A. (2019). Nurdle drifters around South Africa as indicators of ocean structures and dispersion South African *Journal of Science* 115, 1–9. http://dx.doi.org/10.17159/sajs.2019/5372

Sewwandi, M., Hettithanthri, O., Egodage, S.M., Amarathunga, A.A.D. and Vithanage, M. (2022). Unprecedented marine microplastic contamination from the Xpress Pearl container vessel disaster. *Science of the Total Environment* 828, 154374. https://doi.org/10.1016/j.scitotenv.2022.154374

Sewwandi, M., Wijesekara, H., Rajapaksha, A.U., Soysa, S. and Vithanage, M. (2023). Microplastics and plastics-associated contaminants in food and beverages; Global trends, concentrations, and human exposure. *Environmental Pollution* 317, 120747. https://doi.org/10.1016/j.envpol.2022.120747

Sivaramanan, S. and Kotagama, S.W. (2022). Ship disaster threatened environmental security and dwindled down the spirit of maritime Sri Lanka. *Bangladesh Journal of Scientific and Industrial Research* 57(4), 199–206. https://doi.org/10.3329/bjsir.v57i4.59704

Sussarellu, R., Suquet, M., Thomas, Y., Lambert, C., Fabioux, C., Pernet, M.E., Le Goïc, N., Quillien, V., Mingant, C., Epelboin, Y., Corporeau, C., Guyomarch, J., Robbens, J., Paul-Pont, I., Soudant, P. and Huvet, A. (2016a). Oyster reproduction is affected by exposure to polystyrene microplastics. *Proceedings of the National Academy of Sciences of the United States of America* 113(9), 2430–2435. https://doi.org/10.1073/pnas.1519019113

Sussarellu, R., Suquet, M., Thomas, Y., Lambert, C., Fabioux, C., Pernet, M.E.J., Le Goïc, N., Quillien, V., Mingant, C., Epelboin, Y., Corporeau, C., Guyomarch, J., Robbens, J., Paul-Pont, I., Soudant, P. and Huvet, A. (2016b). Oyster reproduction is affected by exposure to polystyrene microplastics. *Proceedings of the National Academy of Sciences* 113(9), 2430–2435. https://doi.org/10.1073/pnas.1519019113

Troost, T.A., Desclaux, T., Leslie, H.A., van Der Meulen, M.D. and Vethaak, A.D. (2018). Do microplastics affect marine ecosystem productivity? *Marine Pollution Bulletin* 135, 17–29. https://doi.org/10.1016/j.marpolbul.2018.05.067

Turner, A., Wallerstein, C., Arnold, R. and Webb, D. (2019). Marine Pollution from Pyroplastics. *Science of the Total Environment* 694, 133610.

Udagedara, S. and Dahanayaka, D.D.G.L. (2020). Current status and checklist of seagrass in Sri Lanka. *International Journal of Aquatic Biology* 8(5), 317–326. https://doi.org/10.22034/ijab.v8i5.619

van Dam, J.W., Negri, A.P., Uthicke, S. and Mueller, J.F. (2011) *Ecological Impacts of Toxic Chemicals*, 187–211.

Van Franeker, J.A. and Law, K.L. (2015). Seabirds, gyres and global trends in plastic pollution. *Environmental Pollution* 203, 89–96. https://doi.org/10.1016/j.envpol.2015.02.034

Ward, J.E., Zhao, S., Holohan, B.A., Mladinich, K.M., Griffin, T.W., Wozniak, J. and Shumway, S.E. (2019). Selective ingestion and egestion of plastic particles by the blue mussel (*Mytilus edulis*) and eastern oyster (*Crassostrea virginica*): Implications for using bivalves as bioindicators of microplastic pollution. *Environmental Science & Technology* 53(15), 8776–8784. https://doi.org/10.1021/acs.est.9b02073

Weber, S. and Schunk, S.A. (2021). Reply to Inoue et al. Comment on "Weber et al. Mayenite-Based Electride C12A7e−: A Reactivity and Stability Study. *Catalysts* 2021, 11, 334". https://www.mdpi.com/2073-4344/11/10/1155/htm

Zhang, Y., Liang, J., Zeng, G., Tang, W., Lu, Y., Luo, Y., Xing, W., Tang, N., Ye, S., Li, X. and Huang, W. (2020). How climate change and eutrophication interact with microplastic pollution and sediment resuspension in shallow lakes: A review. *Science of the Total Environment* 705, 135979. https://doi.org/10.1016/j.scitotenv.2019.135979

Zhao, L., Ru, S., He, J., Zhang, Z., Song, X., Wang, D., Li, X. and Wang, J. (2022). Eelgrass (*Zostera marina*) and its epiphytic bacteria facilitate the sinking of microplastics in the seawater. *Environmental Pollution* 292, 118337. https://doi.org/10.1016/j.envpol.2021.118337

0# 8 Sources and Fate of Plastics into Microplastics
Degradation and Remediation Methods

G.M.S.S. Gunawardhana, U.L.H.P. Perera, and Amila Sandaruwan Ratnayake

8.1 INTRODUCTION

The X-Press Pearl disaster is considered the largest plastic-based marine pollution in the maritime history of the Indian Ocean. A fire broke out on May 20, 2021, in the X-Press Pearl vessel for 13 days due to a leakage of nitric acid from at least one container. The vessel was being towed to deeper waters and eventually sank close to the west coast of Sri Lanka. The ship was carrying 1486 containers including 25 metric tons of nitric acid and several types of plastic pellets (Sewwandi et al., 2022; Ratnayake and Perera, 2022). The spill of plastic pellets along with various hazardous substances has created a significant impact on the sensitive coastal environment, local communities, livelihood, and the economy of Sri Lanka (Figure 8.1). About 1610 metric tons of plastics, other debris, and contaminated sand have been removed from the coastal environment of Sri Lanka soon after the incident. However, there are tons of plastics and other toxic substances left on the beaches and the ocean (Figure 8.1). Plastics can take between 500 to 1000 years to decompose due to its polymeric structure. In addition, plastics can spread across larger distances due to their low density (0.8–1.4 g cm^{-3}). Therefore, the X-Press Pearl disaster would be one of the major sources of microplastics (MPs) in the global marine environments. This book chapter is thus focused on up-to-date facts on (i) plastics production as pollutants with special reference to X-Press Pearl maritime disaster; (ii) sources and toxicity of MPs; (iii) weathering and degradation of plastics; and (iv) remediation methods of MPs.

8.2 PLASTICS PRODUCTION AS POLLUTANTS

8.2.1 PLASTICS

Plastic is a versatile material in many industries such as packaging, transportation, textile, construction, and electronics. The practicality and cheapness of plastic made

FIGURE 8.1 Degradation and marine pollution evidence: (a) six to eight days after the MV X-Press Pearl explosion [where (i) plastic nurdle and pyrolytic debris at Epamulla Beach; (ii) cleaning of polluted Sarakkuwa Beach; (iii) nurdles and plastic debris at the berm; and (iv) white color nurdles six to eight days after the explosion]; and (b) one year after the disaster [where (v); (vi); and (vii) deposition of nurdles along with heavy minerals in various beach sites from Panadura to Marawila areas on the west coast of Sri Lanka; and (viii) well-rounded nurdles one year after the disaster]. All photos were taken by the authors.

it an ideal and essential material in day-to-day life (Table 8.1). The "new plastic era" was inaugurated in the mid-19th century. The utilization and accumulation of plastics in the marine environment increased dramatically from the 1970s onward (Ajith et al., 2020; Malankowska et al., 2021). For example, approximately 60%–80% of worldwide anthropogenic litter is composed of plastics (Derraik, 2002). Although about 10% of waste plastic is recycled (Lamba et al., 2021), the remaining major portion is discharged into the environment regardless of their negative impacts (Hale et al., 2020; Xiang et al., 2022). Moreover, an approximated amount of 4.8–12.7 million tons of plastic waste enters the marine environment each year (Wright et al., 2013; Wang et al., 2015; Sun et al., 2021). These plastic wastes are classified according to sizes as macro (>25 mm), meso (5–25 mm), large micro (1–5 mm), small micro (20 μm to 1 mm), and nano (1–1000 nm) plastics (Anbumani and Kakkar, 2018). Plastic wastes are also classified according to shapes such as fibers, microbeads, fragments, nurdles, and foam. However, nurdles (i.e. the preproduction plastic pellets) are the most dangerous category.

8.2.2 Nurdles

Nurdles are a threat to different organisms and habitats due to their persistence, nearly indestructible morphology, and potential toxicity (Hammer et al., 2012). The density of nurdles varies. For example, polypropylene (PP) and polyethylene (PE) are low-density (float on water), whereas polystyrene (PS), polyvinyl chloride (PVC), polyamide (PA), and polyethylene terephthalate (PET) are high-density (do not float in water) materials (Guo and Wang, 2019; Thushari and Senevirathna, 2020). Nurdle spills can occur during transportation, production, and manufacturing. These nurdles look similar to fish eggs due to their size and color, making them get the attention of turtles, seabirds, fish, and other marine wildlife. Therefore, once these animals ingest nurdles, they tend to trap in an animal's stomach, causing ulceration. This stops them from eating real food and ultimately leads to starvation and death, as observed after the X-Press Pearl maritime disaster. In addition, nurdles have the potential to sorb and accumulate toxic pollutants due to their relatively large surface area/volume ratio (Guo and Wang, 2019).

The X-Press Pearl disaster released white low-density polyethylene (LDPE) plastic nurdles in the microplastic size range (Sewwandi et al., 2022). Consequently, these plastic nurdles are dangerous for marine life (Figure 8.1), and humans can also ingest micro- and nanoplastics via the consumption of seafood and salt.

8.2.3 Polymers

The main ingredient of all plastics is polymers, but all polymers are not plastics. Polymers can be categorized as biopolymers (e.g. wool, cotton, cellulose) and synthetic polymers (e.g. polyurethane, polystyrene, polyvinylchloride). Polymers are composed of hydrocarbons, and C and H atoms are bound together as polymer chains (Sridharan et al., 2022). Polymers can also contain oxygen, chlorine, nitrogen, silicon, phosphorus, and sulfur. Polyethylene terephthalate (PET), polypropylene (PP), polystyrene (PS), polyvinyl chloride (PVC), low- and high-density polyethylene (LDPE and HDPE), and polyurethane are major polymers used for plastic manufacturing (Jones et al., 2020; Scherer et al., 2020). The size, shape, density, and polymer type are varied in a wide range (Andrady, 2011). Some general properties of polymers are chemical resistivity, insulators for heat and electricity, the ability to be processed in different ways, and lightweight (Table 8.1).

TABLE 8.1
Origin of Microplastics Based on Plastic Types, Classes, and Their Properties

Plastic Class		Specific Gravity	Glass Transition (°C)	Density (g/cm³)	Crystallinity (%)	UV/Oxidation Resistance	Strength (psi)	Surface Energy (MJ/m²)	Products and Typical Origin	Recycling Symbols
Low-density polyethylene	LDPE LLDPE	0.91–0.93	−100	0.91–0.925	30–50	Low	600–2300	32.4	Plastic bags, six-pack rings, bottles, netting, drinking straws	♴ LDPE
High-density polyethylene	HDPE	0.94	−80	0.959–0.965	80–90	Low	5000–6000	32.4	Milk and juice jugs	♳ HDPE
Polypropylene	PP	0.85–0.83	−25	0.90	30–50	Low	4500–5500	33	Rope, bottle caps, netting	♷ PP
Polystyrene	PS	1.05	+100	1.04	0	Moderate	5000–7200	40.6	Plastic utensils, food containers	♸ PS
Nylon 6 (Polyamide 6)	PA 6	1.14	+48	1.13–1.15	30–50	Good	6000–24,000	38	Netting and traps	♹ OTHER
Nylon 66	PA-66	1.14	+50	1.13–1.15	30–50	Good	14,000	41.4	Textile fibers, Tennis strings, Fishing lines	♹ OTHER
Thermoplastic Polyester	PET	1.37	+69	1.29–1.40	10–30	Good	7000–10,500	45.1	Plastic beverage bottles	♹ OTHER
Poly(vinyl chloride)	PVC	1.38	+70	1.16–1.20	10	Good			Plastic film, bottles, cups	♵ PVC

Monomers are the starting material to form polymers through polymerization (i.e. a chain reaction which adds new monomer units). Polyethylene (PE), polyvinyl chloride (PVC), acrylics, and polystyrene are made by this method (Table 8.1). In condensation polymerization, monomer units react to form polymers while releasing by-products (condensate). Polyesters and polyamides are made by this process (McKeen, 2016). Polymers can be divided according to their mechanical properties as elastomers and fibers. The elastomer polymers have the weakest attractive forces. These polymers are capable of recovering their original shape after being stretched to great extents. Fibers have high intermolecular attractive forces, which enhances tensile strength. Therefore, fibers are mainly used in the textile industry. Polymers can also be divided according to their thermal properties as thermoplastics and thermoset plastics. Thermoplastics can be melted under heat after curing, allowing them to be molded/reshaped. Thermoplastic plastics are capable of opposing corrosive conditions and are thus useful in electronics, medical, chemical, food and beverage, automotive, and plumbing industries (Table 8.1). Thermoset plastics retain their form during the curing process, but they decompose during overheating (McKeen, 2016). Thermoset plastics are utilized in defense, electrical, adhesives and sealants, automotive, aerospace, energy, and construction due to thermal stability, chemical resistance, easiness to mold, and structural robustness (McKeen, 2016).

8.2.4 Polymers and Plastics Present in the X-Press Pearl Vessel and Their Impacts

The report of the UN Environmental Advisory Mission identified four principal environmental issues, including large quantities of plastic pellets from the vessel washing ashore along the coastline of Sri Lanka (Partow et al., 2021). The X-Press Pearl ship was carrying 87 containers of several types of plastic pellets with an estimated weight of around 1680 tons. Table 8.2 shows the types and quantities of plastics and polymers according to the X-Press Pearl ship manifest. Consequently, epoxy and synthetic resins, high-density polyethylene (HDPE), low-density polyethylene (LDPE), and

TABLE 8.2
Plastic and Polymer Types Onboard in the X-Press Pearl

Plastics and Polymers	Amount (MT)	Plastics and Polymers	Amount (MT)
Epoxy resins	9700.8	Synthetic resins	177.3
HDPE	747.8	LLDPE	245.4
LDPE	574.0	Packaging of PS pellets	31.9
Bare foam pig	1.9	Packaging materials	22.8
Polycarbonate	60.9	Plastic pellets	81.3
Polymer beads	19.3	Polymers of propylene	57.3
PVC Film	20.4	ALKYD resin	19.5
Unsaturated Polyester resin	21.0	Vinyl copolymer	18.6
Vinyl acetate	46.3	Polybutadiene	92.7
Total plastics = 11939.2 (MT)			

linear low-density polyethylene (LLDPE) are the main plastic and polymer materials of the cargo onboard (Table 8.2). These industrial raw materials/plastic pellets and burned plastic fragments have been found along Sri Lankan coastline (but a massive accumulation in certain locations on the west coasts) several days after the outbreak of the major fire. In addition, those plastic pellets reach coastlines in Indonesia, Maldives, and Somalia under the influence of monsoon currents (Partow et al., 2021). However, modeling results suggest that most of the pellets remained at sea.

The plastic pellet bags had been released from a single container that contains low-density polyethylene (LDPE). Therefore, coastline pollution is mainly limited to low-density polyethylene (LDPE)–related pellets. On the other hand, the epoxy resin is the main cargo onboard the vessel (9700 tons loaded in 349 containers), and it is not normally classified as a dangerous good. However, epoxy resin can be dangerous in liquid form. For example, the liquid epoxy can sink and generate a plume close to the seafloor. The plume can drift and interact with suspended particles depending on the characteristics of the oceanic current, and it is toxic to aquatic life (Partow et al., 2021). The ship manifest does not provide adequate details. Consequently, complementary information (e.g. chemical abstracts service number, material safety data sheet) is required to assess the potential impacts of polymers and plastics (Partow et al., 2021). In addition, no analysis has been carried out yet to determine the contamination risks associated with the burned plastics and pellets on the coastline.

8.3 MICROPLASTICS

Microplastics (MPs) are generally defined as small pieces of plastic less than 5 mm in diameter. These MPs can be widely found in marine and coastal environments, including sediments, beaches, seawater, lagoons, and estuaries (Barboza and Gimenez, 2015; Wang et al., 2017; Guo and Wang, 2019). About 80% of marine debris accounts for plastic litter (Auta et al., 2017; Coyle et al., 2020). Therefore, both land-oriented (98%) and marine-oriented (2%) plastic litters are the major sources of marine MPs. The exposure of raw materials during accidents such as the X-Press Pearl disaster (Figure 8.1), illegal plastic disposal of ships and vessels, and fishing and aquaculture operations are examples of marine-oriented plastic litter (Haward, 2018; Sewwandi et al., 2022).

8.3.1 Sources of MPs

Microsized particles can be divided into primary (i.e. enter directly into the environment as nurdles, pellets, and microbeads) and secondary MPs (Jaikumar et al., 2019; Hale et al., 2020; Yang et al., 2020). Primary MPs are manufactured at microscopic sizes as raw materials in plastic products for various industrial and domestic applications (Horton and Dixon, 2018). For example, pellets are granular MPs with a diameter of 2–5 mm (Figure 8.1). Synthetic microfibers are mainly composed of polyester, nylon, and acrylic. Polyester is the most commonly found MPs in marine environments (Mishra et al., 2019). Asia and Africa consume about 63% of synthetic microfibers and thus increase the accumulation of MPs in the Indian Ocean.

Sources and Fate of Plastics into Microplastics 161

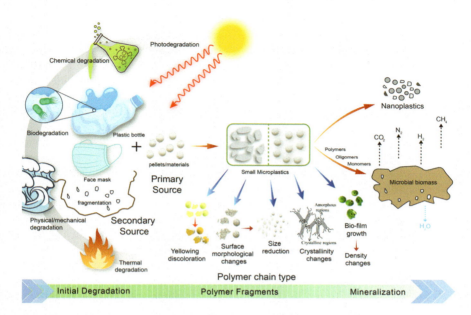

FIGURE 8.2 Degradation of plastics into microplastics.

In addition, primary MPs can be released during the production, transportation, consumption, and maintenance stages in the lifecycle of plastic (Figure 8.2).

Secondary MPs are derived from large plastic debris/macroplastics (e.g. plastic bags, bottles, and packing) under weathering/degradation process over time (Jiang, 2018; Guo and Wang, 2019; Padervand et al., 2020). In contrast, larger plastics such as plastic bags and derelict fishing nets have also become a threat to marine life and the marine environment, especially in coral reef systems and surf zones (Hardesty et al., 2015; Xanthos and Walker, 2017; Pinheiro et al., 2021; McCoy et al., 2022). Wastewater treatment plants also release a significant amount of MPs into the environment, as microsized particles often bypass the treatment units during the cleaning process (Padervand et al., 2020). The world gravitates toward personal protective equipment (PPE) during the past two years after the COVID-19 global pandemic. Surgical face masks (i.e. the most abundant type of PPE) are composed of plastic microfibers such as polypropylene (De-la-Torre et al., 2022) and also liberate both high- and low-density MPs (Abedin et al., 2022). Consequently, face masks would be one of the major sources for marine MPs in the near future (Figure 8.2).

8.3.2 Toxicity of Microplastics

Plastics contain a large variety of chemical additives that are used in improving plasticity, performance, and lifetime. These additives are used as plasticizers, heat stabilizers, antioxidants, flame retardants, colorants, pigments, fillers, and reinforcements. The most common additives are bisphenol A (BPA), phthalates, heavy metals, and polybrominated diphenyl ethers (Figure 8.3). Chemical additives are normally endocrine disruptors and are not chemically bound to the bulk structures during polymer manufacturing. These chemicals can easily be released into the environment along with MPs (Figure 8.3) and are potentially toxic to organisms via three routes: ingestion, dermal absorption, and inhalation (Campanale et al., 2020; Biamis et al., 2021). The accumulation of these additives in organisms could negatively affect them. For example, chemical additives and MPs can cause tissue damage, oxidative stress, changes in immune-related gene expression in humans and neurotoxicity, growth retardation, and behavioral abnormalities in fish (Padervand et al., 2020). The release of volatile compounds (e.g. benzene, toluene, styrene, and methylene chloride) from MPs can also contribute to chronic health issues.

Microplastics possess a relatively large surface area/volume ratio, and thus it is possible to absorb hydrophobic pollution from water (Guo and Wang, 2019; Padervand et al., 2020). In addition, MPs provide substrate to pathogenic bacteria such as *Vibrio parahaemolyticus*. Furthermore, it is found that the interactions between MPs and organic pollutants in aquatic environments can increase toxicity by a factor of 10 (Rubin and Zucker, 2022). Microplastics smaller than 20 μm can penetrate organs, and particles of about 10 μm in size can cross cell membranes and the blood-brain barrier to enter the placenta. The distribution of MPs in secondary tissues (e.g. liver, muscles, and the brain) can cause severe health problems, and it depends on physical and chemical properties and the accumulation concentration of MPs (Campanale et al., 2020). Consequently, MPs are considered one of the prominent health issues in the modern world due to their toxicity effects (Figure 8.3).

8.4 WEATHERING AND DEGRADATION OF PLASTICS

Physical, chemical, and biological processes reduce the structural integrity (Figure 8.2) and then fragment plastics into microplastics (Auta et al., 2017; de Sá et al., 2018). Environmental factors (e.g. sunlight, temperature) and material characteristics (e.g. polymer size, density) also control plastic degradation (Figure 8.2). For example, plastic fragmentation is high in the coastal zone due to high ultraviolet (UV) radiation and physical abrasion by waves (Auta et al., 2017; de Sá et al., 2018). Prolonged exposure of plastic debris to UV radiation weakens its bonds, cracks the surface, discolors, and gradually degrades (Figure 8.2). According to material characterization (e.g. Table 8.1), several synthetic polymers such as polyethylene (PE), polycaprolactone (PCL), polyurethane (PUR), polyhydroxybutyrate (PHB), polyhydroxyalkanoate (PHA), polyvinyl chloride (PVC), polyethylene terephthalate (PET), polybutylene succinate (PBS), polylactic acid or polylactide (PLA), polypropylene (PP), and polystyrene (PS) indicate high resistance against the degradation (Ahmed et al., 2018). Polymers are converted into smaller molecular units such as oligomers, monomers, and/or chemically modified versions during weathering/degradation (Figure 8.2). The most important processes for weathering/degradation are physical/mechanical degradation, photodegradation, chemical degradation, and biodegradation (Bahl et al., 2020).

Sources and Fate of Plastics into Microplastics 163

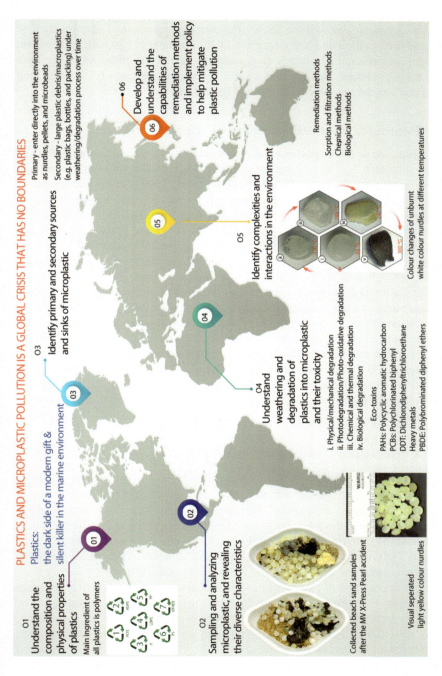

FIGURE 8.3 The outline of the book chapter. Point 1 highlights the nature of plastics. Points 2–5 highlight their distribution in environment and interaction and weathering with time. Finally, point 6 highlights mitigating the global health risks.

8.4.1 Physical/Mechanical Degradation

Physical/mechanical process is an important factor for plastic fragmentation in coastal environments due to wave action (the panning process) and abrasion against beach sand (Fotopoulou and Karapanagioti, 2012; Zhang et al., 2020a). In this case, the weight and volume of plastics (e.g. nurdles) decrease significantly over time (Figure 8.1). Therefore, this process adds a significant amount of MPs and then nanoplastics into the marine environment (Figure 8.2).

8.4.2 Photodegradation/Photooxidative Degradation

Photodegradation is the degradation of plastics under exposure to ultraviolet (UV) radiation (Figure 8.2). The UV radiation in the range of 290–400 nm has enough energy to break several chemical bonds (Gijsman et al., 1999; Cooper and Corcoran, 2012). Therefore, this results in the breaking of the polymer chains (reducing molecular weight) and producing free radicals (Yousif and Haddad, 2013). In addition, photodegradation leads to changes in the physical and optical properties of plastics (Figure 8.2). For example, fading, yellowing, loss of mechanical properties, embrittlement, and changes in molecular weight are the results of photodegradation of plastics (Singh and Sharma, 2008; Cooper and Corcoran, 2012). Furthermore, photodegradation is generally a fast process compared to mechanical and biological processes (Singh and Sharma, 2008; Cooper and Corcoran, 2012). However, the rate of degradation of this process depends on the quantity of antioxidant additives that can prevent the oxidation processes of polymers. Moreover, the photodegradation of plastics in the marine/aquatic environment is slower compared to the terrestrial environment, as UV light is absorbed rapidly by water. Consequently, plastics generally take a longer time to degrade at sea than on land (Andrady, 2011). For this reason, many plastics can remain in marine environments for decades, or even several hundreds of years.

8.4.3 Chemical and Thermal Degradation

The corrosive chemicals, acids (e.g. nitric, sulfuric, and hydrochloric), gases or liquids, and atmospheric pollutants can break and oxidize the polymer chains of plastic, and it is known as chemical degradation (Figure 8.2). For example, in the X-Press Pearl disaster, the white color plastic nurdles are changed to different colors with the reaction of nitric acid and heat (Figure 8.3). Fourier-transform infrared spectroscopy (FTIR) results also show that the chemical structure of the plastic nurdles is also changed during the degradation (Sewwandi et al., 2022).

Thermal degradation of plastic refers to chemical changes affecting heat/temperature but without the simultaneous involvement of any other facts. All polymers can be degraded by heat. However, thermoplastics such as polyethylene (PE), low-density polyethylene (LDPE), linear low-density polyethylene (LLDPE), and polypropylene (PP) are resistant to thermal oxidation (Table 8.1). These plastics are manufactured at high temperatures. In thermal degradation, the depolymerization and statistical fragmentation of chains are considered the two different mechanisms for polymer degradation. After that, polymers follow three major pathways: side-group elimination, random scission, and depolymerization. The side-group elimination takes place in a two-stage process of (i) stripping or elimination of side groups attached to the backbone of the polymer (leaves an unsaturated chain) and (ii) scission into smaller fragments (Cooper and Corcoran, 2012). Random scission results from the formation of free radicals at some point along the backbone of the polymer. Fragments break into smaller fragments during depolymerization (Cooper and Corcoran, 2012). The experimental results show that low-density polyethylene (LDPE) released by the X-Press Pearl disaster altered the polymer structure of plastic nurdles above 60°C (Figure 8.3).

8.4.4 Biological Degradation

Biological degradation refers to the physical or chemical changes induced in plastics by any environmental factor such as light, moisture, heat, and wind, along with biological agents such as bacteria, fungi, yeasts, and their enzymes (Cooper and Corcoran, 2012). Polymers can be potential substrates for microorganisms (Figure 8.2), and factors including polymer characteristics (e.g. mobility, tacticity, crystallinity, molecular weight, and the type of functional groups and additives) and the type of organism control the rate of biodegradation (Bahl et al., 2020). Biodegradation of polymers occurs in four steps: (i) attachment of microorganisms to the polymer surface; (ii) use of the polymer as the carbon source resulting in the growth of the microorganism; (iii) primary polymer degradation; and (iv) ultimate degradation (Cooper and Corcoran, 2012).

In biological degradation, microorganisms colonize the plastic surface and cause the reduction of the molecular weight of plastics, and then the polymer is converted into its monomers. The monomers then become CO_2, H_2O, N_2, H_2, CH_4, salts, minerals, and biomass, and it is known as the mineralization of plastics (Bahl et al., 2020). Other processes such as mechanical and chemical degradation promote microbial reactions (Bahl et al., 2020). However, the pure biodegradation process is rarely successful (Cooper and Corcoran, 2012). In addition, it is difficult to remove MPs from the environment due to their small size and lower density.

8.5 REMOVAL OF MICROPLASTICS

Several methods have been introduced to remove MPs from the environment such as sorption and filtration, chemical, biological removal, and ingestion methods (Padervand et al., 2020; Abuwatfa et al., 2021; Malankowska et al., 2021; Zhang et al., 2021; Shi et al., 2022).

8.5.1 Sorption and Filtration Methods

Several studies focused to investigate the effects of influent flux and particle concentration on the removal efficiency for different membranes. Excellent filtration results (e.g. decrease the influent turbidity from 195 NTU to <1 NTU) were obtained during filtration of the synthetic wastewater using a diatomite platform with 90 μm of supporting mesh (Horton and Dixon, 2018; Li et al., 2018). Advanced treatment technologies like membrane bioreactor, rapid sand filtration, disk filter, and dissolved air floating are applied to remove MPs in wastewater treatment plants. Gurung et al. (2016) compared a conventional activated sludge system and a membrane bioreactor, and the results revealed a better removal of MPs using the membrane bioreactor. Talvitie et al. (2017) also showed that the bioreactors eliminate over 99% of MPs, including the smallest size fractions of 20–100 μm (Talvitie et al., 2017).

Modified filtration methods such as biochar (Wang et al., 2021), membranes (polycarbonate, cellulose acetate, polytetrafluoroethylene) (Pizzichetti et al., 2021), and aluminosilicate filter media (Shen et al., 2021) are used in wastewater treatment plants to enhance the efficiency of MPs removal. Furthermore, Misra et al. (2020) developed a surface binding method to remove MPs using magnetic polyoxometalate–supported ionic liquid phase. Moreover, magnetic nanotubes are another example of removing MPs by sorption method (Tang et al., 2021).

8.5.2 Chemical Methods

Microplastic remediation using chemicals is a cost-effective and widely used method in the industry. These coagulants are either organic or inorganic such as ferric chloride, polyaluminum chloride, polyamine, and polyacrylamide (Rajala et al., 2020; Monira et al., 2021). Oil-based ferrofluid is also used to remove MPs from laundry wastewater (Hamzah et al., 2021).

Coagulation and agglomeration processes (use Fe- and Al-based salts and other coagulants) are widely applied to bind smaller particles in wastewater treatment plants. Experimental results show that Al^{3+} performs better than Fe^{3+} (Ariza-Tarazona et al., 2019). The removal efficiency of MPs is enhanced with the coagulant concentration and pH. Furthermore, anionic polyacrylamide acts better than cationic polyacrylamide during the MPs remediation (Ariza-Tarazona et al., 2019). The electrocoagulation technique has several advantages such as minimum sludge generation, energy efficiency, low operational cost, and flexibility for automation.

The degradative capability of TiO_2-based microdevices and nanodevices for the photocatalytic treatment (Sekino et al., 2012) and an Au-decorated TiO_2-based micromotor under UV illumination are found as efficient photocatalytic removal methods (Wang et al., 2019). A protein-based N-TiO_2 photocatalyst is also applied in degrading the polyethylene MPs in solid and aqueous phases. However, the overall removal efficiency under chemical processes depends on the type of coagulant, pH, chemical composition and salinity, and concentrations.

8.5.3 Biological Methods

Biological methods like biopolymers as flocculation agents (Magalhães et al., 2020), biological filters consisting of anthracite-sand or granular activated carbon (Cherniak et al., 2022), and aerobic and anaerobic processes (Lee and Kim, 2018) are emerging methods to remove MPs. Biodegradation using bacterial, algae, and fungal species is also applied on the experimental scale (Kasmuri et al., 2022; Manzi et al., 2022). These environmentally friendly bioremediation methods show low efficiency compared to chemical and physical methods.

Harrison et al. (2011) discussed the potential of microorganisms, including bacteria and archaea, to facilitate the biological degradation of MPs in coastal sediments. Similarly, Souda consortium shows the potential to remove high-density polyethylene MPs in seawater. Moreover, *Zalerion maritimum* community uses MPs as a nutrient source that triggers biological degradation (Sivan, 2011). Auta et al. (2017) applied *Bacillus* bacterial strains to identify the degradation rates of different MPs. Several investigations show that zooplankton have a high capacity to remove polystyrene MPs through uptake and ingestion, as their intestinal tracts hold MPs for up to 7 days (van Franeker et al., 2011). Arossa et al. (2019) studied Red Sea giant clam, *Tridacna maxima*, to evaluate the removal capacity of polyethylene MPs from wastewater. It shows that the clam shells act as a sorption surface to remove MPs, with larger clams adsorbing higher MPs concentrations. However, the removal efficiency depends on the properties of MPs such as density, shape, and polymer type (Zhang et al., 2020b).

SUMMARY POINT

This book chapter reviewed the properties of plastics, nurdles and polymers, sources and toxicity of MPs, weathering and degradation of plastics into MPs, and their removal methods from the environment (Figure 8.3).

REFERENCES

Abedin, Md. J., Khandaker, M.U., Uddin, Md. R., Karim, Md. R., Ahamad, M.S.U., Islam, Md. A., Arif, A.M., Sulieman, A., Idris, A.M., 2022. PPE pollution in the terrestrial and aquatic environment of the Chittagong city area associated with the COVID-19 pandemic and concomitant health implications. *Environmental Science and Pollution Research* 29, 27521–27533. https://doi.org/10.1007/s11356-021-17859-8

Abuwatfa, W.H., Al-Muqbel, D., Al-Othman, A., Halalsheh, N., Tawalbeh, M., 2021. Insights into the removal of microplastics from water using biochar in the era of COVID-19: a mini review. *Case Studies in Chemical and Environmental Engineering* 4, 100151. https://doi.org/10.1016/j.cscee.2021.100151

Ahmed, T., Shahid, M., Azeem, F., Rasul, I., Shah, A.A., Noman, M., Hameed, A., Manzoor, N., Manzoor, I., Muhammad, S., 2018. Biodegradation of plastics: current scenario and future prospects for environmental safety. *Environmental Science and Pollution Research* 25, 7287–7298. https://doi.org/10.1007/s11356-018-1234-9

Ajith, N., Arumugam, S., Parthasarathy, S., Manupoori, S., Janakiraman, S., 2020. Global distribution of microplastics and its impact on marine environment – a review. *Environmental Science and Pollution Research* 27, 25970–25986. https://doi.org/10.1007/s11356-020-09015-5

Anbumani, S., Kakkar, P., 2018. Ecotoxicological effects of microplastics on biota: a review. *Environmental Science and Pollution Research* 25, 14373–14396. https://doi.org/10.1007/s11356-018-1999-x

Andrady, A.L., 2011. Microplastics in the marine environment. *Marine Pollution Bulletin* 62, 1596–1605. https://doi.org/10.1016/j.marpolbul.2011.05.030

Ariza-Tarazona, M.C., Villarreal-Chiu, J.F., Barbieri, V., Siligardi, C., Cedillo-González, E.I., 2019. New strategy for microplastic degradation: green photocatalysis using a protein-based porous N-TiO$_2$ semiconductor. *Ceramics International* 45, 9618–9624. https://doi.org/10.1016/j.ceramint.2018.10.208

Arossa, S., Martin, C., Rossbach, S., Duarte, C.M., 2019. Microplastic removal by red sea giant clam (*Tridacna maxima*). *Environmental Pollution* 252, 1257–1266. https://doi.org/10.1016/j.envpol.2019.05.149

Auta, H.S., Emenike, C.U., Fauziah, S.H., 2017. Distribution and importance of microplastics in the marine environment: a review of the sources, fate, effects, and potential solutions. *Environment International* 102, 165–176. https://doi.org/10.1016/j.envint.2017.02.013

Bahl, S., Dolma, J., Singh, J.J., Sehgal, S., 2020. Biodegradation of plastics: a state of the art review. *Materials Today* 39, 31–34. https://doi.org/10.1016/j.matpr.2020.06.096

Barboza, L.G.A., Gimenez, B.C.G., 2015. Microplastics in the marine environment: current trends and future perspectives. *Marine Pollution Bulletin* 97, 5–12. https://doi.org/10.1016/j.marpolbul.2015.06.008

Biamis, C., Driscoll, K.O., Hardiman, G., 2021. Microplastic toxicity: a review of the role of marine sentinel species in assessing the environmental and public health impacts. *Case Studies in Chemical and Environmental Engineering* 3, 100073. https://doi.org/10.1016/j.cscee.2020.100073

Campanale, C., Stock, F., Massarelli, C., Kochleus, C., Bagnuolo, G., Reifferscheid, G., Uricchio, V.F., 2020. Microplastics and their possible sources: the example of Ofanto River in southeast Italy. *Environmental Pollution* 258, 113284. https://doi.org/10.1016/j.envpol.2019.113284

Cherniak, S.L., Almuhtaram, H., McKie, M.J., Hermabessiere, L., Yuan, C., Rochman, C.M., Andrews, R.C., 2022. Conventional and biological treatment for the removal of microplastics from drinking water. *Chemosphere* 288, 132587. https://doi.org/10.1016/j.chemosphere.2021.132587

Cooper, D.A., Corcoran, P., 2012. Effects of chemical and mechanical weathering processes on the degradation of plastic debris on marine beaches. Doctor of Philosophy Thesis, the University of Western Ontario, Canada.

Coyle, R., Hardiman, G., Driscoll, K.O., 2020. Microplastics in the marine environment: a review of their sources, distribution processes, uptake and exchange in ecosystems. *Case Studies in Chemical and Environmental Engineering* 2, 100010. https://doi.org/10.1016/j.cscee.2020.100010

de Sá, L.C., Oliveira, M., Ribeiro, F., Rocha, T.L., Futter, M.N., 2018. Studies of the effects of microplastics on aquatic organisms: what do we know and where should we focus our efforts in the future? *Science of the Total Environment* 645, 1029–1039. https://doi.org/10.1016/j.scitotenv.2018.07.207

De-la-Torre, G.E., Dioses-Salinas, D.C., Dobaradaran, S., Spitz, J., Nabipour, I., Keshtkar, M., Javanfekr, F., 2022. Release of phthalate esters (PAEs) and microplastics (MPs) from face masks and gloves during the COVID-19 pandemic. *Environmental Research* 215, 114337. https://doi.org/10.1016/j.envres.2022.114337

Derraik, J.G.B., 2002. The pollution of the marine environment by plastic debris: a review. *Marine Pollution Bulletin* 44, 842–852. https://doi.org/10.1016/S0025-326X(02)00220-5

Fotopoulou, K.N., Karapanagioti, H.K., 2012. Surface properties of beached plastic pellets. *Marine Environmental Research* 81, 70–77. https://doi.org/10.1016/j.marenvres.2012.08.010

Gijsman, P., Meijers, G., Vitarelli, G., 1999. Comparison of the UV-degradation chemistry of polypropylene, polyethylene, polyamide 6 and polybutylene terephthalate. *Polymer Degradation and Stability* 65, 433–441. https://doi.org/10.1016/S0141-3910(99)00033-6

Guo, X., Wang, J., 2019. The chemical behaviors of microplastics in marine environment: a review. *Marine Pollution Bulletin* 142, 1–14. https://doi.org/10.1016/j.marpol bul.2019.03.019

Gurung, K., Ncibi, M.C., Fontmorin, J.M., Särkkä, H., Sillanpää, M., 2016. Incorporating submerged MBR in conventional activated sludge process for municipal wastewater treatment: a feasibility and performance assessment. *Journal of Membrane Science* 6, 3. https://doi.org/10.4172/2155-9589.1000158

Hale, R.C., Seeley, M.E., La Guardia, M.J., Mai, L., Zeng, E.Y., 2020. A global perspective on microplastics. *Journal of Geophysical Research* 125, e2018JC014719. https://doi.org/ 10.1029/2018JC014719

Hammer, J., Kraak, M.H.S., Parsons, J.R., 2012. Plastics in the marine environment: the dark side of a modern gift. Reviews of Environmental Contamination and Toxicology, 1–44. https://doi.org/10.1007/978-1-4614-3414-6_1

Hamzah, S., Ying, L.Y., Azmi, A.A., Abd, R., Razali, N.A., Hairom, N.H.H., Mohamad, N.A., Harun, M.H.C., 2021. Synthesis, characterisation and evaluation on the performance of ferrofluid for microplastic removal from synthetic and actual wastewater. *Journal of Environmental Chemical Engineering* 9, 105894. https://doi.org/10.1016/ j.jece.2021.105894

Hardesty, B.D., Holdsworth, D., Revill, A.T., Wilcox, C., 2015. A biochemical approach for identifying plastics exposure in live wildlife. *Methods in Ecology and Evolution* 6, 92–98. https://doi.org/10.1111/2041-210X.12277

Harrison, J.P., Sapp, M., Schratzberger, M., Osborn, A.M., 2011. Interactions between microorganisms and marine microplastics: a call for research. *Marine Technology Society Journal* 45, 12–20. http://doi.org/10.4031/MTSJ.45.2.2

Haward, M., 2018. Plastic pollution of the world's seas and oceans as a contemporary challenge in ocean governance. *Nature Communications* 9, 667. https://doi.org/10.1038/ s41467-018-03104-3

Horton, A.A., Dixon, S.J., 2018. Microplastics: an introduction to environmental transport processes. *WIREs Water* 5, e1268. https://doi.org/10.1002/wat2.1268

Jaikumar, G., Brun, N.R., Vijver, M.G., Bosker, T., 2019. Reproductive toxicity of primary and secondary microplastics to three cladocerans during chronic exposure. *Environmental Pollution* 249, 638–646. https://doi.org/10.1016/j.envpol.2019.03.085

Jiang, J.Q., 2018. Occurrence of microplastics and its pollution in the environment: a review. *Sustainable Production and Consumption* 13, 16–23. https://doi.org/10.1016/ j.spc.2017.11.003

Jones, K.L., Hartl, M.G.J., Bell, M.C., Capper, A., 2020. Microplastic accumulation in a *Zostera marina* L. bed at Deerness Sound, Orkney, Scotland. *Marine Pollution Bulletin* 152, 110883. https://doi.org/10.1016/j.marpolbul.2020.110883

Kasmuri, N., Tarmizi, N.A.A., Mojiri, A., 2022. Occurrence, impact, toxicity, and degradation methods of microplastics in environment – a review. Environmental Science and Pollution Research 29, 30820–30836. https://doi.org/10.1007/s11356-021-18268-7

Lamba, P., Kaur, D.P., Raj, S., Sorout, J., 2021. Recycling/reuse of plastic waste as construction material for sustainable development: a review. *Environmental Science and Pollution Research*. https://doi.org/10.1007/s11356-021-16980-y

Lee, H., Kim, Y., 2018. Treatment characteristics of microplastics at biological sewage treatment facilities in Korea. *Marine Pollution Bulletin* 137, 1–8. https://doi.org/10.1016/j.marpol bul.2018.09.050

Li, L., Xu, G., Yu, H., Xing, J., 2018. Dynamic membrane for micro-particle removal in wastewater treatment: performance and influencing factors. *Science of the Total Environment* 627, 332–340. https://doi.org/10.1016/j.scitotenv.2018.01.239

Magalhães, S., Alves, L., Medronho, B., Romano, A., Rasteiro, M., da G., 2020. Microplastics in ecosystems: from current trends to bio-based removal strategies. *Molecules* 25, 3954. https://doi.org/10.3390/molecules25173954

Malankowska, M., Echaide-Gorriz, C., Coronas, J., 2021. Microplastics in marine environment: a review on sources, classification, and potential remediation by membrane technology. *Environmental Science: Water Research & Technology* 7, 243–258. https://doi.org/10.1039/D0EW00802H

Manzi, H.P., Zhang, M., Salama, E.S., 2022. Extensive investigation and beyond the removal of micro-polyvinyl chloride by microalgae to promote environmental health. *Chemosphere* 300, 134530. https://doi.org/10.1016/j.chemosphere.2022.134530

McCoy, K.S., Huntington, B., Kindinger, T.L., Morioka, J., O'Brien, K., 2022. Movement and retention of derelict fishing nets in Northwestern Hawaiian Island reefs. *Marine Pollution Bulletin* 174, 113261. https://doi.org/10.1016/j.marpolbul.2021.113261

McKeen, L. W., (ed.). 2016. Polyether Plastics. *Fatigue and Tribological Properties of Plastics and Elastomers (Second Edition)*, William Andrew Publishing, 73–98.

Mishra, S., Rath, C.C., Das, A.P., 2019. Marine microfiber pollution: a review on present status and future challenges. *Marine Pollution Bulletin* 140, 188–197. https://doi.org/10.1016/j.marpolbul.2019.01.039

Misra, A., Zambrzycki, C., Kloker, G., Kotyrba, A., Anjass, M.H., Franco, C.I., Mitchell, S.G., Güttel, R., Streb, C., 2020. Water purification and microplastics removal using magnetic polyoxometalate-supported ionic liquid phases (magPOM-SILPs). *Angewandte Chemie International Edition* 59, 1601–1605. https://doi.org/10.1002/anie.201912111

Monira, S., Bhuiyan, M.A., Haque, N., Pramanik, B.K., 2021. Assess the performance of chemical coagulation process for microplastics removal from stormwater. *Process Safety and Environmental Protection* 155, 11–16. https://doi.org/10.1016/j.psep.2021.09.002

Padervand, M., Lichtfouse, E., Robert, D., Wang, C., 2020. Removal of microplastics from the environment: a review. *Environmental Chemistry Letters* 18, 807–828. https://doi.org/10.1007/s10311-020-00983-1

Partow, H., Lacroix, C., Le Floch S., Alcaro, L., 2021. X-Press Pearl Maritime Disaster. Report of the UN environmental advisory mission, Sri Lanka.

Pinheiro, L.M., Lupchinski, J.E., Denuncio, P., Machado, R., 2021. Fishing plastics: a high occurrence of marine litter in surf-zone trammel nets of Southern Brazil. *Marine Pollution Bulletin* 173, 112946. https://doi.org/10.1016/j.marpolbul.2021.112946

Pizzichetti, A.R.P., Pablos, C., Álvarez-Fernández, C., Reynolds, K., Stanley, S., Marugán, J., 2021. Evaluation of membranes performance for microplastic removal in a simple and low-cost filtration system. *Case Studies in Chemical and Environmental Engineering* 3, 100075. https://doi.org/10.1016/j.cscee.2020.100075

Rajala, K., Grönfors, O., Hesampour, M., Mikola, A., 2020. Removal of microplastics from secondary wastewater treatment plant effluent by coagulation/flocculation with iron, aluminum and polyamine-based chemicals. *Water Research* 183, 116045. https://doi.org/10.1016/j.watres.2020.116045

Ratnayake, A.S., Perera, U.L.H.P., 2022. Coastal zone management in Sri Lanka: a lesson after recent naval accidents. *Marine Pollution Bulletin* 182, 113994. https://doi.org/10.1016/j.marpolbul.2022.113994

Rubin, A.E., Zucker, I., 2022. Interactions of microplastics and organic compounds in aquatic environments: a case study of augmented joint toxicity. *Chemosphere* 289, 133212. https://doi.org/10.1016/j.chemosphere.2021.133212

Scherer, C., Weber, A., Stock, F., Vurusic, S., Egerci, H., Kochleus, C., Arendt, N., Foeldi, C., Dierkes, G., Wagner, M., Brennholt, N., Reifferscheid, G., 2020. Comparative assessment of microplastics in water and sediment of a large European river. *Science of the Total Environment* 738, 139866. https://doi.org/10.1016/j.scitotenv.2020.139866

Sekino, T., Takahashi, S., Takamasu, K., 2012. Fundamental study on nanoremoval processing method for microplastic structures using photocatalyzed oxidation. *Key Engineering Materials* 523, 610–614. https://doi.org/10.4028/www.scientific.net/KEM.523-524.610

Sewwandi, M., Hettithanthri, O., Egodage, S.M., Amarathunga, A.A.D., Vithanage, M., 2022. Unprecedented marine microplastic contamination from the X-Press Pearl container vessel disaster. *Science of the Total Environment* 828. https://doi.org/10.1016/j.scitotenv.2022.154374

Shen, M., Hu, T., Huang, W., Song, B., Zeng, G., Zhang, Y., 2021. Removal of microplastics from wastewater with aluminosilicate filter media and their surfactant-modified products: performance, mechanism and utilization. *Chemical Engineering Journal* 421, 129918. https://doi.org/10.1016/j.cej.2021.129918

Shi, X., Zhang, X., Gao, W., Zhang, Y., He, D., 2022. Removal of microplastics from water by magnetic nano-Fe$_3$O$_4$. *Science of the Total Environment* 802, 149838. https://doi.org/10.1016/j.scitotenv.2021.149838

Singh, B., Sharma, N., 2008. Mechanistic implications of plastic degradation. *Polymer Degradation and Stability* 93, 561–584. https://doi.org/10.1016/j.polymdegradstab.2007.11.008

Sivan, A., 2011. New perspectives in plastic biodegradation. *Current Opinion in Biotechnology* 22, 422–426. https://doi.org/10.1016/j.copbio.2011.01.013

Sridharan, S., Kumar, M., Saha, M., Kirkham, M.B., Singh, L., Bolan, N.S., 2022. The polymers and their additives in particulate plastics: what makes them hazardous to the fauna? *Science of the Total Environment* 824, 153828. https://doi.org/10.1016/j.scitotenv.2022.153828

Sun, J., Zhu, Z.R., Li, W.H., Yan, X., Wang, L.K., Zhang, L., Jin, J., Dai, X., Ni, B.J., 2021. Revisiting microplastics in landfill leachate: unnoticed tiny microplastics and their fate in treatment works. *Water Research* 190, 116784. https://doi.org/10.1016/j.watres.2020.116784

Talvitie, J., Mikola, A., Koistinen, A., Setälä, O., 2017. Solutions to microplastic pollution – removal of microplastics from wastewater effluent with advanced wastewater treatment technologies. *Water Research* 123, 401–407. https://doi.org/10.1016/j.watres.2017.07.005

Tang, Y., Zhang, S., Su, Y., Wu, D., Zhao, Y., Xie, B., 2021. Removal of microplastics from aqueous solutions by magnetic carbon nanotubes. *Chemical Engineering Journal* 406, 126804. https://doi.org/10.1016/j.cej.2020.126804

Thushari, G.G.N., Senevirathna, J.D.M., 2020. Plastic pollution in the marine environment. *Heliyon* 6, e04709. https://doi.org/10.1016/j.heliyon.2020.e04709

van Franeker, J.A., Blaize, C., Danielsen, J., Fairclough, K., Gollan, J., Guse, N., Hansen, P.L., Heubeck, M., Jensen, J.K., le Guillou, G., Olsen, B., Olsen, K.O., Pedersen, J., Stienen, E.W.M., Turner, D.M., 2011. Monitoring plastic ingestion by the northern fulmar fulmarus glacialis in the North Sea. *Environmental Pollution* 159, 2609–2615. https://doi.org/10.1016/j.envpol.2011.06.008

Wang, C., Wang, H., Fu, J., Liu, Y., 2015. Flotation separation of waste plastics for recycling – a review. *Waste Management* 41, 28–38. https://doi.org/10.1016/j.wasman.2015.03.027

Wang, C., Zhao, J., Xing, B., 2021. Environmental source, fate, and toxicity of microplastics. *Journal of Hazardous Materials* 407, 124357.https://doi.org/10.1016/j.jhazmat.2020.124357

Wang, J., Peng, J., Tan, Z., Gao, Y., Zhan, Z., Chen, Q., Cai, L., 2017. Microplastics in the surface sediments from the Beijiang River littoral zone: composition, abundance, surface textures and interaction with heavy metals. *Chemosphere* 171, 248–258. https://doi.org/10.1016/j.chemosphere.2016.12.074

Wang, L., Kaeppler, A., Fischer, D., Simmchen, J., 2019. Photocatalytic TiO_2 micromotors for removal of microplastics and suspended matter. *ACS Applied Materials & Interfaces* 11, 32937–32944. https://doi.org/10.1021/acsami.9b06128

Wright, S.L., Thompson, R.C., Galloway, T.S., 2013. The physical impacts of microplastics on marine organisms: a review. *Environmental Pollution* 178, 483–492. https://doi.org/10.1016/j.envpol.2013.02.031

Xanthos, D., Walker, T.R., 2017. International policies to reduce plastic marine pollution from single-use plastics (plastic bags and microbeads): a review. *Marine Pollution Bulletin* 118, 17–26. https://doi.org/10.1016/j.marpolbul.2017.02.048

Xiang, Y., Jiang, L., Zhou, Y., Luo, Z., Zhi, D., Yang, J., Lam, S.S., 2022. Microplastics and environmental pollutants: key interaction and toxicology in aquatic and soil environments. *Journal of Hazardous Materials* 422, 126843. https://doi.org/10.1016/j.jhazmat.2021.126843

Yang, Y., Wang, J., Xia, M., 2020. Biodegradation and mineralization of polystyrene by plastic-eating superworms *Zophobas atratus*. *Science of the Total Environment* 708, 135233. https://doi.org/10.1016/j.scitotenv.2019.135233

Yousif, E., Haddad, R., 2013. Photodegradation and photostabilization of polymers, especially polystyrene: review. *SpringerPlus* 2, 398. https://doi.org/10.1186/2193-1801-2-398

Zhang, B., Chen, L., Chao, J., Yang, X., Wang, Q., 2020a. Research progress of microplastics in freshwater sediments in China. *Environmental Science and Pollution Research* 27, 31046–31060. https://doi.org/10.1007/s11356-020-09473-x

Zhang, Y., Kang, S., Allen, S., Allen, D., Gao, T., Sillanpää, M., 2020b. Atmospheric microplastics: a review on current status and perspectives. *Earth-Science Reviews* 203, 103118. https://doi.org/10.1016/j.earscirev.2020.103118

Zhang, Y., Zhou, G., Yue, J., Xing, X., Yang, Z., Wang, X., Wang, Q., Zhang, J., 2021. Enhanced removal of polyethylene terephthalate microplastics through polyaluminum chloride coagulation with three typical coagulant aids. *Science of the Total Environment* 800, 149589. https://doi.org/10.1016/j.scitotenv.2021.149589

9 X-Press Pearl Ship Impact on Sea Turtles

E.M. Lalith Ekanayake, U.S.P.K. Liyanage, and M.S.O.M. Amararathna

9.1 SEA TURTLES OF SRI LANKA

Sea turtles are reptiles that belong to the order Testudines, suborder Cryptodira and spend their entire lives in marine or estuarine habitats. They use terrestrial habitats only for nesting and some restricted cases of basking (Musick and Limpus, 1997). Today there are seven species of sea turtles in the world representing two families, Cheloniidae and Dermochelyidae. The seven species are green turtle (*Chelonia mydas*), olive ridley turtle (*Lepidochelys olivacea*), leatherback turtle (*Dermochelys coriacea*), loggerhead turtle (*Caretta caretta*), hawksbill turtle (*Eretmochelys imbricata*), Kemp's ridley turtle (*Lepidochelys kempii*), and flatback turtle (*Natator depressus*). All seven species of sea turtles are classified as follows according to the Red List of Threatened Species by the International Union for the Conservation of Nature (IUCN) (IUCN-MTSG). The hawksbill turtle and Kemp's ridley turtle are listed as critically endangered, while green turtle is listed as endangered. The loggerhead turtle, olive ridley turtle, and leatherback turtle are listed as vulnerable, while flatback back turtle is listed as data deficient. Moreover, all sea turtles are listed in Appendix I of CITES (Convention on International Trade in Endangered Species of Wild Fauna and Flora), and all species, except the flatback, are included in Appendix I and II of CMS (Convention on the Conservation of Migratory Species of Wild Animals).

Of the seven species of sea turtles in the world, five come ashore to nest in Sri Lanka, while their feeding habitats and migratory routes are located around the island (Figure 9.1). They are the green turtle, loggerhead turtle, hawksbill turtle, olive ridley turtle, and leatherback turtle. Sea turtles have been protected in Sri Lanka under government legislation since 1972 by Fauna and Flora Protection Ordinance (FFPO, 1972; Amendments 1993, 2009, and 2022). Moreover, Sri Lanka also signed the CITES in 1979, CMS Convention in 1991, and IOSEA Marine Turtles MOU in 2001. The turtle-nesting beaches are mainly located from Mount Lavinia on the western coast to Arugambay on the eastern coast. All five species are found nesting in few places, while two or three species are found nesting on most of these beaches. High nesting abundance was observed in Rekawa, Kosgoda, Kahandamodara, and Bundala, while scattered nesting was observed in the other beaches such as Balapitiya

FIGURE 9.1 Five species of sea turtles in Sri Lanka.

and Walawemodara (Amarasooriya, 2000; Ekanayake et al., 2002a; Ekanayake et al., 2012).

The green turtle is the most frequent nesting turtle in Sri Lanka (Deraniyagala, 1939: 1953) and it was observed that about 96% of nests at Rekawa (Ekanayake et al., 2002a) and 90% at Kosgoda (Ekanayake et al., 2010) are of green turtles. Olive ridley turtle is the second most frequent nester on Sri Lankan beaches, with small number of nests from other three species. When considering all the nesting beaches in Sri Lanka approximately 70% of the nests were green turtles and 18% belonged to olive ridley turtles (Rajakaruna et al., 2021). Although sea turtles nest throughout the year in Sri Lanka a clear peak was observed from March to June (Ekanayake et al., 2002a, Ekanayake et al., 2010; Rajakaruna et al., 2021). This nesting pattern was observed at Rekawa (the largest sea turtle rookery in Sri Lanka) from 1996 to 2000, as well as in 2020 and at Kosgoda rookery from 2003 to 2008.

There were about 832 average nests per year recorded at Rekawa rookery for five species of turtles from 1996 to 2000 (Ekanayake et al., 2002a). However, it increased up to 1730, 1537, and 1770 nests per year in 2020, 2021, and 2022, respectively (Figure 9.2). High frequency of leatherback nesting was observed in Godawaya Beach in southern Sri Lanka (Ekanayake et al., 2002b). Amarasooriya (2000) recorded about 4000 nests for 5 species of sea turtles in 21 sites from Benthota to Bundala. All the sea turtle hatcheries in Sri Lanka are located between Mount Lavinia and Koggala. In total, 1752 nests were recorded in 2014 for 5 species of sea turtles in 13 nesting sites from Mount Lavinia to Koggala (Jayathilake et al., 2016).

9.2 FORAGING POPULATION

While it was observed that green turtles are the most frequent nesters, many studies confirmed that olive ridley turtles are the most abundant species in the coastal waters of Sri Lanka. Oliver (1946) and Deraniyagala (1953) have reported large

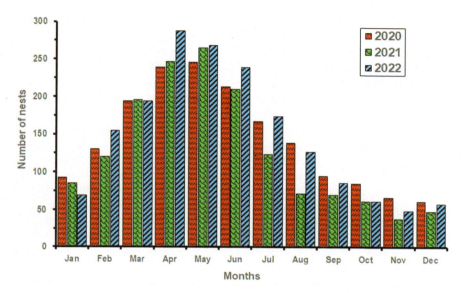

FIGURE 9.2 Number of sea turtle nests at Rekawa rookery – 2020–2022.

concentrations of olive ridley turtles in the coastal waters of Sri Lanka, migrating northward during September and November. During Arribadas in February and March, every year, hundreds of thousands of olive ridleys nests along the coast of Gahirmatha and Rushikulya in Odisha state, India. In 2020, over 300,000 olive ridley nests were recorded in the Arribada, while in 2021 their number was recorded to be 350,000 (B.C. Choudry, personal communication). In India, certain postnesting female turtles migrated to the coasts of Sri Lanka and the Gulf of Mannar, following the gyres within the Bay of Bengal (Shanket et al., 2021). Satellite tagging studies revealed that some of this nesting population foraging around Sri Lankan coastal waters (Kuppusamy et al., 2012; Behera et al., 2018)

A sea turtle disentanglement program was conducted focusing on rescuing turtles caught accidentally in fishing nets off the western coast of Sri Lanka during the flying fish season from November 2006 to May 2007. In total, 46 sea turtles were rescued during the six-month period, with olive ridleys accounting for a large proportion (96%, n = 44; Ramanathan et al., 2010). Out of these 44 olive ridley turtles 54% were adults and also 52% were males (Ramanathan et al., 2010). These studies confirm that there is foraging olive ridley population in the coastal waters of Sri Lanka.

9.3 THREATS TO SEA TURTLES

Sea turtles face many threats all around the world, both in their nesting habitats and in their marine habitats. Most of these threats are man induced, including direct and indirect harvesting of adults, juveniles, and eggs. People kill turtles for meat and shells. Egg collection of sea turtles is taking place in many turtle nesting beaches around the world (Miller, 1997). Nesting females and eggs buried in sand are more accessible to man than other coastal resources. Females are highly vulnerable during their nesting phase as they spend a long time on the beach. Sea turtle eggs and meat are consumed by the coastal communities for nutrition and for purported aphrodisiac qualities (Vargas et al., 2009; Gakuo, 2009). The market demand for green turtle meat is high in some countries, especially for green turtle soup. The largest green turtle slaughter in the world took place in Bali, Indonesia, where about 30,000 turtles were slaughtered in a year (Limpus, 1994).

Loss of feeding habitats due to coral mining is another problem for sea turtles. Turtle-nesting beaches are also constantly being lost due to beach erosion. In areas where erosion is high, beach armoring such as vertical or inclined sea walls, wooden walls, rock revetments, and sandbags are used to protect the beach. These activities result in the drop of the nesting beach areas. Beachfront development for tourism and other purposes is directly affecting the nesting females of their nesting sites. Exotic vegetation replantation programs such as *Casuarina* plantation (after the 2004 tsunami in the Godawaya turtle sanctuary, Hambantota, Sri Lanka – personnel observation) may lead to modification of the natural vegetation profile behind the beach, affecting features such as amount of shade available or underground root mass. Lighting on nesting beaches disrupts critical behavior including the nest site selection of nesting females and the nocturnal sea-finding behavior of both hatchlings and nesting females (Raymond, 1984). The artificial lighting on a beach could be a threat and nesting turtles may be forced to select less appropriate nesting sites when lighting deters sea turtles from nesting beaches (Witherington and Martin, 1996).

Some developed countries use mechanical pumping of sand for beach nourishment, which changes the quality and properties of the nesting habitats such as sorting, moisture content, reflection, and conduction. This could affect the shape of the egg chamber, incubation temperature, water uptake for the developing clutches, and gas exchange (Lutcavage et al., 1997). In some beaches large quantities of sand are dug from the beaches to use as a filling object in various construction activities. Sand mining destroys the beach, disturbs the nesting habitats, and hence becomes a threat to nesting turtles (Sella, 1982). Some countries use heavy vehicles for beach cleaning which could result in crushing the developing eggs and preemergent hatchlings (Lutcavage et al., 1997). Marine pollution due to oil and gas exploration, transportation, and development also are considerable threats to sea turtles. The mode of respiration of the sea turtle is rapid inhalation of surface layer air before diving. Presence of petroleum vapor in the environment could introduce toxins into their lungs. The oil deposits could interfere with normal development of the embryos in the egg clutch as well as present a lethal hazard to newly emerged hatchlings (Lutcavage et al., 1997). The turtle population in the Gulf of Mexico was badly affected by the "Deepwater Horizon" oil spill (Anonymous, 2010; Bjorndal et al., 2011). Sea turtles consume a lot of nondegradable debris such as plastic bags, pellets, lines, ropes, latex balloons, aluminum, paper, cardboard, Styrofoam, and rubber (Mrosovsky, 1981; Lutcavage et al., 1997; Mrosovsky, Ryan and James, 2009).

However, incidental capture in fisheries (bycatch) has been considered a major source of mortality and subsequent population declines for many sea turtle species around the world (Chan et al., 1988; Kapurusinghe and Coorey, 2002; Lewison et al., 2004; Peckham et al., 2007). Sea turtle bycatch occurs in large-scale as well as small-scale fishing fleet, in gear types such as trawls, longlines, gillnets, pound nets, dredges, and also in pots and traps (Chuenpagdee et al., 2003; Moore et al., 2009; Wallace et al., 2010; Finkbeiner et al., 2011). Approximately 40% of all animals caught in fisheries are discarded as trash. Marine mammals, sea turtles, seabirds, and other species are caught and discarded, usually dead. For those animals that are caught and released injured but still alive, their fate after being released is unknown. The likelihood that a turtle is accidentally captured in fisheries is influenced by the interaction between its behavior, ecology, and life history and its susceptibility due to spatial or temporal overlap with fishing gear. Some of these bycatch turtles' carcasses wash to beaches while others float to the deep sea or drown. But there was a question about the percentage of carcasses washed on the beaches. So, there were several studies conducted by researchers to estimate the percentage of carcasses beaching after bycatch dead. According to Hart et al. (2006) drift bottle return rates suggest an upper limit of about 20% for the proportion of sea turtle carcasses that strand and the timeframe for stranding is within two weeks. During another study conducted in Baja California Sur, Mexico, a total of 4752 individually marked drifters were deployed during 9 trials and the overall recovery rate was 22% (Koch et al., 2013).

9.4 THREATS IN SRI LANKA

The most widespread form of sea turtle exploitation in Sri Lanka is the illegal poaching of turtle eggs. Almost a hundred percent of the sea turtle nests occurring on the south and southwest coasts of Sri Lanka had been robbed by egg poachers before 1996 (Ekanayake et al., 2002a). Due to the in situ and ex situ conservation programs protection of nests has increased since 1996 (Ekanayake et al., 2002a, Ekanayake et al., 2010, Rajakaruna et al., 2021). One of the least understood and possibly most serious threats that face marine turtle populations in Sri Lanka is bycatch in fishing gear. Turtles often get entangled in the sea and also in the lagoon, causing damage for each entanglement for fishing nets. In response, fishers either beat the turtles' heads until they are rendered unconscious or hack off the turtles' body parts to make disentanglement easier. Many turtles had been slaughtered for meat especially in areas such as Beruwela, Trincomalee, Jaffna, Negombo, Chilaw, and Kalpitiya (Parsons, 1962; Currey and Matthew, 1995). There was a high demand for turtle meat in Jaffna (Anonymous, 1973). The fishermen of the south and particularly those along the east coast sent turtles that were accidentally caught in their nets to the collecting centers, and from those places the turtles were used to be periodically transported to Jaffna in large lorry consignments and in a most cruel manner (Anonymous, 1973). However, this transportation gradually decreased after the amendment of the FFPO in 1972 (personal communication). Kalpitiya had been the main location for turtle slaughtering since the civil war began in Jaffna in 1983 (Thilakasena, 1998).

During 1999 and 2000 a turtle rescue program was conducted at Kandakkuliya in Kalpitiya, where the bycatch turtles trapped in gill nets were released with the support from fishermen (Kapurusinghe and Cooray, 2002). Several surveys in

Kalpitiya area have reported that some fishermen are willing to rescue the entangled turtles while some are not (Ekanayake et al., 2015). However, due to the continuous awareness programs by various institutions and also law enforcement by government authorities, many fishers release the entangled turtles without harm (Ekanayake & Karunarathna, 2020).

9.5 FLIPPER AND SATELLITE TAGGING IN SRI LANKA AND THE REGIONAL TAGGING PROGRAMS

All sea turtles migrate long distances between feeding sites and nesting sites. Navigational ability of sea turtles has been attributed to a variety of potential mechanisms including a chemical trail or odor plume in the currents (Luschi et al., 1998; 2001), geomagnetic parameters such as magnetic inclination angle and field intensity (Lohmann and Lohmann, 1996), region-specific magnetic cues (Lohmann et al., 2001), landmark-based orientation along the coast (Luschi et al., 1998), or physical and chemical features provided by winds and currents (Papi et al., 2000). Mating of sea turtles usually takes place offshore of the nesting beaches or in the nearshore internesting habitats which are called courtship areas. Males mate in their natal region near female natal beaches and migrate back to their feeding grounds after mating (FitzSimmons et al., 1996; 1997). Mating may also occur along the migration corridors or at the feeding grounds. Flipper tagging and satellite telemetry of nesting females have provided insights into distribution and movements. Conventional flipper tags were used to track renesting movements and postnesting migrations of sea turtles nesting in the Rekawa and Kosgoda rookeries in Sri Lanka. Over 1500 turtles were tagged at both sites from 1996 to 2008, and renesting frequencies of green turtles were 78.9% at Rekawa and 54.3% at Kosgoda (Ekanayake et al., 2013). Many females were found renesting on these two beaches and some on other beaches that were in close proximity to the original tagging sites. Migratory patterns were determined for two female green turtles which were tracked away from the coastal waters of Sri Lanka. One turtle was found dead at the northern tip of Agatti Island, Lakshadweep, India, four years after tagging and the other turtle was found dead on the beach north of Phuket in Phang-nga Province, Thailand, one year after tagging (Ekanayake et al., 2013).

In recent decades, the study of adult marine turtle behavior at sea has been revolutionized by the application of satellite telemetry, usually deployed at nesting beaches where females are most accessible (Godley et al., 2008). This technology has provided intriguing insights into the use of breeding, migration, and feeding habitats and informs conservation and management by assisting in the identification of threats to individuals and habitats throughout their range (James et al., 2005; Peckham et al., 2007; Witt et al., 2011).

First satellite telemetry study for sea turtles in Sri Lanka was conducted at Rekawa Sanctuary in 2006 and 2007. Out of ten green turtles tagged five went along the west coast toward Gulf of Mannar (Richardson et al., 2013). So, this study confirmed that green turtles nesting in southern Sri Lanka migrate toward Gulf of Mannar and Lakshadweep Islands along the western coast of Sri Lanka, passing Colombo, where the X-Press Pearl Accident occurred (Figure 9.3). Moreover, another satellite

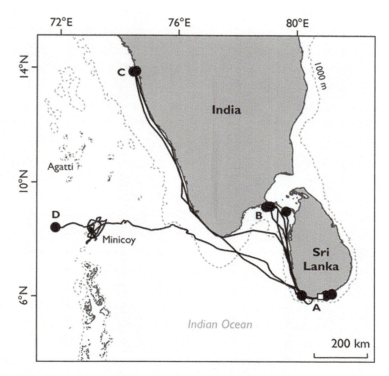

FIGURE 9.3 Migratory path of satellite-tagged green turtles from Rekawa.
Source: Richardson et al., 2013.

telemetry study conducted for olive ridley turtles confirmed that nesting olive ridley in southern Sri Lanka moved toward western sea and Gulf of Mannar (Sivakumar et al., 2010). Furthermore, several studies conducted in India confirmed that olive ridley turtles nesting in Indian beaches migrating toward Sri Lanka and foraging in Sri Lankan waters (Pandav and Choudhury, 1998; Shanker et al., 2002; Behera et al., 2018).

9.6 LARGER NUMBER OF SEA TURTLE STRANDINGS FROM 27 MAY TO 30 SEPTEMBER 2021

The X-Press Pearl ship fire began on 21 May 2021 and continued until 1 June 2021. The first turtle carcass stranding was recorded by the DWC on 27 May 2021 on Negombo Beach. Since then, 372 turtle carcasses have been stranded on the beach until 30 September 2021 along the coast of Sri Lanka from Mannar in the Northern coast to Batticaloa in the eastern coast (Figure 9.4). The carcasses of all five sea turtle species were recorded which is nesting and foraging around Sri Lanka, and also juvenile, subadult, and adult turtle carcasses in both sexes. However, about 83% of the carcasses were olive ridley turtles.

FIGURE 9.4 Sea turtle stranding after X-Press Pearl ship accident. Photo Credit: Sumith Ranawake.

9.7 SEA TURTLE STRANDING IN SRI LANKA

Every year some sea turtle strandings are recorded along the coast of Sri Lanka, mostly due to bycatch in fishing gear (Figure 9.5). From 2014 to 2019 less than 40 strandings were recorded per year, and in 2020 about 50 strandings were recorded (Ekanayake, 2015; unpublished data from DWC and BCSL). In 2020 some turtle deaths were recorded due to the New Diamond oil tanker accident. This could be the reason for the increase in turtle strandings compared to other years from 2014. However, 372 sea turtle strandings were recorded from 27 May to 30 September 2021 from Mannar in the Northern coast to Batticaloa in the eastern coast. This is a 744% increase in sea turtle strandings compared to 2020 (Figure 9.6). This sudden and unusual increase of turtle strandings began a couple of days after the X-Press Pearl ship accident. According to the available sources from Department of Fisheries and Department of Wildlife Conservation, there were no unusual or illegal fisheries activities in the coastal zone of Sri Lanka during these periods which affected sea turtle mortality. Hence, there should be a direct link between ship accident and this sudden increase in sea turtle deaths.

X-Press Pearl Ship Impact on Sea Turtles

FIGURE 9.5 Sea turtle bycatch in fishing gear and stranding.

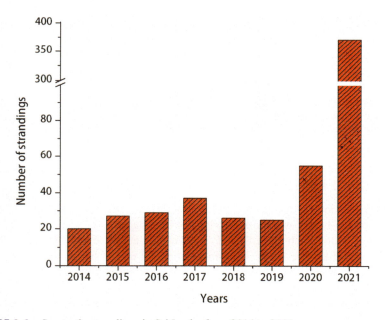

FIGURE 9.6 Sea turtle strandings in Sri Lanka from 2014 to 2021.

9.8 POSSIBLE CAUSES FOR SUDDEN SEA TURTLE DEATHS AFTER THE X-PRESS PEARL SHIP ACCIDENT

Sea turtles are air-breathing reptiles that have lungs, so they regularly surface to breathe. So, turtles could be exposed to the various noxious fumes released from the burning ship while they surface for breathing. Moreover, they keep air from one breath for 15–30 minutes or more and about 3–4 hours while sleeping on sea bed (Lutcavage & Lutz, 1991). Further, adult leatherback turtles can dive deeply, holding their breath for up to 70 minutes. Hence, most of the noxious contents could be absorbed into the lungs and body. There was a continuous fire in the ship for about 10 days. So the heat of the flames and also heat of the water surrounding the ship could be increased. Moreover, burning containers and other substances fall into water with high heat. So there is a higher possibility for heat-induced burning to the sea turtles moving by ship during the incident.

Observation of peculiarities in morphological and anatomical features, such as injuries, damages, and color changes, in sea turtle carcasses found on the beaches can provide valuable insights into the causes of their deaths and the overall health of marine ecosystems. It was known that the ship transported acids and some chemicals (www.bbc.com/news/world-asia-57395693, www.npr.org/2021/06/02/1002484499/sri-lanka-faces-environmental-disaster-as-ship-full-of-chemicals-starts-sinking). According to the chemical list transported by the X-Press Pearl ship there were several toxic chemicals released to the sea during the accident (Rubesinghe et al., 2022). So these chemicals could badly impact sea turtles. The analysis of some sea turtle biosamples is ongoing to confirm the cause of death related to chemicals in the ship. Moreover, there is a possibility that turtles swimming closer to the ship get burned due to higher concentrations of acids and some chemicals.

It is a well-known factor that plastic debris highly impacts sea turtle health. There were larger quantities of plastic nurdles released from the X-Press Pearl ship while some nurdles were burned and debris was released. There is a higher possibility that sea turtles swallow these nurdles or debris while they are swimming and feeding. These pieces of plastic debris could have been stuck and caused blockage in turtle intestines. Moreover, burned bunches of plastic nurdles could even pierce the intestinal wall, causing internal bleeding which is lethal to the turtle.

9.9 IMPACT ON THE SEA TURTLE POPULATION

As discussed above possibly only about 20% of the turtle carcasses could be found stranded on the beach, while others could be floating into the deep sea or drowning (Hart et al., 2006; Koch et al., 2013). So, it can be assumed that the actual number of deaths could be five times the number stranded on the beach. Hence, it can be estimated that about 1860 turtles (372 × 5) from the wild population died due to the impact of the ship.

Sea turtles are better characterized by a type III survivorship curve, with mortality rates inversely related to age (Iverson, 1991). Few sea turtles survive from egg to adulthood, with various models yielding an often-cited 1:1000 survival rate (Frazer, 1986). It is assumed that most mortalities occur in the hatchling and juvenile stages (Figure 9.7).

In total, 372 juvenile and adult turtle carcasses were observed after the ship accident. According to the survivorship curve, generally few juveniles and adults will be surviving in a natural turtle population. Hence, loss of juveniles and adult turtles could have a significant negative impact on the wild population.

After the ship accident plastic debris and nurdles moved long distances with the ocean current in the coastal waters of Sri Lanka. Many studies confirmed that plastic debris is lethal for sea turtles including both hatchlings and adults (Barreiro's and Raykov, 2014; Mendes et al., 2015; Jerdy et al., 2017; Duncan et al., 2021). The ship accident happened during the peak sea turtle nesting season in Sri Lanka (Ekanayake et al., 2002a, Ekanayake et al., 2010; Rajakaruna et al., 2021). So thousands of sea

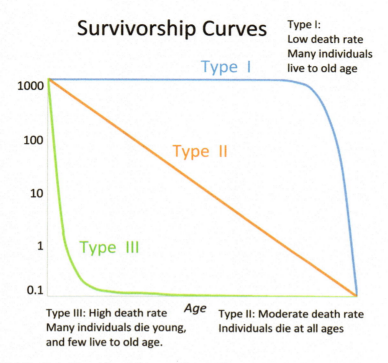

FIGURE 9.7 Type III survivorship curve of sea turtles.

Source: Redrawn from https://greenseaturtlesendangered.weebly.com/general-information-its-life-history.html

turtle hatchings moving along the coast of Sri Lanka could be trapped in the plastic debris and nurdles released to the sea during the ship accident. According to the study conducted by Duncan et al. (2021) smaller stranded turtles displayed higher body burdens of plastic than large bycaught individuals. Hence, release of plastic debris due to the X-Press Pearl ship accident could have a long-lasting impact on the nesting and foraging sea turtle population in Sri Lanka. In addition to sea turtles, plastic debris could be a threat to many other marine megafauna (Senko et al., 2020).

Chemical pollution has been identified as a major threat to sea turtle populations around the world, although level of impact could vary (Finlayson et al., 2019a). The impact level could vary with both organic and inorganic contaminants as well as level of concentration (Finlayson et al., 2019b; Dogruer et al., 2021). Further there is a possibility that chemicals released from the ship will remain for a long time in the coastal zone of Sri Lanka including nesting beaches and impact sea turtles in future.

The plastic debris could be deposited on the beaches around the island including turtle nesting beaches. This may impact the hatching of sea turtle eggs on the nesting beaches. Plastic nurdles, fragments of burned plastic, and microplastics can alter the specific conditions needed for beaches to provide a suitable nesting environment. In sea turtles the sex ratio and hatchling success are influenced by temperature, and hence, alterations to the incubating environment could negatively affect the development of hatchlings, the proportion of males to females produced, and mortality rates (Mrosovsky and Yntema, 1980; Godfrey, Barreto and Mrosovsky, 1996; Cole et al., 2011; Beckwith and Fuentes, 2018). In sea turtles the incubation of eggs at some temperatures yields 100% phenotypic males, whereas 100% phenotypic females are produced at other temperatures (Godfrey, Barreto and Mrosovsky, 1996) while both sexes may be produced only in a 2–3°C range of temperature, often narrow, between these male and female producing temperatures (Mrosovsky and Pieau, 1991: Godfrey, Barreto and Mrosovsky, 1997). The pivotal temperature is the constant incubation temperature that results in a 50:50 sex ratio among individuals. (Mrosovsky and Yntema, 1980: Mrosovsky, 2000). The pivotal temperature cannot be determined for a single egg but for a single clutch (Mrosovsky and Pieau, 1991). For marine turtles the lower temperature produces more males, while the temperature higher than the pivotal temperature produces more females (Godfrey, Barreto, and Mrosovsky, 1997). Plastics, particularly those containing a dark pigment, warm up when exposed to heat, and their presence within the sand may increase the nest temperature, potentially leading to a higher proportion of female hatchlings being produced. So burned nurdles and lumps washed and deposited on the nesting beaches will have a long-lasting negative impact on sea turtles (see Chapter 6 – Figures 6.7 and 6.8).

9.10 CHALLENGES, LIMITATIONS, AND FUTURE PERSPECTIVES

According to the available information, the X-Press Pearl ship accident, which released tons of chemicals and plastic together into the ocean, was one of the largest ship accidents in the world and the first of its kind in Sri Lanka. Although Sri Lanka is located in a maritime hub in the Indian Ocean, the preparedness and facilities available to react are minimal during a ship accident. Apart from the administrative problems faced by government institutions, it was also a major challenge to monitor the impact on sea turtles and other animals as well as on the entire coastal biodiversity due to the accident. The COVID-19 pandemic and subsequent national lockdown and travel restrictions directly influenced the deployment of people to monitor the impact on coastal biodiversity. Moreover, the lack of necessary equipment and facilities and previous experience and deficiencies in updated technology management were other barriers to timely monitoring activities. The beaching of an unusually high number of dead sea turtles was reported after 27 May 2021. Inadequate numbers of trained persons for autopsies and laboratory analysis, poor storage facilities for the preservation of carcasses, difficulty in conducting time sample collection for analysis, lack of reputed laboratory facilities, and poor coordination among responsible authorities were major drawbacks and challenges faced after the incident.

It is a must to initiate a central government body for coordination and immediate action during such a disaster in the future. The necessary resource persons should be equipped with proper practical training as well as required equipment to be mobilized in the event of any kind of maritime disaster in Sri Lankan waters. As Sri Lanka is located in the maritime hub in the Indian Ocean and the Sri Lankan economy depends on the maritime trade, a 24/7-ready monitoring and research team should be available to face any kind of future disaster and to collect proper evidence on time. Further, it should be established in conformity with internationally acceptable affiliated laboratories to provide analytical reports to local and international courts and other required agencies. Deficiencies in policies and legal frameworks in Sri Lanka should be updated and strengthened to conduct internationally acceptable research and studies to challenge any international arbitration or court case.

REFERENCES

Amarasooriya, K.D. (2000). Classification of marine turtle nesting beaches of southern Sri Lanka. In: N. Pilchar and G. Ismail (Editors), Sea turtles of the Indo Pacific, Research, Management and Conservation. *Proceedings of the second ASEAN Symposium and workshop on sea turtle biology and conservation.* 228–237.

Anonymous (1973). Editorial in Loris. The turtles and us. *Loris* 13(1).8.

Anonymous (2010). Sea turtle strandings and the deepwater oil spill. National Oceanic and Atmospheric Administration, U.S. Department of Commerce, Accessed on 25th June 2011 from: http://sero.nmfs.noaa.gov/sf/deepwater_horizon/sea_turtles.pdf

Barreiros JP, Raykov VS (2014) Lethal lesions and amputation caused by plastic debris and fishing gear on the loggerhead turtle Caretta caretta (Linnaeus, 1758). Three case reports from Terceira Island, Azores (NE Atlantic). Mar Pollut Bull 86: 518–522.

Beckwith, V. K., & Fuentes, M. M. P. B. (2018). Microplastic at nesting grounds used by the northern Gulf of Mexico loggerhead recovery unit. *Marine Pollution Bulletin*, *131*(January), 32–37. doi:10.1016/j.marpolbul.2018.04.001

Behera, S., B. Tripathy, B.C. Choudhury and K. Sivakumar (2018). Movements of Olive Ridley Turtles (*Lepidochelys olivacea*) in the Bay of Bengal, India, Determined via Satellite Telemetry. *Chelonian Conservation and Biology*, 17(1):44–53.doi:10.2744/CCB-1245.1

Bjorndal, K.A., Bowen, B.W., Chaloupka, M., Crowder, L.B., Heppell, S.S., Jones, C.M., Lutcavage, M.E., Policansky, D., Solow, A.R. and Witherington, B.E. (2011). Better science needed for restoration in the Gulf of Mexico. *Science* **331**, 537–538.

Chan, E.H., Liew, H.C. and Mazlan, A.G. (1988). The incidental capture of sea turtles in fishing gear in Terengganu, Malaysia. *Biological Conservation* **43**, 1–7.

Chuenpagdee, R., Morgan, L.E., Maxwell, S.M., Norse, E.A. and Pauly, D. (2003). Shifting gears: assessing collateral impacts of fishing methods in US waters. *Frontiers in Ecology and the Environment.* **1**, 517–524.

Cole, M., Lindeque, P., Halsband, C., Galloway, T.S., 2011. Microplastics as contaminants in the marine environment: a review. *Mar. Pollut. Bull.* 62 (12), 2588–2597.

Currey, D. and Matthew, E. (1995). Report on an investigation into threats to marine turtles in Sri Lanka and Maldives. *Environmental Investigation Agency report.*

Deraniyagala, P.E.P. (1939). *The Tetrapod reptiles of Ceylon. Colombo Museum, Colombo,* **1**, 412p.

Deraniyagala, P.E.P. (1953). *A coloured atlas of some vertebrates from Ceylon. Tetrapod Reptiles, Colombo Museum, Colombo,* **2**, 101p.

Dogruer, G., Kramer, N. I., Schaap, I. L., Hollert, H., Gaus, C., van de Merwe, J. P. An integrative approach to define chemical exposure threshold limits for endangered sea turtles. *Journal of Hazardous Materials* 420 (2021) 126512.

Duncan EM, Broderick AC, Critchell K, Galloway TS, Hamann M, Limpus CJ, Lindeque PK, Santillo D, Tucker AD, Whiting S, Young EJ and Godley BJ (2021) Plastic Pollution and Small Juvenile Marine Turtles: A Potential Evolutionary Trap. *Frontiers in Marine Science* 8:699521. doi: 10.3389/fmars.2021.699521

Ekanayake, Lalith (2015). A survey of marine turtle by-catch and fisherfolk attitude at Kalpitiya, Sri Lanka. *Indian Ocean Turtle Newsletter* 22: 11–12.

Ekanayake, Lalith & Y.K. Karunarathna (2020). Marine turtle bycatch, the turtle-human conflict at Kalpitiya Peninsula, Sri Lanka. 6th *International Marine Conservation Congress*, Kiel, Germany (Online Conference).

Ekanayake, E.M.L., Kapurusinghe, T., Saman, M.M. and Premakumara, M.G.C. (2002b). Estimation of the number of leatherback (*Dermochelys coriacea*) nesting at the Godavaya turtle rookery in southern Sri Lanka during the nesting season in the year 2001. *Kachhapa* 6, 11–12.

Ekanayake, E.M.L., Kapurusinghe, T., Saman, M.M. and Rathnakumara, A.M.D.S. (2012). A declining trend of nesting frequency of sea turtles at the largest rookery in Sri Lanka. In: T.T. Jones and B.P. Wallace (Compilers), *Proceedings of the thirty-first Annual Symposium on Sea Turtle Biology and Conservation*. NOAA Technical Memorandum, NOAA NMFS-SEFSC-631, 201.

Ekanayake, E.M.L., R.S. Rajakaruna, T. Kapurusinghe, M.M. Saman, D.S. Rathnakumara, P. Samaraweera & K.B. Ranawana (2010). Nesting Behaviour of the Green Turtle at Kosgoda Rookery, Sri Lanka. *Ceylon Journal of Science (Biological Science)*, 39 (2): 109–120.

Ekanayake, E.M.L., Ranawana, K.B., Kapurusinghe, T., Premakumara, M.G.C. and Saman, M.M. (2002a). Marine turtle conservation in Rekawa turtle rookery in southern Sri Lanka. *Ceylon Journal of Science (Biological Science)* 30, 79–88.

Ekanayake, E.M.L., Thushan Kapurusinghe, M.M. Saman, A.M.D.S. Rathnakumara, R.S. Rajakaruna, P. Samaraweera & K.B. Ranawana (2013). Re-nesting movements and post-nesting migrations of green turtles tagged in two turtle rookeries in Sri Lanka. J. Blumenthal, A. Panagopoulou, and A.F. Rees, Compilers. *Proceedings of the Thirtieth Annual Symposium on Sea Turtle Biology and Conservation*. NOAA Technical Memorandum NMFS-SEFSC-640, 123.

Finkbeiner, E.M., Wallace, B.P., Moore, J.E., Lewison, R.L., Crowder, L.B. and Read, A. J. (2011). Cumulative estimates of sea turtle bycatch and mortality in USA fisheries between 1990 and 2007. *Biological Conservation* doi:10.1016/j.biocon.2011.07.033.

Fitz Simmons, N.N., Moritz, C., Limpus, C.J., Miller, J.D., Parmenter, C.J. and Prince, R. (1996). Comparative genetic structure of green, loggerhead and flatback populations in Australia based on variable mtDNA and nDNA regions. In: B. Bowen and W. Witzell (Compilers), *Proceedings of the International Symposium on Sea Turtle Conservation Genetics*. NOAA Technical Memorandum, NMFS-SEFSC-396, 25–32.

Fitzsimmons, N.N., Limpus, C.J., Norman, J.A., Goldizen, A.R., Miller, J.D. and Moritz, C. (1997). Philopatry of male marine turtles inferred from mitochondrial DNA markers. *Proceedings of the National Academy of Sciences USA* 94, 8912–8917.

Flora and Fauna Protection Ordinance (FFPO), Government of Sri Lanka (1972).

Frazer, N.B. and Ladner, R.C. (1986). A growth curve for green sea turtles, *Chelonia mydas*, in the U.S. Virgin Islands, 1913–14. *Copeia* 1986(3), 798–802.

Gakuo, A.M. (2009). Advances in sea turtle conservation in Kenya. *Indian Ocean Turtle Newsletter* 9, 10–13.

Godfrey, M.H., Barreto, R. and Mrosovsky, N. (1996). Estimating past and present sex ratio of marine turtles in Suriname. *Can.J.Zool*, 74, 267–277.

Godfrey, M.H., Barreto, R. and Mrosovsky, N. (1997). Metabolically-generated Heat of Developing Eggs and its Potential Effect on Sex Ratio of Marine turtle Hatchlings. *Journal of Herpetology.*, **31**(4), 616–619.

Godley, B.J., Blumenthal, J.M., Broderick, A.C., Coyne, M.S., Godfrey, M., Hawkes, L., Witt, M. (2008). Satellite tracking of sea turtles: where have we been and where do we go next? *Endanger. Species Res.* 4, 3–22. http://dx.doi.org/10.3354/esr00060.

Gulko, D. and Eckert, K.L. (2004). Sea Turtles: An Ecological Guide. Mutual Publishing. Honolulu, Hawaii, USA. 122 pp.

Hart KM, Mooreside P, Crowder LB (2006) Interpreting the spatio-temporal patterns of sea turtle strandings: going with the flow. *Biol Conserv* 129:283–290.

IUCN-MTSG. Red List Assessments | IUCN-SSC Marine Turtle Specialist Group (iucn-mtsg.org)

Iverson, J. B. (1991). Patterns of survivorship in turtles (order Testudines). *Can. J. Zool.* 69: 385–391.

James, M.C., S.A. Eckert & R.A. Myers (2005). Migratory and reproductive movements of male leatherback turtles (*Dermochelys coriacea*). *Marine biology* 147: 845–853.

Jayathilake, R.A.M., Maldeniya, R. and Kumara, M.D.I.C. (2016). A Study on Temporal and Spatial Distribution of Sea Turtle Nesting on the Southwest Coast of Sri Lanka. In: *Proceedings of the National Aquatic Research and Development Agency (NARA), Scientific Session*. Colombo, Sri Lanka, 143–146.

Jerdy, H., Werneck, M. R., da Silva, M. A., Ribeiro, R. B., Bianchi, M., Shimoda, E., & de Carvalho, E. C. Q. (2017). Pathologies of the digestive system caused by marine debris in *Chelonia mydas*. *Marine Pollution Bulletin*, 116(1–2), 192–195. https://doi.org/10.1016/j.marpolbul.2017.01.009.

Kapurusinghe, T. and Cooray. R. (2002). Marine turtle by-catch in Sri Lanka: Survey report. Turtle Conservation Project (TCP) Publications, Sri Lanka.

Kimberly A. Finlayson, Frederic D.L. Leusch, Jason P. van de Merwe. Cytotoxicity of organic and inorganic compounds to primary cell cultures established from internal tissues of Chelonia mydas Science of the Total Environment 664 (2019a) 958–967.

Kimberly A. Finlayson, Frederic D.L. Leusch, Jason P. van de Merwe. Primary green turtle (Chelonia mydas) skin fibroblasts as an in vitro model for assessing genotoxicity and oxidative stress Aquatic Toxicology 207 (2019b) 13–18.

Koch V, Peckham H, Mancini A, Eguchi T (2013) Estimating At-Sea Mortality of Marine Turtles from Stranding Frequencies and Drifter Experiments. *PLoS ONE* 8(2): e56776. doi:10.1371/journal.pone.0056776

Lewison, R.L., Freeman, S.A. and Crowder, L.B. (2004). Quantifying the effects of fisheries on threatened species: the impact of pelagic longlines on loggerhead and leatherback sea turtles. *Ecology Letters* **7**, 221–231.

Limpus, C.J. (1994). Current declines in Southeast Asian turtle populations. In: B.A. Schroeder and B.E. Witherington (Compilers), *Proceedings of the Thirteenth Annual Symposium on Sea Turtle Biology and Conservation*. NOAA Technical Memorandum, NMFS-SEFSC-341, 89-91.

Lohmann, K.J. and Lohmann, C.M.F. (1996). Detection of magnetic field intensity by sea turtles. *Nature* **380**, 59–61.

Lohmann, K.J., Cain, S.D., Dodge, S.A. and Lohmann, C.M. F. (2001). Regional magnetic fields as navigational markers for sea turtles. *Science* **294**, 364–366.

Luschi, P., Akesson, S., Broderick, A.C., Fiona, G., Godley, B.J., Papi, F. and Hays, G.C. (2001). Testing the navigational abilities of oceanic migrants: displacement experiments on green sea turtles (*Chelonia mydas*). *Behavioral Ecology and Sociobiology* **50**, 528–534.

Luschi, P., Hays, G.C., Seppia, C.D., Marsh, R. and Papi, F. (1998). The navigational feats of green sea turtles migrating from Ascension Island investigated by satellite telemetry. *Proceedings of the Royal Society of London Biological Sciences* **265**, 2279–2284.

Lutcavage, M.E. & P.L. Lutz (1991). Voluntary diving metabolism and ventilation in the loggerhead. *Journal of Experimental Marine Biology and Ecology* 147: 287–296.

Lutcavage, M.E., Plotkin, P., Witherington, B. and Lutz, P.L. (1997). Human impacts on marine turtle survival. In *The biology of sea turtles*, eds. P. L. Lutz and J. A. Musick. CRC Press, Washington D.C., Chapter 15, 387–409.

Mendes, S.S., Carvalho, R.H., Faria, A.F., Sousa, B.M. (2015): Marine debris ingestion by *Chelonia mydas* (Testudines: Cheloniidae) on the Brazilian coast. *Marine Pollution Bulletin* 92: 8–10.

Miller, J.D. (1997). Reproduction in Marine turtles. In *The biology of marine turtles*, eds. P. L. Lutz and J. A. Musick. CRC Press, Washington D.C., Chapter 3, 51–81.

Moore, J.E., Wallace, B.P., Lewison, R.L., Zydelis, R., Cox, T.M. and Crowder, L.B. (2009). A review of marine mammal, sea turtle and seabird bycatch in USA fisheries and the role of policy in shaping management. *Marine Policy* 33(3), 435–451.

Mrosovsky, N. (1981). Plastic jellyfish. *Marine Turtle Newsletter* **17**, 5–7.

Mrosovsky, N. (2000). A simple method of estimating sex ratio of populations of sea turtle hatchlings. In: Kalb, H.J. and T. Wibbels (Compilers), *Proceedings of the Nineteenth Annual Symposium on Marine turtle Biology and Conservation*. NOAA Technical Memorandum, NMFS-SEFSC-443, 19–20.

Mrosovsky, N. and Yntema, C.L. (1980). Temperature dependence of sexual differentiation in marine turtles: implication for conservation practices. *Biological Conservation* **18**, 271–280.

Mrosovsky, N., Ryan, G.D. and James. M.C. (2009). Leatherback turtles: The menace of plastic. *Marine Pollution Bulletin* **58**, 287–289.

Mrosovsky. N., and Pieau, C.(1991). Transitional range of temperature, pivotal temperatures and thermosensitive stages for sex determination in reptiles. *Amphibia-Reptilia.,* **12**, 169–179.

Oliver, J.A. (1946). *An aggregation of Pacific sea turtles. Copeia 1946: 103.*

Pandav, B. & B.C. Choudhury (1998). Olive ridley tagged in Orissa recovered in the coastal waters of East Sri Lanka. *Marine Turtle Newsletter* 82: 9–10.

Papi, F., Luschi, P., Akesson, S., Capogrossi, S. and Hays, G.C. (2000). Open-sea migration of magnetically disturbed sea turtles. *The Journal of Experimental Biology* **203**, 3435–3443.

Parsons, J.J. (1962). The Green turtle and man. Gainesville: University of Florida, 125p.

Peckham, S.H., Diaz, D.M., Walli, A., Ruiz, G., Crowder, L.B. and Nichols, W.J. (2007). Small-scale fisheries bycatch jeopardizes endangered Pacific loggerhead turtles. *PLoS ONE* **2**(10), e1041. doi:10.1371/journal.pone.0001041.

Rajakaruna, R.S., Ekanayake, E.M. L and Suraweera, P.A.C.N.B. (2021). Sri Lanka. In: Phillott, A.D. and Rees, A.F. (eds.), *Sea Turtles in the Middle East and South Asia Region. MTSG Annual Regional Report 2021*. Draft Report to the IUCN SSC Marine Turtle Specialist Group.

Ramanathan, A., Mallapur, A., Rathnakumara, S., Ekanayake, L. and Kapurusinghe, T. (2010). Untangling the tangled: Knowledge, attitudes and perceptions of fishermen to the rescue and the disentanglement of sea turtles in Kalpitiya, Sri Lanka. In: K. Dean and M. C. Lopez-Castro (Compilers), *Proceedings of the Twenty-eighth Annual Symposium on Sea Turtle Biology and Conservation*. NOAA Technical Memorandum, NMFS-SEFSC-602, 118.

Raymond, P.W. (1984). Marine turtle hatchling disorientation and artificial beachfront lighting. A report for the Centre for Environmental Education sea turtle rescue fund.72p.

Richardson, P.B., Annette C. Broderick, Michael S. Coyne, Lalith Ekanayake, Thushan Kapurusinghe, Chandralal Premakumara, Susan Ranger, M. M. Saman, Matthew J. Witt & Brendan J. Godley (2013). Satellite tracking suggests size-related differences in behaviour and range of female green turtles nesting at Rekawa Wildlife Sanctuary, Sri Lanka. *Marine Biology* 160(6), 1415–1426.

Rubesinghe, C., Brosché, S., Withanage, H., Pathragoda, D., Karlsson, T. X-Press Pearl, a "new kind of oil spill" consisting of a toxic mix of plastics and invisible chemicals. *International Pollutants Elimination Network (IPEN)*

Satyaranjan Behera, Sajan John and Ved Prakash Ola (2010). Applications of satellite telemetry technique in sea turtle research in India. *Telemetry in Wildlife Science*. Vol.13, No.1.

Sella, I. (1982). Sea turtles in the eastern Mediterranean and northern Red Sea. In *Biology and conservation of marine turtles*, ed. K.A. Bjorndal. *Proceedings of the World Conference on Sea Turtle Conservation*. Smithsonian Institution Press, Washington D.C., 417–422.

Senko, J.F., Nelms, S.E., Reavis, J.L., Witherington, B., Godley, B.J., Wallace, B.P., (2020). Understanding individual and population-level effects of plastic pollution on marine megafauna. *Endangered Species Res.* 43, 234–252.

Shanker, K., A. Abreu-Grobois, V. Bezy, R. Briseño, L. Colman, A. Girard, M. Girondot, M. Jensen, M. Manoharakrishnan, J. M. Rguez-Baron, R. A. Valverde and L. West (2021). Olive Ridleys, The quirky turtles that conquered the world. *SWOT Report – State of the World's Sea Turtles*, vol. 16, 24–33.

Shanker, K., Choudhury, B.C., Pandav, B., Tripathy, B., Kar, C.S., Kar, S.K., Gupta, N.K., and Frazier, J.G. (2002). Tracking olive ridley turtles from Odisha. In: Seminoff, J. (Ed.). *Proceedings of the 22nd Annual Symposium on Sea Turtle Biology and Conservation*. NOAA Tech. Memor. NMFS-SEFSC-503, pp. 50–51.

Sivakumar, K., Choudhury, B.C., Kumar, R.S., Behera, S.K., Behera, S., John, S., Ola, V.P. and Tripathy, B. (2010). *Application of satellite telemetry techniques in sea turtle research in India*. Telemetry in Wildlife Science, pp. 140–145.

Thilakasena, S. (1998). Viya siduren ahasa balannange kedavachakaya, kesbavunge jivithaya anature. (Turtles are in danger). *Desathiya* 21(20–21), 3–9. (In Sinhala medium)

Vargas, S.M., Araujo, F.C.F. and Santos, F.R. (2009). DNA barcoding of Brazilian sea turtles (Testudines). *Genetics and Molecular Biology* 32(3), 608–612.

Wallace, B.P., Lewison, R.L., McDonald, S., McDonald, R.T., Bjorkland, R.K., Kelez, S., Kot, C., Finkbeiner, E.M., Helmbrecht, S. and Crowder, L.B. (2010). Global patterns of marine turtle bycatch. *Conservation Letters* 3, 131–142.

Witherington, B.E. and Martin, R.E. (1996). Understanding, assessing, and resolving light pollution problems on marine turtle nesting beaches. Florida Marine Research Institute, Technical Report TR-2.,73p.

Witt MJ, Augowet Bonguno E, Broderick AC, Coyne MS, Formia A, et al. (2011) Tracking leatherback turtles from the world's largest rookery: assessing threats across the South Atlantic. *Proceedings of the Royal Society B: Biological Sciences* 278: 2338–2347.

10 Impacts of the MV X-Press Pearl Ship Disaster on Marine Mammals in Sri Lankan Marine Waters

U.S.P.K. Liyanage, P.R.T. Cumaranatunga,
E.M. Lalith Ekanayaka, and G.A. Tharaka Prasad

10.1 INTRODUCTION

Maritime accidents are undesired events faced by maritime and coastal communities. (Ceyhun, 2014; Luo & Shing, 2019; Dominguez-Péry et al., 2021). However, technological developments including modification of ship design, communication and safety equipment, updated shipping legislations, and training have resulted in a decline in shipping accidents by 70% within the past decade (Dominguez-Péry et al., 2021). Even though the number of accidents is small, damage caused by a single event raises long-term ecological consequences due to large vessel sizes and huge volumes of cargo carried by them (Roberts & Hansen 2002). Human errors attribute to more than 80% of ship accidents (Chan et al., 2016; Acejo et al. 2018; Galieriková, 2019; Sánchez-Beaskoetxea et al., 2021) and the outcomes might be collisions or being stuck by other ships, groundings, fire, explosions, and physical damages due to heavy weather or ultimate sinking (Rømer et al., 1996; HELCOM, 2018).

Sri Lanka is an island in the Indian Ocean neighboring the east-west shipping lane, one of the busiest shipping routes in the world (Priyadarshana et al., 2015), although it has experienced very little number of maritime accidents related to shipping in its history. Unexpectedly, two merchant ships named the MT New Diamond and the MV X-Press Pearl caught fire respectively within the contiguous zone off the south eastern coast in September 2020 and within the territorial waters off the western coast of Sri Lanka in May 2021. MV X-Press Pearl is a newly built (in 2021) container ship registered under the flag of Singapore, having a capacity to carry 2700 containers with volume of twenty-foot equivalent units (TEU). At the time of the accident, it had 1486 TEU, out of which approximately 81 were loaded with dangerous goods including 25 MT of nitric acid, 1040 MT of caustic soda, and 1880 MT of urea. Furthermore, it was reported to carry 15 kinds of hazardous and noxious items categorized in accordance with the International Maritime Dangerous

Goods code. In addition, about 9700 MT of potentially harmful epoxy resins and about 1686 MT of microplastic nurdles and pellets were also listed among loaded cargo (Partow et al., 2021; de Vos et al., 2021). On 20 May 2021, an orangish brown smoke was detected from the nitric acid stored containers and it was followed by fire, which lasted for 13 days. The MV X-Press Pearl ship burning incident is considered the worst shipping disaster that occurred in the Sri Lankan maritime zone and globally and it created a huge socioeconomic loss and ecological consequences to the coastal resource users in the western, southwestern and northwestern provinces and many species of marine living resources. Coastal and marine sensitive ecosystems affected by the effects associated and noxious and hazardous material released by the accident include beaches, lagoons, estuaries, river mouths, coral reefs, nursery and feeding grounds of fish and other marine organisms, and nesting beaches of sea turtles. Plastic pellets or nurdles, residues from fire and explosions, and polluted soot are the most common hazardous materials or pollutants released. Enormous damage has occurred to the highly protected Endangered, Threatened and Protected (ETP) species such as marine mammals (whales, dolphins and porpoises) and sea turtles. Among the 31 species of marine mammals inhabiting the coastal and deep sea marine waters of Sri Lanka, carcasses of 10 species of coastal and oceanic dolphins, baleen whales, and two small sperm whale species were recovered mainly from the southern, western and northwestern coasts immediately after the ship disaster. Aim of this chapter is to document the possible short-term, medium-term, and long-term negative impacts on the marine mammals (cetaceans) encountered in the coastal and marine waters around Sri Lanka due to the pollution caused by the incident/pollution associated with the ship accident, especially due to smoke emitted, explosions, chemicals released, and the oil spills following the MV X-Press Pearlship disaster and also to highlight the research gaps related to marine disasters around Sri Lanka and globally.

10.2 MARINE MAMMALS RECORDED IN MARINE WATERS OF SRI LANKA

Marine waters of Sri Lanka have been identified as a well-known home for 31 species of marine mammals including 7 species of baleen whales, 23 species of toothed whales, and 1Dugong species (Ilangakoon, 2002; Martenstyne, 2019; de Vos, 2017) and the number of species in the list, which have been updated with time and identification of new species with current details of taxonomic classification, are provided in Table 10.1. Distribution and abundance of some marine mammal species have shown a seasonal variation with the monsoonal current pattern, although many species are sighted in Sri Lankan waters throughout the year (de Vos et al., 2012; Kirumbara et al., 2022). During the period from May to August, whales and dolphins aggregated in the south and west coastal regions of Sri Lanka to take the feeding advantages from the southwest monsoonal wind-induced upwelling (Krakstad et al., 2018). Therefore, the damage caused to marine mammals due to MV X-Press Pearlship accident might have been severe when compared to similar accidents that may possibly occur during other times of the year.

TABLE 10.1
List of the Marine Mammals Recorded in Marine Waters of Sri Lanka with their Taxonomic Classification

Order	Suborder	Family	No. of Species	Common names	Scientific nomenclature
Cetacea	Mysticeti (Baleen whales)	Balaenopteridae	07	Blue whale	*Balaenoptera musculus*
				Fin whale	*B. physalus*
				Bryde's whale	*B. brydei*
				Ormura's whale	*B. omurai*
				Eden's whale	*B. edeni*
				Humpback whale	*Megaptera novaeangliae*
				Minke whale	*B. acutorostrata*
	Odontoceti (Toothed whales)	Physeteridae	01	Sperm whale	*Physeter macrocephalus*
		Kogiidae	02	Pygmy sperm whale	*Kogia breviceps*
				Dwarf sperm whale	*Kogia sima*
		Ziphiidae	05	Longman's Beaked Whale	*Indopacetus pacificus*
				Blainville's beaked Whale	*Mesoplodon densirostris*
				Cuvier's beaked Whale	*Zpihius cavirostris*
				Deraniyagala's beaked whale	*M. hotaula deraniyagala*
				Ginko toothed beaked whale	*Mesoplodon ginkgodens*
		Phocoenidae	01	Finless porpoise	*Neophocaena phocaenoides*
		Delphinidae	14	Spinner dolphin	*Stenella longirostris*
				Bottlenose dolphin	*Tursiops truncates*
				Stripped dolphin	*Stenella coeruleoalba*
				Spotted dolphin	*Stenella attenuate*
				Common dolphin	*Delphinus delphis*
				Frasers dolphin	*Lagenodelphis hosei*
				Risso's dolphin	*Grampus griseus*
				False killer whale	*Pseudorca crassidens*
				Pygmy killer whale	*Feresa attenuate*
				Killer whale	*Orcinus orca*
				Melon-headed whale	*Peponocephala electra*
				Short-finned pilot whale	*Globicephala macrorhyncus*
				Indo-Pacific humpback dolphin	*Sousa chinensis*
				Rough-toothed dolphin	*Steno bredanensis*
Sirenia		Dugongidae	01	Dugong	*Dugung dugong*

10.3 HABITAT TYPES OCCUPIED BY MARINE MAMMALS

Marine mammals are distributed among a variety of habitats, horizontally from coastal lagoons to high seas and vertically from surface to depths of 3000 m. Curviour's beaked whales and sperm whales are the deepest divers, whichdive to depths below 2000 m where upwelling occurs, although they prefer to rest at the surface waters. Members of the family Balaenopteridae (e.g. blue whale, fin whale, Bryde's whale, minke, humpback and Ormura's whales) are widely distributed cetaceans in the upwelling regions of the ocean. Blue whale, *Balaenoptera musculus,* belongs to family Balaenopteridae and is the largest animal that has ever lived on the earth. They are frequently found in submarine canyons and along the continental shelf edge, where prey aggregate (Figure 10.1). These species live in epipelagic zones above 200 m of the water column. Marine waters off Trincomalee, Mirissa, and Kalpitiya are well-known habitats of marine mammals in Sri Lanka,where blue whales are found throughout the year with 1–12 individuals per pod. Panadura Canyon on the west coast of Sri Lanka, the newly identified blue whale feeding ground (Kirumbara et al., 2022), might be the most affected by the MV X-Press Pearl disaster due to its close proximity to the location of the ship accident. Coastal dolphins are the largest group of marine mammals, most abundant and widely distributed throughout the coastal waters of Sri Lanka from surface down to bottom of shallow coastal waters, bays, and shelf edges. Larger dolphins, known as whales, belong to order Cetacea (Table 10.1), are the top-level predators in marine ecosystems, prefer to live in offshore waters, and rarely move to productive coastal waters. Dugong is the only marine mammal species belonging to order Serenia recorded from shallow coastal seagrass beds in the Kalpitiya and Gulf of Mannar in Sri Lanka.

10.4 FOOD AND FEEDING HABITS OF MARINE MAMMALS

Feeding habits of marine mammals show a greater diversity from herbivores to planktivores to carnivores (which include top-level predators) and main source of contaminants found in the bodies of marine mammals is considered to be ingested with their food. Further, levels of bioaccumulation are believed to be due to form of food, tropic level, type of food chains, and longevity of the species (Pierce et al., 2013).

Dugongs are a unique species of marine mammals, among all marine mammals living in ocean ecosystems, due to their herbivorous feeding habit, and also their distribution is limited to shallow coastal regions, where dense seagrass beds occur. Dugong is the only primary consumer group of marine mammals on earth. Seagrass beds are found in different parts of coastal waters of Sri Lanka but most of them are not large enough to provide the nutritional requirements of Dugongs; hence they aggregate in Kalpitiya and Gulf of Mannar, where sufficient forage is available for feeding. Since they are bottom feeders, pollutants that settle in the sediments among seagrasses and dissolve in water can get into their body during feeding on seagrasses directly or indirectly through ingestion of pollutant contaminated seagrasses, sediments or sea water. Baleen whales are another unique group of mammals found in marine ecosystems, and due to their skimming or filter feeding pattern, they consume huge quantities of zooplankton consisting of fish larvae, copepods, crustaceans (krill), etc.,

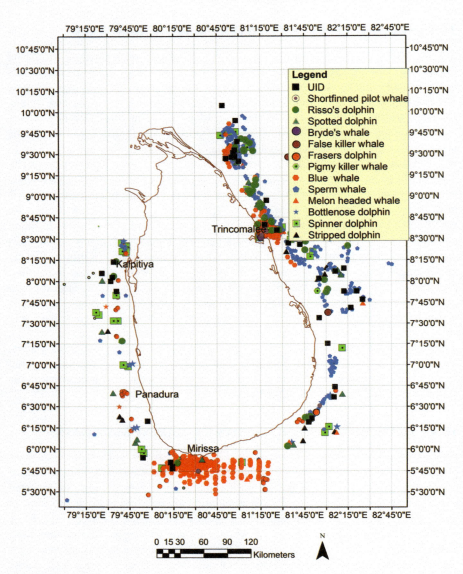

FIGURE 10.1 Some of the historical and recent marine mammal sighting records in Sri Lanka (Alling et al., 1991; Alling et al., 1986; Bröker and Ilangakoon, 2008; de Vos et al., 2012; Priyadarshana et al., 2015; Nanayakkara et al., 2014; Kirumbara et al., 2022).

and hence are more vulnerable to ingesting plastic nodules which are suspended in the surface waters after the ship accident. Baleen whales make feeding aggregates in areas with submarine canyons and steep continental slopes where upwelling occurs. Mirissa-Dondra in South Coast, Panadura Canyon in the west coast, and Kalpitiya area in the northwest coast are famous baleen whale habitats and they are in close proximity to the location where the ship accident occurred.

Toothed whales (family Delphinidae) are the most affected group of marine mammals from the MV X-Press Pearlship accident. All members of the toothed whales are top-level predators in the marine food chain, playing vital roles in the marine ecosystems. Prey selection of these dolphins depends on their living environment, although coastal dolphins feed on small fish schools such as *Sardinella*, scads, herrings, anchovies, and squids as well as bottom-living invertebrates. Therefore, long-term predation on contaminated prey species living among the bottom sediments in the vicinity of the location of the ship wreck, which released/is releasing hazardous pollutants that are deposited among the marine sediments, could be harmful to coastal dolphin species. Although large offshore dolphins, the top-level predators, are predating on large fish species like yellowfin or bigeye tuna and sharks and also on other cetaceans, they also can face harmful effects of the ship disaster, due to bioaccumulation of the toxic compounds and heavy metals released from the shipwreck along food chains.

Members of the family Physeteridae (sperm whales) and two species of the family Kogiidae (pygmy and dwarf sperm whales), which spend most of their time in deeper waters and prefer to feed on crustaceans, cephalopods (squid and octopus), fish, skates, and sharks, can accidentally ingest sediments contaminated with hazardous material, including burned plastic nurdles, released from the shipwreck while feeding at the sea floor areas.

As mentioned, the MV X-Press Pearl disaster is considered the worst ever maritime disaster that occurred in the coastal waters of Sri Lanka and caused significant damage to local economy and coastal and marine environment. This incident was first reported on 20 May 2021, about 17 km northwest of the port of Colombo, Sri Lanka (Figure 10.2). Among the various kinds of marine life that became victims of the above ship disaster, marine mammals could be considered a group of very important top-level predators notably affected.

10.5 SHIPPING INDUSTRY AND MARINE POLLUTION WITH SPECIAL REFERENCE TO X-PRESS PEARL DISASTER

The shipping industry plays a significant role in the global economy and trade while handling more than 90% of the transport. Further, seaborne trade handled 10.7 billion tons of goods in volume during the year 2020 (UNCTAD, 2020). As a consequence of global trade expansion, shipping industry has developed at a massive scale, although it has been having a number of negative impacts on the ocean and coastal ecosystems (Freedman et al., 2017), with special reference to acoustic pollution (Celi et al., 2015; Camerlenghi, 2021), whale-ship collision (Vanderlaan & Taggart, 2007; Redfern et al., 2013;Peltier et al., 2019), invasive species contamination (Jing et al., 2012; Balaji et al., 2014; Gollasch et al., 2015), chemical pollution (Rømer et al., 1996), and oil pollution (Botello et al., 1997: Rogowska & Namieśnik, 2010) etc.

Moreover, the majority of maritime accidents create adverse effects on properties, coastal activities, fisheries, and activities aboard the ships. Chemical spills including nitric acid, caustic soda, urea, and many other hazardous chemicals; emission of smoke with noxious gasses and soot due to burning of fuel, plastic, chemicals, and other flammable material on board; explosions; spills of different kinds of oil; release

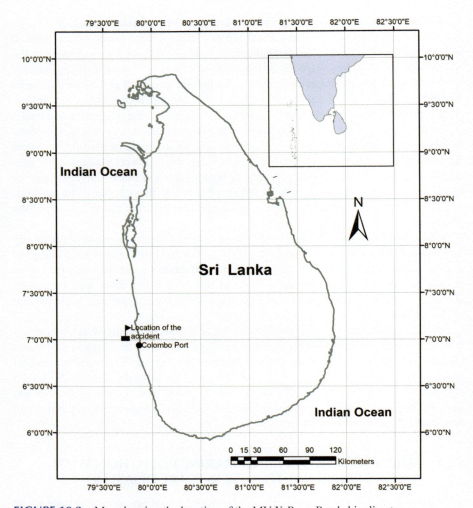

FIGURE 10.2 Map showing the location of the MV X-Press Pearl ship disaster.

of plastic pellets and resins; accumulation of microplastic; ash and other burned and partially burned plastic within sediments and on the beaches, etc., are some of the dangerous and unpleasant results of the disaster. As a result of the incident, unusually large numbers of deaths and strandings of turtles and mammals (whales and dolphins) were encountered on the beaches in close proximity to the location of the burning ship within a short period after the incident.

The severity of marine accidents including oil spills is determined based on the quantity of marine mammals and sea birds killed; although, recovery of the carcasses is challenging (Williams et al., 2011) due to poor detection following sinking and floating away with the currents or scavenging and due to many more reasons (Peltier et al., 2012; Reisdorf et al., 2012; Carretta et al., 2016; Moor et al., 2020). Studies on the behavior of marine mammal carcasses which are floating and sinking are rare, although 18% of net-entangled, dead common dolphins have been observed to be buoyant when released and 6% of them were found to be beached (Peltier et al., 2012; Peltier et al., 2016). Some previous studies revealed that release distance was negatively correlated with carcass recovery (Peltier et al., 2012).Carcasses are more easily discoverable when stranding on sandy beaches in populated areas than on rocky shores (Evans et al., 2005), forests, or unpopulated areas. Furthermore, salinity, temperature, and density of the animals are considered some of the factors determining the carcass sink or float (Reisdorf et al., 2012; Wells et al., 2015; Moore et al., 2020). High scavenging damages are recorded on the deep sunken carcasses due to lack of enough food on the deep-sea floor, and hence refloating possibility is comparatively less. In shallow waters physical, chemical, and microbial decompositions are higher than the deeper waters and hence sunken carcasses are bloated and refloat with the microbial decomposition (Smith & Baco, 2003; Anderson & Hobischak, 2004; Reisdorf et al., 2012). The stranding recovery rate could vary with the time and space also. Previous stranding data in Sri Lanka revealed that highest strandings were recorded in the south, west, and northwest coastal regions during the southwest monsoon (Liyanage et al., in press), especially where the southwest monsoonal winds interface.

Novel information available on dead cetacean (Odontocetes) recovery revealed that the fraction of stranding is reliably lower than the real number of deaths. The preliminary study on cetacean carcass recovery in Mexico after the *Deepwater Horizon* oil spill estimated the carcass recovery rate to be 2% (CI 0–6%) (Williams at al., 2011). Consequently, the true death count could be 50 times the number of carcasses recovered. In the study done on the California coastal bottlenose dolphin population, the estimated carcass recovery fraction was 0.25 (Carretta et al., 2016), cetacean carcass in the Gulf of Mexico was 0–0.062 (Williams et al., 2011), harbor porpoise recovery fraction in North Atlantic was <0.01 (Moore & Read, 2008), and residence common dolphin recovery fraction in Sarasota Bay, Florida, was 0.08 (Wells et al., 2015). This indicates that after the MV X-Press Pearl ship accident, about fifty carcasses of marine mammals were recovered, although actual number of marine mammal deaths could be much higher and this matter should be seriously considered during damage estimates and designing plans for long-term restoration.

Among all the families of affected marine mammals, spinner dolphin, *Stenella longirostris,* is the most affected species, followed by striped dolphin, *Stenella coeruleoalba*. These two species are very common in coastal waters around Sri Lanka and are usually found in small to big pods (Alling et al, 1991). In addition to the common species, a number of rare species including Fraser's dolphin (*Lagenodelphis hosei*), Risso's dolphin (*Grampus griseus*), and false killer whale (*Pseudorca crassidens*), were also stranded. The total number of affected dolphins seems to be several-fold when the numbers of recovered strandings are considered. Damaged populations will take a long time to recover, due to their late maturity, long lifespan, long gestation period, and litter constituted of one calf (K selected species). Therefore, in-depth consideration of life history information is important for assessment of the risk and economic losses caused by the deaths of mammalian species, although such information is limited for many species of marine mammals (Davidson et al., 2012; IUCN 2020). Average life span of toothed whales is closely similar and estimated as

FIGURE 10.3 Condition of dolphins stranded in different locations. (Photo credit: M.M. Morathanna, Department of Wildlife conservation.)

25–30 years (Robeck et al., 2015; Jaakkola & Willis, 2019), and it is 80–110 years for baleen whales such as blue whales (Buddhachat et al., 2021).

10.6 POTENTIAL IMPACTS ON MARINE MAMMALS FROM THE INCIDENTS THAT OCCURRED DURING AND AFTER THE FIRE IN THE MV X-PRESS PEARL

10.6.1 Noxious Gasses Emitted During Fire on Board the Ship

The majority of aquatic animals respire using gills or skin or lungs, and marine mammals use lungs for respiration. Especially cetaceans (whales and dolphins) are a specific group of marine mammals descending from terrestrial mammals, which have been adapted to aquatic life without many functional and structural modifications in their respiratory systems. Marine mammals belong to orders Cetacea and Sirenia, which are fully aquatic mammals that respire atmospheric air; therefore, they frequently come to the surface for breathing. Moreover, cetaceans are adapted to life at sea, where air pollution is very small when compared to land; hence they would be more susceptible to the noxious gasses and dust released into the air during the outbreak of fire in the MV X-Press Pearl, which persisted for several days (Figure 10.4).

Further, cetaceans are highly prone to lung injuries caused by smoke emitted during fire outbreaks, because of the absence of turbinates and cilia, apneustic breathing, and high lung air exchange (Venn-Watson et al., 2013). Anatomy and physiology of respiratory system of cetaceans are adapted to gulp large volumes of atmospheric air when they surface for respiration and they can easily inhale any noxious gasses or airborne smoke particles. Inhalation of the smoke originating from chemical and

FIGURE 10.4 Emission of black smoke which may have consigned noxious gases during the fire outbreak in the MV X-Press Pearl.(Photo credit: Sri Lanka Air Force.)

oil fire can lead to harmful damage to the respiratory system, which could be due to carbon monoxide poisoning, hypoxia, and damage to airways and epithelia caused by toxic compounds, and ultimately suffer from bronchopneumonia (Stone & Martin, 1969; Herndon et al., 1987; Venn-Watson et al., 2013). This is possible because, five years after the wildfire in the USA, those who were exposed to noxious gasses released during the fire frequently suffered from lung diseases (Venn-Watson et al., 2013). Meanwhile, marine mammals exposed to toxic gasses will attempt to escape from that environment and immediately move to deeper waters, where they may die before anybody canobserve the marine mammals in distress. Therefore, total number of marine mammals affected will not be known, and only the number of carcasses washed off onto shores of Sri Lanka can be taken into account and others may get carried to other destinies around the Indian Ocean or sink to the bottom.

10.6.2 Polyaromatic Hydrocarbons

Polyaromatic hydrocarbons (PAHs) dare considered environmental pollutants emitted through incomplete combustion of organic materials, such as coal, oil, wood, gas, and plastic, which can contaminate the body especially through inhalation (Verheyen et al., 2021). They enter aquatic systems directly due to oil spills, some industries such as petroleum industry, and due to natural seepage and usage of petroleum products (Ylitalo et al., 2017; Honda and Suzuki, 2020). PAHs can cause harmful impacts on marine protected species including marine mammals by affecting their inhalation during surfacing for breathing, ingestion of contaminated foods, or absorption through the skin (Schwacke et al., 2014; Venn-Watson et al., 2015; Ruberg et al., 2021). When PAHs are released into water some parts evaporate and dissolve in seawater and the major part is adsorbed by suspended materials. Therefore, planktivorous baleen whale species are more vulnerable to engulfing contaminated plankton and oil droplets suspended in the water column (Geraci and St. Aubin 1990). High lipid solubility nature of PAHs accelerates the fast absorption through the gastrointestinal tract and distribution with different kinds of body tissues immediately (Abdel-Shafy & Mansure, 2016). Scientific evidence suggests that environmental exposure to PAHs leads to long-term immunosuppression, inflammation, cancers, oxidative stresses, abortions, and abnormalities in the endocrine system (Moorthy et al., 2015; Rengarajan et al., 2015; Farzan et al., 2016; Wang et al., 2017) and contribute to changes in movement and vocalization of marine mammals (Sanderfoot et al., 2022). Furthermore, many PAHs have toxic, mutagenic, and carcinogenic properties (Patel et al., 2020; Wetzel et al., 2006). Huge quantity of PAHs was generated during the period when MV X Press Pearlship was under fire and an oil spill was coupled with the incident.

Population-level damages to the marine mammals caused by the noxious gasses released during the ship fire cannot be restored easily since the marine mammals may have died due to contact poisoning or acute toxicity and they may have disappeared unnoticed because all carcasses would not settle on beaches of Sri Lanka. Others may have faced the impacts of chronic toxicity of the gases released and would suffer from chronic lung diseases, and diagnosis of such diseases is not possible due to practical difficulties in carrying out investigations based on their internal anatomy. However, if mammals who were in distress were found prior to their death or sinking to deeper waters, there would have been a possibility of saving them from death by providing some medical assistance, at least to smaller mammalian species. Affected marine mammalian populations should be allowed to be restored naturally on their own by declaring the natural habitats of those species as marine protected areas (MPAs), which is also a difficult task because EEZ of Sri Lanka has important ship routes.

10.6.3 Dissolved Hazardous Chemicals and Heavy Metals

From the moment the concentrated nitric acid leak appeared on the MV X-Press Pearl, it reacted with the methanol cargo and continued to react throughout the fire on board. This resulted in the release of a massive amount of chemicals, some of which may have undergone further reactions to form even more toxic substances that could have harmed marine life. Release of toxic substances will continue until the wreck removal, and also there is a strong possibility that resuspension of toxic substances settled among the bottom sediments would occur with the disturbances caused by monsoonal currents for many more years even after the removal of the wreck.

Various types of chemicals released at different phases may behave differently within marine and coastal ecosystems, causing a vast array of toxic effects on marine biota. Certain chemicals released are very stable and persistent, in general, and many of them are resistant to metabolic degradation; this subject will be discussed in a separate chapter. The chronic toxicity of chemical pollution is reflected indifferent forms of health impacts, mortalities, and strandings in marine biota including marine mammals. Some studies have revealed that morbillivirus infections are the primary causes of disease outbreaks and mass strandings of marine mammals, affected by immunosuppression associated with environmental pollution (Hall, 1995; Osterhaus et al.,1995; Barrett et al., 1995). Bennett et al, (2001) studied the liver concentrations of Hg and selenium (Se), the Hg:Se molar ratio, and zinc (Zn) of the dead harbor porpoise—*Phocoena phocoena*—with infectious disease caused by parasites and pathogens (most frequently pneumonia caused by lungworm and bacterial infections) against the porpoises that died due to fisheries interactions. This study revealed that mean liver concentrations of heavy metals especially mercury (Hg), selenium (Se), and (Zn) were exclusively higher in the porpoises that died due to infectious disease than trauma, and also lead (Pb), cadmium (Cd), copper (Cu), and nickel (Ni) concentrations were not much different among the groups.

Heavy metals are persistent and cause toxic effects on aquatic ecosystems (Ali et al., 2019). Biomagnification of the heavy metals in body tissues of the toothed whales (suborder Odontoceti) is subsequently elevated since more or less all the species are included long-living top-level predators (Muir et al., 1992; Bouquegneau and Joiris, 1988). Furthermore, concentrations in the various body parts vary with age, sex, and reproductive stage of the animal (Beck et al., 1997; Dietz et al., 1996; Bowles, 1999). The study on stranded bottlenose dolphins on Cannery Island revealed that selenium concentrations of the blubber and liver were 7.29 and 68.63 µg/g dry weight (dw), respectively, and mercury concentrations were consequently 80.83 and 223.77 µg/g dw (García-Alvarez, et al., 2015).

Baleen whales (suborder Mysticeti) are a well-known group of filter feeders with shorter food chains; hence, accumulated concentrations are comparatively lower than in the toothed whales (Bowles, 1999; Honda et al., 1987). Among thousands of baleen whale tissue samples analyzed throughout the world, maximum concentrations of accumulated pollutants are generally lower than in other marine mammal species (O'Shea & Brownell, 1994).

Among the different kinds of heavy metals, mercury, cadmium, selenium, zinc, and copper are the major elements accumulated in the body tissues of dolphins (Bowles, 1999). Mercury is one of the most toxic heavy metals encountered in them. The long-term bioaccumulation of mercury can result in extensive deposits of a granular pigment in the liver, causing active liver diseases including necrosis and fat globules (Rawson et al., 1993). Presence of fat globules reveals that animal fat metabolism is affected and it can lead to cell death. Previous studies revealed that accumulation of heavy metal varies across species and body parts. Andre et al. (1990) report that mercury concentrations in the striped dolphin (*S. coeruleoalba*) declined according to the following order of organs: liver > spleen > blubber > kidney > pancreas > stomach > lungs. Combustion has been identified as one of the anthropogenic sources of mercury in aquatic ecosystems, from which it is first released into the atmosphere and later into the sea (Pacyna et al., 2010). Mercury accumulation in tissues of marine mammals can cause renal and hepatic damage and it also causes neurotoxic, genotoxic, and immunotoxic effects (Frederick et al., 2018; Kershaw & Hall, 2019), which may ultimately cause pathological infections that lead to the stranding and deaths (Esposito et al., 2020).

The impact of cadmium on dolphins is not yet clearly understood, but it also has mammalian toxicities even under minor concentrations. In dolphins cadmium is mainly accumulated in kidneys, and the levels in the kidneys were 3-fold higher than in the liver and 26-fold higher than in the muscle (Long et al.,1998). The accumulation of cadmium in the body causes cancers, genetic mutations, and birth defects ((Jarup et al., 2009; Eisler et al., 1985). Therefore, the samples of liver, kidney, and some other types of tissues collected by the Department of Wildlife Conservation (DWC) officials during postmortems conducted on the stranded mammals after the X-Press Pearl ship accident have been sent to the government analyst department and other local and foreign laboratories to be tested for heavy metal contamination and to Veterinary Research Institute for histopathological examinations.

Container loads of lithium batteries have been released into the environment during the MV X-Press Pearl ship disaster. Due to its long oceanic residence time (~1.2 million years) and its weak capacity to adsorb onto marine particles (Decarreau et al., 2012), Li is reported to be homogeneously distributed throughout the water column (Misra & Froelich, 2012). However, no studies have been conducted on the impacts of Li on marine mammals. Thibon et al. (2021) have studied the Li accumulation in different trophic groups (filterfeeders to mesopredators) and habitats (benthic, demersal, and pelagic) from three contrasting marine biogeographic areas, i.e. temperate (Bay of Biscay, northeast Atlantic Ocean), tropical (New Caledonia, Pacific Ocean), and subpolar climates (Kerguelen Islands, southern Indian Ocean), and have found that highest Li concentrations are found in filter feeders and lowest in predatory fish and strong variations have been found among organs. Biochemical similarity has been observed between Na and Li during transport in the brain and in osmoregulatory organs. Relatively high Li concentrations have been observed in fish gills and kidneys (0.26 and 0.15 µg/g, respectively), and a large range of Li contents (up to 0.34 µg/g) in fish brains, while depleted levels have been observed in fish liver and muscles (0.07 ± 0.03 and 0.06 ± 0.08 µg/g, respectively). Marine mammals also represent different tropic groups and similar trends should be seen among them, although Li toxicity has not been studied on marine mammals. Hereford, 2021 studied the human toxic effects caused by Li; hence, chronic toxicity of Li on marine mammals cannot be ruled out and studies on Li concentrations in different organs of stranded marine mammals and their food items would reveal the impacts of Li on marine mammals. Chronic symptoms of Li toxicity in the central nervous system and heart consequently include tremor, ataxia, confusion, agitation, neuromuscular excitability, seizures, coma, and myocarditis (Hereford, 2021).

If the mammals were in the vicinity of the ship at the time of spilling of strong acids (conc. nitric acid) and release of caustic soda from the ship cargo, and if they came in contact with those chemicals, there is a possibility that skin damage may have occurred. Wounds and lesions found on marine mammals stranded on the beaches could be attributed to effects of hazardous chemicals released in large quantities to the environment.

10.6.4 Plastic Nurdles, Burned Plastic Floating or in Suspension within Marine Waters

Plastics are considered widely used organic polymer products with strong and persistent characteristics (Laist, 1987). Unburned and burned plastic pellets have been identified as one of the long-lasting major pollutants released to the marine ecosystem with the X-Press Pearl disaster and plastic nurdles were observed on beaches of Sri Lanka even more than a year after the ship accident. The amount of plastic spilled during the disaster was about 1680 MT and burned plastics were chemically complex and larger in size than the unburned (de Vos et al., 2021); hence, this pollution was more critical than observed. Microplastics are the small plastic particles directly disposed of or produced through breaking down of large plastic material into small pieces by the physical degradation within the environment (Barnes et al., 2009; Jahnke et al., 2017). All the plastic nurdles in the ship were below 5 mm in dimension and hence can be considered in the category of microplastics (MPs). MP-contaminated marine debris is considered one of the major threats for cetaceans feeding in different layers of the marine ecosystem (Baulch and Perry, 2014), and 60–80% of the marine plastics are found in marine debris (Derraik, 2002). Even though defragmentation occurred MPs in marine debris are persistent in the ecosystem 100–1000 years without biodegradation (Boren et al., 2006; Derraik, 2002).

MP has been found in early stages of the marine foodchains including bivalves, planktivorous fish, and zooplanktons especially copepods and krill (Foekema et al., 2013; Van Cauwenberghe & Janssen, 2014; Dawson et al., 2018), and also, they are transferred via planktonic organisms from a lower trophic level (mesozooplankton) to higher levels (Cole et al., 2011; Setälä et al., 2014; Nelms et al., 2019). Filter-feeders in marine ecosystems are heavily exposed to microplastic contamination because of their selection of small particles as food sources (Bessling et al., 2015; Farrell & Nelson 2013). Baleen whales are the largest filter feeders in marine ecosystems that consume thousands of kilograms of krill, copepods, and other mesozooplankton (small fish species and fish larvae) on a daily basis. Coastal waters of Sri Lanka have been identified as a well-known home for the Indian Ocean blue whale (*Balaenoptera musculus indica*), the largest baleen whale species in the region inhabiting marine waters off Sri Lanka throughout the year (Kirumbara et al., 2022). In addition, six more baleen whale species, namelyfin whale(*Balaenoptera physalus*), Bryde's whale(*Balaenoptera brydei*), minke whale(*Balaenoptera acutorostrata*), Ormura's whale (*Balaenoptera omurai*), Humpback whale (*Megaptera novaeangliae*), and Eden's whale(*Balaenoptera edeni*), were also recorded. Therefore, all these species might get highly exposed to the microplastic pollution that occurred due to heavy plastic load released into the sea after the ship accident (Figure 10.5).

MP contamination of the different prey species of dolphins and whales varies with the location where they live. Detailed studies on MP-contaminated prey species are rare, although some data is available as a percentage of prey species contaminated (percentage of number of contaminated to number of sampled individuals). Family Scombridae 32.2% contaminated in English Channel 32.2% (Nelms et al., 2018), in North Sea 4.1% (Foekema et al.,2013; Rummel et al., 2016), and in Mediterranean Sea 71.4% (Guven et al., 2017), and the contaminations of family Clupeidae in North Sea, Baltic Sea offshore, and Baltic Sea coastal regions are respectively 1.5%, 1.8%, and 19.7% (Foekema et al., 2013; Rummel et al., 2016; Hermsen et al., 2017; Budimir et al., 2018). Cetaceans feeding on families Scombridae and Engraulidae are highly vulnerable to MP contamination (Burkhardt-Holm & N'Guyen, 2019). Records are rare on presence of microplastic in digestive tract of the cetaceans (Lusher et al., 2018; Burkhardt-Holm & N'Guyen, 2019), although Denuncio et al. (2011) examined 28% of the incidentally captured Franciscana dolphins (*Pontoporia blainvillei*) in Argentina having plastic in their stomach. Moreover, postmortem of the plastic debris ingested by Blainville's beaked whale, *Mesoplodon densirostris,* stranded ashore in

FIGURE 10.5 Plastic-contaminated water and beach. (Photo credit: Marine Environmental Pollution Prevention Authority.)

Brazil revealed that the cause of death was starvation due to intestinal obstruction (Secchi and Zarzur, 1999; De Stephanis et al., 2013). Live pygmy sperm whale and Minke whale stranded in Texas, USA (Tarpley & Marwitz, 1993), and another fin whale stranded in Korea also died due to the same reason listed above.

MPs consisted with substantial concentrations of persistent organic pollutants and toxins can transfer to the organisms through ingestion and can cause biomagnification over decades in long-living filter feeders, leading to interruption of biological processes (Germanov et al., 2018). Nelms et al. (2019) investigated stranded marine mammal carcasses with various causes and revealed that harbor porpoises and common dolphins that died with the infectious disease had slightly elevated MP levels than the animals that died due to other reasons. Furthermore, MP in marine organisms causes starvation, poor digestion, malnutrition, immune system impairment, oxidative stress, induction and modulation in fitness, poor survival, and fecundity (Van Cauwenberghe et al., 2014; Guzzetti et al., 2018; Strungaru et al., 2019; Katsanevakis, 2008; Fossi et al., 2018).

10.6.5 EXPLOSIONS FOLLOWED BY FALLING OF HOT METAL, BURNING PLASTIC, CONTAINERS AND HARD DEBRIS INTO MARINE WATERS

Small dolphin species especially bottlenose and spinner dolphins are frequently found allover the coastal waters of Sri Lanka. The hard noises generated from offshore and marine activities have the capacity to directly cause disturbance to marine mammals such as whales, dolphins, and porpoises. However, the effects of heat and noise can be harmful to marine mammals and their intraspecific communication using sound.

After the ship accident few possible net-entangled dolphin carcasses were stranded on the west coast of Sri Lanka, especially close to the site of the ship disaster. This might be due to nonlethal influences of the ship accident such as panic caused by incidents that occurred during the ship fire (explosions, heat generation, etc.) or unconsciousness caused by noxious gases and/or other substances released into marine environment during fire and explosions. Unconscious marine mammals are more vulnerable to net entanglement or are prone to other accidents such as ship or boat strikes.

Marine mammals are normally very friendly animals, and because of that they usually flock around boats and ships because of their friendly nature toward humans. Bottlenose dolphins are the most popular and friendly of all marine species, and also they are the most intelligent and happy creatures in the marine environment. Spinner dolphins are also frequently seen in the coastal waters. Therefore, there is strong possibility that they flocked around the MV X-Press Pearl ship in distress and also after that around the boats which were involved in rescue and fire control operations.

10.7 RECOMMENDATIONS

Marine mammals are highly protected species in Sri Lankan waters with both local and international legislation due to scarcity and threats faced by them. After the burning incident around 50 marine mammals (including a fetus) were stranded on the Sri Lankan coast especially at the southwest monsoonal wind interface. Counts of the strandings recorded during the disaster period have been significantly higher compared to that of early month of the same year and also when compared to those that occurred during the same months of the previous years. Considering the ecological and economic importance and severity of the current issue, investigations on the short-term and long-term impacts on marine mammals and their habitat have been launched and they are identified as priority areas. Large burned plastic pieces and unburned pellets which are buried under the sediments may subjected to chemical and mechanical degradation, and suspension of micro- and nanoplastics will be a continuous process for several years, and it will be a problem which may not have a clear solution.

Following investigations on parameters related to marine mammals are included in post–ship accident monitoring of living marine mammalian populations and also when carrying out autopsies on the carcasses of stranded marine mammals

- Taxonomic identification
- Morphometric characterization
- Identification of sex and maturity stages
- Parasitic and other pathogenic disease conditions
- External abnormalities including wounds and other physical changes resulting from the accident, net entanglement marks, and marks on the bodies of mammals due to predation, scavenging, etc.
- Conditions of stranding (live and released, died after stranding, fresh, moderately decomposed, advanced decomposed or mummified, single or mass stranding)
- Anatomical, histological, and histopathological changes
- Gut contents (quality, condition and contaminants)
- Chemical contamination in the environment

FIGURE 10.6 (A and B) MV X-Press Pearl on fire with thick black smoke and debris falling into the surrounding environment due to explosions. (Photo credit: Sri Lanka Air Force.)

REFERENCES

Abdel-Shafy, H.I. and Mansour, M.S., 2016. A review on polycyclic aromatic hydrocarbons: source, environmental impact, effect on human health and remediation. *Egyptian Journal of Petroleum*, 25(1):107–123.

Acejo, I., Sampson, H., Turgo, N., Ellis, N. and Tang, L., 2018. The causes of maritime accidents in the period 2002–2016, Seafarers International Research Centre (SIRC), Cardiff University, United Kingdom. Available from http://orca.cf.ac.uk/17481/1/Sampson_The%20causes%20of%20mar itime %20accidents%20in%20the%20period%20 2002-2016.pdf.

Ali, H., Khan, E. and Ilahi, I., 2019. Environmental chemistry and ecotoxicology of hazardous heavy metals: environmental persistence, toxicity, and bioaccumulation. *Journal of Chemistry*, 2019: 6730305.

Alling, A., 1986. Records of odontocetes in the northern Indian Ocean (1981–1982) and off the coast of Sri Lanka (1982–1984). *Journal of the Bombay Natural History Society. Bombay*, 83(2):376–394.

Alling, A.K., Dorsey, E.M., Gordon, J.C.D., 1991. Blue whales Balaenoptera musculus off the northeast coast of Sri Lanka: Distribution, feeding and individual identification. In S. Leatherwood, G.P. Donovan (Eds.). *Cetaceans and Cetacean Research in the Indian Ocean Sanctuary*, Technical Report No. 3, United Nations Environment Programme Marine Mammal: Nairobi, Kenya, pp. 247–258.

Anderson, G.S. and Hobischak, N.R., 2004. Decomposition of carrion in the marine environment in British Columbia, Canada. *International Journal of Legal Medicine*, 118(4):206–209.

André, J.M., Ribeyre, F. and Boudou, A., 1990. Mercury contamination levels and distribution in tissues and organs of delphinids (*Stenella attenuata*) from the eastern tropical Pacific, in relation to biological and ecological factors. *Marine Environmental Research*, 30(1):43–72.

Balaji, R., Yaakob, O. and Koh, K.K., 2014. A review of developments in ballast water management. *Environmental Reviews*, 22(3):298–310.

Barnes, D.K.A., Galgani, F., Thompson, R.C., Barlaz, M., 2009. Accumulation and fragmentation of plastic debris in global environments. *Philosophical Transaction of the Royal Society, London, B. Biological Science*, 364:1985–1998. http://dx.doi.org/10.1098/rstb.2008.0205

Barrett, T., Blixenkrone-Møller, M., Di Guardo, G., Domingo, M., Duignan, P., Hall, A., Mamaev, L. and Osterhaus, A.D.M.E., 1995. Morbilliviruses in aquatic mammals: report on round table discussion. *Veterinary Microbiology*, 44 (2–4):261–265.

Baulch, S. and Perry, C., 2014. Evaluating the impacts of marine debris on cetaceans. *Marine Pollution Bulletin*, 80(1–2):210–221.

Beck, K.M., Fair, P., McFee, W. and Wolf, D., 1997. Heavy metals in livers of bottlenose dolphins stranded along the South Carolina coast. *Marine Pollution Bulletin*, 34(9):734–739.

Bennett, P.M., Jepson, P.D., Law, R.J., Jones, B.R., Kuiken, T., Baker, J.R., Rogan, E. and Kirkwood, J.K., 2001. Exposure to heavy metals and infectious disease mortality in harbour porpoises from England and Wales. *Environmental Pollution*, 112(1):33–40.

Besseling, E., Foekema, E.M., Van Franeker, J.A., Leopold, M.F., Kühn, S., Rebolledo, E.B., Heße, E., Mielke, L.J.I.J., IJzer, J., Kamminga, P. and Koelmans, A.A., 2015. Microplastic in a macro filter feeder: humpback whale *Megaptera novaeangliae*. *Marine Pollution Bulletin*, 95(1):248–252.

Boren, L.J., Morrissey, M., Muller, C.G., Gemmell, N.J., 2006. Entanglement of New Zealand fur seals in man-made debris at Kaikoura. *New Zealand Marine Pollution Bulletin*, 52:442–446.

Botello, A.V., Susana Villanueva, F. and Gilberto Diaz, G., 1997. Petroleum pollution in the Gulf of Mexico and Caribbean Sea. *Reviews of Environmental Contamination and Toxicology*, 153:91–118.

Bouquegneau, J.M., Joiris, C., 1988. The fate of stable pollutants—heavy metals and organochlorines—in marine organisms. In S.H. Wright (ed.), *Advances in Comparative and Environmental Physiology*. Springer Berlin Heidelberg: Berlin, Heidelberg, pp. 219–247.

Bowles, D., 1999. An overview of the concentrations and effects of metals in cetacean species. *Journal of Cetacean Research and Management*, 1999: 125–148.

Bröker, K.C.A. and Ilangakoon, A., 2008. Occurrence and conservation needs of cetaceans in and around the Bar Reef Marine Sanctuary, Sri Lanka. *Oryx*, 42(2):286–291.

Buddhachat, K., Brown, J.L., Kaewkool, M., Poommouang, A., Kaewmong, P., Kittiwattanawong, K. and Nganvongpanit, K., 2021. Life expectancy in marine mammals is unrelated to telomere length but is associated with body size. *Frontiers in Genetics*, 12: 1792.

Budimir, S., Setälä, O. and Lehtiniemi, M., 2018. Effective and easy to use extraction method shows low numbers of microplastics in offshore planktivorous fish from the northern Baltic Sea. *Marine Pollution Bulletin*. 127:586–592. https://doi.org/10.1016/j.marpolbul.2017.12.054

Burkhardt-Holm, P. and N'Guyen, A., 2019. Ingestion of microplastics by fish and other prey organisms of cetaceans, exemplified for two large baleen whale species. *Marine Pollution Bulletin*, 144:24–234.

Camerlenghi, A., 2021. The future challenge of decreasing underwater acoustic pollution. *Bulletin of Geophysics and Oceanography*, 62: 91.

Carretta, J.V., Danil, K., Chivers, S.J., Weller, D.W., Janiger, D.S., Berman-Kowalewski, M., Hernandez, K.M., Harvey, J.T., Dunkin, R.C., Casper, D.R. and Stoudt, S., 2016. Recovery rates of bottlenose dolphin (*Tursiops truncatus*) carcasses estimated from stranding and survival rate data. *Marine Mammal Science*, 32(1):349–362.

Celi, M., Filiciotto, F., Vazzana, M., Arizza, V., Maccarrone, V., Ceraulo, M., Mazzola, S. and Buscaino, G., 2015. Shipping noise affecting immune responses of European spiny lobster (*Palinurus elephas*). *Canadian Journal of Zoology*, 93(2):113–121.

Ceyhun, G.C., 2014. The impact of shipping accidents on marine environment: A study of Turkish seas. *European Scientific Journal*, 10(23):10–23.

Chan, S.R., Hamid, N.A. and Mokhtar, K., 2016. A theoretical review of human error in maritime accidents. *Advanced Science Letters*, 22(9):2109–2112.

Cole, M., Lindeque, P., Halsband, C., Galloway, T.S., 2011. Microplastics as contaminants in the marine environment: A review. *Marine Pollution Bulletin*, 62:2588–2597. http://dx.doi.org/10.1016/j.marpolbul.2011.09.025

Davidson, A. D., Boyer, A. G., Kim, H., et al., 2012. Drivers and hotspots of extinction risk in marine mammals. *Proceedings of National Academy of Science, U. S. A.* 109:3395–3400. doi: 10.1073/pnas.1121469109

Dawson, A.L., Kawaguchi, S., King, C.K., Townsend, K.A., King, R., Huston, W.M. and Nash, S.M.B., 2018. Turning microplastics into nanoplastics through digestive fragmentation by Antarctic krill. *Nature Communications*, 9(1):1–8.

de Vos, A., 2017. First record of Omura's whale, Balaenoptera omurai, in Sri Lankan waters. *Marine Biodiversity Records*, 10(1):1–4.

de Vos, A., Aluwihare, L., Youngs, S., DiBenedetto, M.H., Ward, C.P., Michel, A.P., Colson, B.C., Mazzotta, M.G., Walsh, A.N., Nelson, R.K. and Reddy, C.M., 2021. The M/V X-Press Pearl nurdle spill: Contamination of burnt plastic and unburnt nurdles along Sri Lanka's beaches. *ACS Environmental Au*, 2(2):128–135.

de Vos, A., Clark, R., Johnson, G., Johnson, C., Kerr, I., Payne, R., and Madsen, P. T., 2012. Cetacean sightings and acoustic detections in the offshore waters of Sri Lanka: March–June 2003. *Journal of Cetacean Research and Management*, 12(1):185–193.

Decarreau, A., Vigier, N., Pálková, H., Petit, S., Vieillard, P., Fontaine, C., 2012. Partitioning of lithium between smectite and solution: An experimental approach. *Geochimca Cosmochimica Acta*, 85:314–325.

Denuncio, P., Bastida, R., Dassis, M., Giardino, G., Gerpe, M. and Rodríguez, D., 2011. Plastic ingestion in Franciscana dolphins, *Pontoporia blainvillei* (Gervais and d'Orbigny, 1844), from Argentina. *Marine Pollution Bulletin*, 62(8):1836–1841.

Derraik, J.G., 2002. The pollution of the marine environment by plastic debris: a review. *Marine Pollution Bulletin*, 44(9):842–852.

De Stephanis, R., Giménez, J., Carpinelli, E., Gutierrez-Exposito, C. and Cañadas, A., 2013. As main meal for sperm whales: Plastics debris. *Marine Pollution Bulletin*, 69(1–2):206–214.

Dietz, R., Riget, F. and Johansen, P., 1996. Lead, cadmium, mercury and selenium in Greenland marine animals. *Science of the Total Environment*, 186(1–2):67–93.

Dominguez-Péry, C., Vuddaraju, L.N.R., Corbett-Etchevers, I. and Tassabehji, R., 2021. Reducing maritime accidents in ships by tackling human error: A bibliometric review and research agenda. *Journal of Shipping and Trade*, 6(1):1–32.

Eisler, R., 1985. Cadmium hazards to fish, wildlife, and invertebrates: A synoptic review. U.S. Fish and Wildlife Service Biological Report, 85 (1.2), Contaminant Hazard Reviews Report 2:46.

Esposito, M., Capozzo, D., Sansone, D., Lucifora, G., La Nucara, R., Picazio, G., Riverso, C. and Gallo, P., 2020. Mercury and cadmium in striped dolphins (*Stenella coeruleoalba*) stranded along the Southern Tyrrhenian and Western Ionian coasts. *Mediterranean Marine Science*, 21(3):519–526.

Evans, K., Thresher, R., Warneke, R.M., Bradshaw, C.J., Pook, M., Thiele, D. and Hindell, M.A., 2005. Periodic variability in cetacean strandings: Links to large-scale climate events. *Biology Letters*, 1(2):147–150.

Farrell, P. and Nelson, K., 2013. Trophic level transfer of microplastic: *Mytilus edulis* (L.) to *Carcinus maenas* (L.). *Environmental Pollution*, 177:1–3.

Farzan, S.F., Chen, Y., Trachtman, H. and Trasande, L., 2016. Urinary polycyclic aromatic hydrocarbons and measures of oxidative stress, inflammation and renal function in adolescents: NHANES 2003–2008. *Environmental Research*, 144:149–157.

Foekema, E.M., De Gruijter, C., Mergia, M.T., van Franeker, J.A., Murk, A.J., Koelmans, A.A., 2013. Plastic in North Sea fish. *Environmental Science and Technology*, 47:8818–8824. https:// doi.org/10.1021/es400931b

Fossi, M.C., Panti, C., Baini, M. and Lavers, J.L., 2018. A review of plastic-associated pressures: Cetaceans of the Mediterranean Sea and eastern Australian shearwaters as case studies. *Frontiers in Marine Science*, 5:173.

Freedman, R., Herron, S., Byrd, M., Birney, K., Morten, J., Shafritz, B., Caldow, C. and Hastings, S., 2017. The effectiveness of incentivized and non-incentivized vessel speed reduction programs: Case study in the Santa Barbara channel. *Ocean & Coastal Management*, 148:31–39.

Galieriková, A., 2019. The human factor and maritime safety. *Transportation Research Procedia*, 40:1319–1326.

García-Alvarez, N., Fernández, A., Boada, L.D., Zumbado, M., Zaccaroni, A., Arbelo, M., Sierra, E., Almunia, J. and Luzardo, O.P., 2015. Mercury and selenium status of bottlenose dolphins (*Tursiops truncatus*): A study in stranded animals on the Canary Islands. *Science of the Total Environment*, 536:489–498.

Geraci, J.R. and St Aubin, D.J. 1990. Summary and conclusions. In: Marine pelagic ecosystem future 229 Sea Mammals and Oil: Confronting the Risks, ed. J.R. Geraci and D.J. St Aubin, pp. 253–256. New York: Academic Press, Inc.

Germanov, E.S., Marshall, A.D., Bejder, L., Fossi, M.C. and Loneragan, N.R., 2018. Microplastics: No small problem for filter-feeding megafauna. *Trends in Ecology & Evolution*, 33(4):227–232.

Gollasch, S., Minchin, D. and David, M., 2015. The Transfer of Harmful Aquatic Organisms and Pathogens with Ballast Water and Their Impacts. Global Maritime Transport and Ballast Water Management, Springer, Dordrecht, pp. 35–58.

Güven, O., Gökdağ, K., Jovanović, B. and Kıdeyş, A.E., 2017. Microplastic litter composition of the Turkish territorial waters of the Mediterranean Sea, and its occurrence in the gastrointestinal tract of fish. *Environmental Pollution*, 223:286–294. https://doi.org/10.1016/j.envpol.2017.01.025

Guzzetti, E., Sureda, A., Tejada, S. and Faggio, C., 2018. Microplastic in marine organism: Environmental and toxicological effects. *Environmental Toxicology and Pharmacology*, 64:164–171.

Hall, A.J., 1995. Morbilliviruses in marine mammals. *Trends in Microbiology*, 3(1):4–9.

HELCOM, 2018. HELCOM Assessment on maritime activities in the Baltic Sea 2018. Baltic Sea Environment Proceedings No.152. Helsinki Commission, Helsinki, pp. 1–253. http://helcom.fi/Lists/Publications/BSEP152.pdf

Hereford, A., 2021. Lithium toxicity: A battery of problems. Emergency Medicine Board Review. www.emboardbombs.com/papers/2021/6/13/lithium-toxicity-a-battery-of-problems-9c45s

Hermsen, E., Pompe, R., Besseling, E. and Koelmans, A.A., 2017. Detection of low numbers of microplastics in North Sea fish using strict quality assurance criteria. *Marine Pollution Bulletin*, 122:253–258. https://doi.org/10.1016/j.marpolbul.2017.06.051

Herndon, D.N., Langner, F., Thompson, P., et al. 1987. Pulmonary injury in burned patients. *Surgical Clinic of North America*, 67:31–46.

Honda, K., Yamamoto, Y., Kato, H. and Tatsukawa, R., 1987. Heavy metal accumulations and their recent changes in southern minke whales *Balaenoptera acutorostrata*. *Archives of Environmental Contamination and Toxicology*, 16(2):209–216.

Honda, M. and Suzuki, N., 2020. Toxicities of polycyclic aromatic hydrocarbons for aquatic animals. *International Journal of Environmental Research and Public Health*, 17(4):1363.

Ilangakoon, A.D., 2002. Whales and Dolphins, Sri Lanka: A Guide to Cetaceans in the Waters around Sri Lanka. WHT Publications Ltd., Colombo, Sri Lanka.

IUCN., 2020. The IUCN Red List of Threatened Species. Version 2020-2. IUCN, Switzerland.

Jaakkola, K. and Willis, K., 2019. How long do dolphins live? Survival rates and life expectancies for bottlenose dolphins in zoological facilities vs. wild populations. *Marine Mammal Science*, 35(4):1418–1437.

Jahnke, A., Arp, H.P.H., Escher, B.I., Gewert, B., Gorokhova, E., Kühnel, D., Ogonowski, M., Potthoff, A., Rummel, C., Schmitt-Jansen, M. and Toorman, E., 2017. Reducing uncertainty and confronting ignorance about the possible impacts of weathering plastic in the marine environment. *Environmental Science & Technology Letters*, 4(3):85–90.

Jarup, L. and Åkesson, A., 2009. Current status of cadmium as an environmental health problem. *Toxicology and Applied Pharmacology*, 238:201–208.

Jing, L., Chen, B., Zhang, B. and Peng, H., 2012. A review of ballast water management practices and challenges in harsh and arctic environments. *Environmental Reviews*, 20(2):83–108.

Katsanevakis, S., 2008. Marine debris, a growing problem: Sources, distribution, composition and impacts. In: Hofer, T.N. (Ed.), *Science*. New Research, Nova Science Publishers: New York, pp. 53–100.

Kershaw, J.L. and Hall, A.J., 2019. Mercury in cetaceans: Exposure, bioaccumulation and toxicity. *Science of the Total Environment*, 694, p. 133683.

Kirumbara, L.U., Krishantha, J.R., Jens-Otto, K. and Kanapathipillai, A., 2022. Distribution and abundance of the blue whale (*Balaenoptera musculus* indica) off Sri Lanka during the Southwest Monsoon 2018. *Journal of Marine Science and Engineering*, 10(11):1626.

Krakstad, J.O., Jayasinghe, P., Totland, A., Dalpadado, P., Søiland, H., Cervantes, D., Gunasekara, S., Liyanage, U., Haputhantri, S., Rathnasuriya, I., Wimalasiri, U., Weerakoon, A., Nirbadha, S., Harischandra, A., Wanigatunga, R. and Kithsiri, P., 2018. Survey of regional resources and ecosystems of the Bay of Bengal: Part 1 Sri Lanka, 24 June to 16 July 2018. NORAD-FAO programme GCP/GLO/690/NOR, Cruise reports Dr. Fridtjof Nansen, EAF-Nansen/CR/2018/8, pp. 162.

Laist, D.W., 1987. Overview of the biological effects of lost and discarded plastic debris in the marine environment. *Marine Pollution Bulletin*, 18:319–326.

Long, M., Reid, R.J. and Kemper C.M.,1998. Cadmium accumulation and toxicity in the bottlenose dolphin *Tursiops truncatus*, the common dolphin *Delphinus delphis*, and some dolphin prey species in South Australia. *Australian Mammalogy*, 20:25–33.

Luo, M. and Shin, S.H., 2019. Half-century research developments in maritime accidents: Future directions. *Accident Analysis & Prevention*, 123:448–460.

Lusher, A.L., Hernandez-Milian, G., Berrow, S., Rogan, E. and O'Connor, I., 2018. Incidence of marine debris in cetaceans stranded and bycaught in Ireland: Recent findings and a review of historical knowledge. *Environmental Pollution*, 232:467–476. doi: https://doi.org/ 10.1016/j.envpol.2017.09.070

Martenstyn, H., 2019. *Sri Lanka Marine Mammals Research and Conservation 1560–2019, Technical Report-Vol. 01*. Sri Lanka: Author.

Misra, S. and Froelich, P. N., 2012. Lithium isotope history of Cenozoic seawater: Changes in silicate weathering and reverse weathering. *Science*, 335(6070):818–823.

Moore, J.E. and Read, A.J., 2008. A Bayesian uncertainty analysis of cetacean demography and bycatch mortality using age-at-death data. *Ecological Applications*, 18(8):1914–1931.

Moore, M.J., Mitchell, G.H., Rowles, T.K. and Early, G., 2020. Dead cetacean? Beach, bloat, float, sink. *Frontiers in Marine Science,* 7: 7–33.

Moorthy, B., Chu, C. and Carlin, D.J., 2015. Polycyclic aromatic hydrocarbons: From metabolism to lung cancer. *Toxicological Sciences*, 145(1):5–15.

Muir, D.C.G., Wagemann, R., Hargrave, B.T., Thomas, D.J., Peakall, D.B. and Norstrom, RJ., 1992. Arctic marine ecosystem contamination. *Science of the Total Environment*, 122:75–134.

Nanayakkara, R.P., Herath, J. and de Mel, R.K., 2014. Cetacean presence in the Trincomalee Bay and adjacent waters, Sri Lanka. *Journal of Marine Biology*, 2014: 819263.

Nelms, S.E., Barnett, J., Brownlow, A., Davison, N.J., Deaville, R., Galloway, T.S., Lindeque, P.K., Santillo, D. and Godley, B.J., 2019. Microplastics in marine mammals stranded around the British coast: Ubiquitous but transitory?. *Scientific Reports,* 9:1075.https://doi.org/10.1038/s41598-018-37428-3

Nelms, S.E., Galloway, T.S., Godley, B.J., Jarvis, D.S. and Lindeque, P.K., 2018. Investigating microplastic trophic transfer in marine top predators. *Environmental Pollution.* https://doi.org/10.1016/j.envpol.2018.02.016

O'Shea, T.J. and Brownell Jr, R.L., 1994. Organochlorine and metal contaminants in baleen whales: A review and evaluation of conservation implications. *Science of the Total Environment*, 154(2–3):179–200.

Osterhaus, A.D., de Swart, R.L., Vos, H.W., Ross, P.S., Kenter, M.J. and Barrett, T., 1995. Morbillivirus infections of aquatic mammals: Newly identified members of the genus. *Veterinary Microbiology*, 44 (2–4):219–227.

Pacyna, E.G., Pacyna, J.M., Sundseth, K., et al., 2010. Global emission of mercury to the atmosphere from anthropogenic sources in 2005 and projections to 2020. *Atmospheric Environment*, 44:2487–2499.

Partow, H., Lacroix, C., Le Floch, S., Alcaro, L. 2021. X-Press Pearl Maritime disaster Sri Lanka. Report of the UN environmental advisory mission July 2021. *United Nations Environment Programme*, 1–50.

Patel, A.B., Shaikh, S., Jain, K.R., Desai, C. and Madamwar, D., 2020. Polycyclic aromatic hydrocarbons: Sources, toxicity, and remediation approaches. *Frontiers in Microbiology*, 11, p. 562813.

Peltier, H., Authier, M., Deaville, R., et al., 2016. Small cetacean bycatch as estimated from stranding schemes: The common dolphin case in the northeast Atlantic. *Environmental Science & Policy*, 63:7–18.

Peltier, H., Beaufils, A., Cesarini, C., Dabin, W., Dars, C., Demaret, F., Dhermain, F., Doremus, G., Labach, H., Van Canneyt, O. and Spitz, J., 2019. Monitoring of marine mammal strandings along French coasts reveals the importance of ship strikes on large cetaceans: A challenge for the European Marine Strategy Framework Directive. *Frontiers in Marine Science*, 6: 486.

Peltier, H., Dabin, W., Daniel, P., et al., 2012. The significance of stranding data as indicators of cetacean populations at sea: Modelling the drift of cetacean carcasses. *Ecological Indicators*, 18:278–290. https://doi.org/10.1371/journal.pone.0062180

Pierce, G.J., Caurant, F. and Law, R.J., 2013. Bioaccumulation of POPs and toxic elements in small cetaceans along European Atlantic coasts. *Chemical Pollution and Marine Mammals*, 71–84.

Priyadarshana, T., Randage, S. M., Alling, A., Calderan, S., Gordon, J., Leaper, R., and Porter, L. 2015. Distribution patterns of blue whale (*Balaenoptera musculus*) and shipping off southern Sri Lanka. *Regional Studies in Marine Science*, 3:181–188. https://doi.org/10.1016/j.rsma.2015.08.002

RAwsoN, A.J., Patton, G.W., Hofmann, S.U.Z.A.N.N.E., Pietra, G.G. and Johns, L., 1993. Liver abnormalities associated with chronic mercury accumulation in stranded Atlantic bottlenosed dolphins. *Ecotoxicology and Environmental Safety*, 25(1):41–47.

Redfern, J.V., McKenna, M.F., Moore, T.J., Calambokidis, J., Deangelis, M.L., Becker, E.A., Barlow, J., Forney, K.A., Fiedler, P.C. and Chivers, S.J., 2013. Assessing the risk of ships striking large whales in marine spatial planning. *Conservation Biology*, 27(2):292–302.

Reisdorf, A., Bux, R., Wyler, D., et al. 2012. Float, explode or sink: postmortem fate of lung-breathing marine vertebrates. *Palaeobiodiversity and Palaeoenvironment*, 92:67–81. https://doi.org/10.1007/s12549-011-0067-z

Rengarajan, T., Rajendran, P., Nandakumar, N., Lokeshkumar, B., Rajendran, P. and Nishigaki, I., 2015. Exposure to polycyclic aromatic hydrocarbons with special focus on cancer. *Asian Pacific Journal of Tropical Biomedicine*, 5(3):182–189.

Robeck, T.R., Willis, K., Scarpuzzi, M.R. and O'Brien, J.K., 2015. Comparisons of life-history parameters between free-ranging and captive killer whale (*Orcinus orca*) populations for application toward species management. *Journal of Mammalogy*, 96(5):1055–1070.

Roberts, S.E. and Hansen, H.L., 2002. An analysis of the causes of mortality among seafarers in the British merchant fleet (1986–1995) and recommendations for their reduction. *Occupational Medicine*, 52(4):195–202.

Rogowska, J. and Namieśnik, J., 2010. Environmental implications of oil spills from shipping accidents. *Reviews of Environmental Contamination and Toxicology*, 206:95–114.

Rømer, H., Haastrup, P. and Petersen, H.J., 1996. Exploring environmental effects of accidents during marine transport of dangerous goods by use of accident descriptions. *Environmental Management*, 20(5):753–766.

Ruberg, E.J., Elliott, J.E. and Williams, T.D., 2021. Review of petroleum toxicity and identifying common endpoints for future research on diluted bitumen toxicity in marine mammals. *Ecotoxicology*, 30(4):537–551.

Rummel, C.D., Löder, M.G.J., Fricke, N.F., Lang, T., Griebeler, E.-M., Janke, M. and Gerdts, G., 2016. Plastic ingestion by pelagic and demersal fish from the North Sea and Baltic Sea. *Marine Pollution Bulletin*, 102:134–141. https://doi.org/10.1016/j.marpolbul.2015.11.043

Sánchez-Beaskoetxea, J., Basterretxea-Iribar, I., Sotés, I. and Machado, M.D.L.M.M., 2021. Human error in marine accidents: Is the crew normally to blame?. *Maritime Transport Research*, 2:100016.

Sanderfoot, O.V., Bassing, S.B., Brusa, J.L., Emmet, R.L., Gillman, S.J., Swift, K. and Gardner, B., 2022. A review of the effects of wildfire smoke on the health and behavior of wildlife. *Environmental Research Letters*, 16(12):123003.

Schwacke, L.H., Smith, C.R., Townsend, F.I., Wells, R.S., Hart, L.B., Balmer, B.C., Collier, T.K., De Guise, S., Fry, M.M., Guillette Jr, L.J. and Lamb, S.V., 2014. Health of common bottlenose dolphins (*Tursiops truncatus*) in Barataria Bay, Louisiana, following the Deepwater Horizon oil spill. *Environmental Science & Technology*, 48(1):93–103.

Secchi, E.R., Zarzur, S., 1999. Plastic debris ingested by a Blainville's beaked whale, *Mesoplodon densirostris*, washed ashore in Brazil. *Aquatic Mammals*, 25:21–24.

Setälä, O., Fleming-Lehtinen, V. and Lehtiniemi, M., 2014. Ingestion and transfer of microplastics in the planktonic food web. *Environmental Pollution*, 185:77–83.

Smith, C.R. and Baco, A.R., 2003. Ecology of whale falls at the deep-sea floor. In *Oceanography and Marine Biology, An Annual Review*. 41:319–333.

Stone H. H., Martin, J. D., 1969. Pulmonary injury associated with thermal burns. *Surgery Gynecology and Obstetrics*, 129:1242–6.

Strungaru, S.A., Jijie, R., Nicoara, M., Plavan, G. and Faggio, C., 2019. Micro-(nano) plastics in freshwater ecosystems: Abundance, toxicological impact and quantification methodology. *TrAC Trends in Analytical Chemistry*, 110:116–128.

Tarpley, R. J. and Marwitz, S., 1993. Plastic debris ingestion by cetaceans along the Texas coast: Two case reports. *Aquatic Mammals*, 19(2):93–98.

Thibon, F., Weppe, L., Vigier, N., Churlaud, C., Lacoue-Labarthe, T., Metian, M., Cherel, Y. and Bustamante, P., 2021. Large-scale survey of lithium concentrations in marine organisms. *Science of the Total Environment*, 751:141453. http://dx.doi.org/10.1016/j.scitotenv.2020.141453

UNCTAD, 2020. Review of Maritime Transport 2020. United Nations Publications, New York, New York.

Van Cauwenberghe, L., Janssen, C.R., 2014. Microplastics in bivalves cultured for human consumption. *Environmental Pollution*, 193:65–70. http://dx.doi.org/10.1016/j.envpol.2014.06.010

Vanderlaan, A.S.M. and Taggart, C.T., 2007. Vessel collisions with Whales: the probability of lethal injury based on vessel speed. *Marine Mammal Science*, 23:144–156. http://dx.doi.org/10.1111/j.1748-7692.2006.00098.x

Venn-Watson, S., Colegrove, K.M., Litz, J., Kinsel, M., Terio, K., Saliki, J., Fire, S., Carmichael, R., Chevis, C., Hatchett, W. and Pitchford, J., 2015. Adrenal gland and lung lesions in

Gulf of Mexico common bottlenose dolphins (*Tursiops truncatus*) found dead following the Deepwater Horizon oil spill. *PLoS One*, 10(5):e0126538.

Venn-Watson, S., Smith, C.R., Jensen, E.D. and Rowles, T., 2013. Assessing the potential health impacts of the 2003 and 2007 firestorms on bottlenose dolphins (*Tursiops truncatus*) in San Diego Bay. *Inhalation Toxicology*, 25(9):481–491.

Verheyen, V.J., Remy, S., Govarts, E., Colles, A., Rodriguez Martin, L., Koppen, G., Voorspoels, S., Bruckers, L., Bijnens, E.M., Vos, S. and Morrens, B., 2021. Urinary polycyclic aromatic hydrocarbon metabolites are associated with biomarkers of chronic endocrine stress, oxidative stress, and inflammation in adolescents: FLEHS-4 (2016–2020). *Toxics,* 9(10):245.

Wang, I.J., Karmaus, W.J. and Yang, C.C., 2017. Polycyclic aromatic hydrocarbons exposure, oxidative stress, and asthma in children. *International Archives of Occupational and Environmental Health*, 90(3):297–303.

Wells, R.S., Allen, J.B., Lovewell, G., et al., 2015. Carcass-recovery rates for resident bottlenose dolphins in Sarasota Bay, Florida. *Marine Mammal Science,* 31:355–368.

Wetzel, D.L., J.E. Reynolds, III, and Morales-Vela, B., 2006. Possible impacts of chemical contaminants on manatees in Mesoamerica. Proceedings Congreso de la Societdad Mesoamericana para la Biologia y la Conservacion. October, Antigua Guatemala, Guatemala.

Williams, R., Gero, S., Bejder, L., et al., 2011. Underestimating the damage: Interpreting cetacean carcass recoveries in the context of the Deepwater Horizon/BP incident. *Conservative Letters,* 4:228–233.

Ylitalo, G.M., Collier, T.K., Anulacion, B.F., Juaire, K., Boyer, R.H., da Silva, D.A., Keene, J.L. and Stacy, B.A., 2017. Determining oil and dispersant exposure in sea turtles from the northern Gulf of Mexico resulting from the Deepwater Horizon oil spill. *Endangered Species Research*, 33:9–24.

11 Impact of MV X-Press Pearl Disaster on Coastal and Marine Fish and Fisheries

*P.R.T. Cumaranatunga, K.R. Dalpathadu,
H.M.U. Ayeshya, K.A.W.S. Weerasekera,
S.S.K. Haputhantri, W.N.C. Priyadarshani,
G.K.A.W. Fernando, L.D. Gayathry, and
H.B.U.G.M. Wimalasiri*

11.1 INTRODUCTION TO MV X-PRESS PEARL SHIP DISASTER

The MV X-Press Pearl cargo ship accident, which occurred on May 21, 2021, was the worst ship accident globally recorded to date. Ship first released a brown-colored smoke which later turned into a fire emitting a black-colored, thick smoke, followed by an explosion. The accident occurred within the territorial waters, in the Western part of Sri Lanka, when the South-West monsoons were strongly active, which made it very difficult for the fire fighters to douse the fire, although the Sri Lanka Navy worked hard together with the Sri Lanka Air Force and the Indian Coast Guard.

Accident occurred in close proximity to the Colombo harbor. Release of large volumes of plastic nurdles from the ship cargo was the most significant catastrophe, because not only did the nurdles cover the ocean surface but they also got carried long distances with strong currents that prevailed at that time, covering a significant stretch of the beaches from North-Western coast through West coast to South-Western coast, with virgin and burned chunks of plastic of various sizes (Figure 11.1). Together with the above unmanageable load of plastic entering the marine environment, release of thick black smoke which carried noxious gases and leakage of concentrated nitric acid from one of the containers on the upper deck, which reacted with methanol to form hazardous chemicals, and release of urea, fertilizer, and other organic matter containing nutrients, different kinds of oil including fuel oil, caustic soda, lithium batteries, etc., from damaged containers, possibly together with other hazardous chemicals released into the environment, polluted the atmosphere, coastal waters, and the sandy shores of the North-Western, Western, and South-Western parts of the country.

Few days after the outbreak of fire in the ship, which created a thick black smoke released into the atmosphere and debris resulting from the fire and the explosions

FIGURE 11.1 Virgin plastic nurdles and burned plastic particles distributed on beaches of Sri Lanka.

and sinking of the ship wreck (Figure 11.2), carcasses of different species of fish (Figure 11.3) together with carcasses of turtles and mammals, some of which have not been commonly encountered on the shores of Sri Lanka, prior to the accident, washed off on to the beaches around Sri Lanka, indicating a possible serious impact on the marine and coastal fish fauna and other marine biodiversity.

Among the carcasses of fish found on the beaches off the Western and North-Western coasts of Sri Lanka, there were carcasses of economically and ecologically valuable species (Figure 11.3A–J) such as giant trevally and rock fish species (important food fish), physically damaged or deteriorated whale shark (a nontargeted food fish species in capture fisheries, because it is a considered a threatened species), reef fish such as parrot fish (as a food and of ornamental value), lion fish (valuable ornamental fish), and a moray eel (which lives inside cracks and crevices and is considered an ornamental fish) (Figure 11.3). There was a headless carcass of an anguillid eel, possibly *Anguilla bicolor*, at silver eel stage of its life cycle (Figure 11.3k), which is found in freshwater and migrates to oceans for spawning and completing its life cycle. These Asian freshwater anguillid eels (*A. bicolor* and *Anguilla nebulosa*) which spend their growth phase within freshwater of Sri Lanka are known to migrate to deep oceans for spawning when the strong monsoon currents are in operation within the coastal belt of Sri Lanka (Cumaranatunga et al., 1997). The presence of a segment of a silver eel, part of the Anguilla eel life cycle, suggests that the ship accident may have negatively affected these eels during their migration for spawning in the deep oceans.

During and after the ship accident, debris from the shipwreck and various items included in the cargo were released into the environment, creating complex alterations within the sensitive coastal marine ecosystems, which should be seriously considered as they may result in irreversible short-term, medium-term, and long-term impacts on coastal and marine ecosystems and biodiversity, especially on fish species that are of high economic importance to Sri Lanka. Most of the carcasses of fish found on the coastal belt were either physically damaged or in a highly deteriorated condition. Most fish have delicate bodies and their dead bodies can be subjected to degeneration due to the action of lytic enzymes (autolysis), microbial degeneration, and physical damages caused by strong monsoonal currents and wave action prevailing at the time of the ship accident. All dead fish will not float or stay suspended in water, especially fish affected by the explosions, which will receive damage to their air bladder and lose their ability to maintain buoyancy; therefore their carcasses will sink to the bottom of the sea. Therefore, it is difficult to estimate the damage caused to fish fauna only by considering the number of carcasses washed off on to the shores of Sri Lanka.

Impact on Coastal and Marine Fish and Fisheries 217

FIGURE 11.2 MV X-Press Pearl ship on fire releasing a thick black smoke (top), shipwreck shortly after sinking (middle), and the status of the wreck off the Western coastal belt of Sri Lanka at the time of preparation of this chapter (bottom).

FIGURE 11.3 (A–J) Carcasses of fish found on the Western coastline of Sri Lanka few days after the ship accident, among which there were few rarely encountered species. A) Giant Trevally, B) Whale shark, C) Indo-Pacific tarpon, D) Rockfish, E) Parrot fish, F) Pufferfish, G) Morey eel, H) Lionfish, I) Porcupine fish and J) part of an anguillid eel at silver eel stage. Source for D, E, F, G: Environmental Science Division, NARA.

Most carcasses may have been carried to long distances with the strong currents moving northward parallel to the Western and North-Western coast of Sri Lanka, and prior to deposition of the carcasses on the beaches, they may have been subjected to mechanical degradation and scavenging by other organisms. To evaluate the total ecosystem damage and socioeconomic impact on fish resources and fishery industry, it is important to conduct a multifaceted survey. Such a monitoring program should especially include a survey on diversity, density, and histopathology of economically important fish species which spend the whole or part of their life history within the area affected by the incidents that occurred from the time of the ship accident until the removal of the wreck and at least for two years after the removal of the wreck. It is important to carry out a comprehensive analysis of the socioeconomic impacts on fishery industry in the South-Western, Western, and North-Western coastal districts of Sri Lanka, which were subjected to severe disturbances due to the ship disaster and due to prohibition of fishing imposed by the government within the area affected by the MV X-Press Pearl ship accident, as it is considered the worst marine disaster that has occurred globally to date.

11.2 MARITIME ZONES OF SRI LANKA AFFECTED BY THE SHIP ACCIDENT

Sri Lanka is an island in the Indian Ocean and it has an Exclusive Economic Zone (EEZ) that extends outward up to 370 km (200 nautical miles [nm]) from its shores

Impact on Coastal and Marine Fish and Fisheries

and covers an area of about 510,000 km². Sri Lanka also has the exclusive right to authorize, regulate, and control scientific research within this zone, besides having sovereign rights to resources in the water column, seabed, and subsurface (Martenstyn, 2019). EEZ of Sri Lanka is equivalent to eight times its land area. The territorial waters of Sri Lanka extend to 22 km (12 nm) beyond the coastline and cover an area of about 21,500 km². The contiguous zone is the band of water extending from the outer edge of the territorial sea to up to 24 nm (Figure 11.4).

MV X-Press Pearl ship accident occurred in the territorial waters off the Western coast of Sri Lanka. According to the vertical classification of oceanic zones (Figure 11.5), the zone where accident occurred is classified as the neritic zone, which is within the light-penetrating, photic zone. The neritic zone is the most productive ocean region, as it supports a rich biodiversity. It has been estimated that 90% of the world's finfish and shellfish harvest comes from the neritic zone. The stable environment of this zone provides light, oxygen, and nutrients contributed by runoff from nearby land and upwelling from the continental shelf, as well as suitable salinity and temperature to

FIGURE 11.4 Map indicating the maritime zones (Exclusive Economic zone [EEZ], Contiguous zone and the Territorial waters). Modified from Admiralty Charts & Sri Lanka Survey Department.

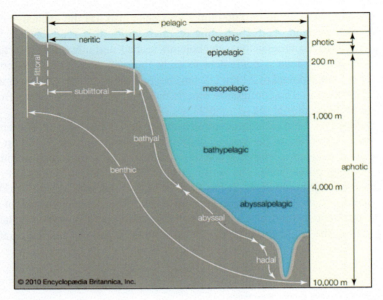

FIGURE 11.5 Zonation of the ocean. The open ocean, the pelagic zone, includes all marine waters throughout the globe beyond the continental shelf, as well as the benthic, or bottom, environment on the ocean floor.

Source: Encyclopedia Britanica, Inc.

support a wide range of marine life. Among the important organisms occupying the neritic zone, most abundant are the photosynthetic protists called phytoplankton that support marine ecosystems by forming the basis of the food web. Phytoplankton are microscopic algae which use light from the sun to generate their own food and are themselves food for filter feeders and zooplankton. Marine animals such as fish feed on zooplankton, and fish in turn become food for other fish, marine mammals, birds, and humans. Marine bacteria also play an important role in the flow of trophic energy by decomposing organisms and recycling nutrients in the marine environment.

11.3 MOVEMENT OF WATER CURRENT WITHIN THE AREA AFFECTED BY THE SHIP ACCIDENT

The ocean around Sri Lanka is under the effect of two monsoons: the South-West monsoon (May to September) and the North-East monsoon (December to February) (Figure 11.6a and b). In between there are two periods referred to as intermonsoons: First Intermonsoon (March/April) and Second Intermonsoon (October/November). Figure 11.6c indicates the quarterly variation of current pattern around Sri Lanka, due to the effect of two monsoons. Monsoons affect the weather pattern in Sri Lanka and drastically change the directions of ocean currents (both horizontal and vertical) around Sri Lanka, resulting in seasonal variations in upwelling and downwelling processes, which influence the nutrient dynamics that directly affect the productivity of marine waters around Sri Lanka. Seasonal variations of the current

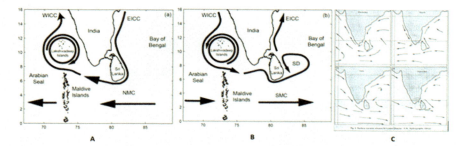

FIGURE 11.6 Current pattern around Sri Lanka during North-East (a) and South-West (b) monsoons (Source: de Vos et al., 2014) and variations in current pattern around Sri Lanka in December, March, June, and September (c). Source: HM Hydrographic Office.

pattern and the nutrient dynamics as well as seasonal variations in the diversity and density of different fish species are major factors affecting the fishery industry.

At the time of the ship accident the Western coastal and marine waters were quite turbulent and strong currents that were moving toward the Western coast of Sri Lanka were traveling northward along the Western coastal belt. These monsoonal currents are used by marine organisms, including fish, to move long distances in order to fulfill their biological needs such as reproduction and feeding. At the time of the ship disaster these strong currents carried all the debris, and a variety of materials were released from the ship cargo. Most prominent out of them were large metal containers, broken pieces of metal from the containers, barrels/containers of oil, chemicals, plastic bags containing plastic pellets of different sizes, virgin and burned plastic nurdles, and synthetic yarn which were entangled in fishing nets (Figure 11.7). All or most of the material released from the ship disturbed or polluted the marine and coastal ecosystems, most probably causing severe impacts on the biodiversity including fish and, in turn, on the coastal capture fishery industry. Due to the effect of these currents most of the pollutants which entered the marine environment were distributed to a vast area, including both coastal and offshore areas. Synthetic yarn entangled in fishing nets caused irreparable damage to fishing nets.

11.4 EFFECTS OF THE SHIP DISASTER ON MIGRATORY FISH SPECIES AND STAGES OF THEIR LIFE HISTORY

Many species of fish are known to annually move within a particular area of the ocean and regularly travel over great distances, especially to complete their life history from eggs through larval and juvenile stages to adult stage and move to spawning grounds where they meet individuals of opposite sex and release gametes (spawning), which produce fertilized eggs. After spawning they return to the feeding grounds. When eggs hatch out, larvae which emerge usually leave the spawning grounds for areas where they develop into juveniles, before joining the adult stock at the feeding grounds (Figure 11.8).

Fish migrate to fulfill different biological needs, such as in search of feeding grounds (alimentary or feeding migration), spawning grounds (genetic or spawning migration), and suitable climatic conditions (climatic or seasonal migration), and for water

FIGURE 11.7 Fishing nets entangled with synthetic yarn released from ship cargo.

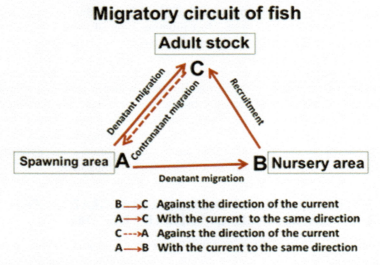

FIGURE 11.8 Cushing's triangle on fish migration showing the directions of movement of fish during different stages of their life history, with respect to the direction of ocean currents. Adopted from Paul et al. (2016).

and electrolytes balance from sea to freshwater and vice versa (osmoregulatory migration). Migratory patterns of fish are related to oceanographic factors and ocean currents (Dorst, 2022). Eggs, larvae, and young are moved passively with the current (drifting or contranatant migration), although migration of adult fish toward breeding grounds is an active swimming movement against the direction of the current (denatant migration)

(Figure 11.8). Adult movements thus are usually directional or oriented movements, where fish respond to environmental conditions and fish swim either toward or away from a source of stimulus. They also involve a random locomotory movement from a uniform habitat to diverse directions. Two major categories of migration that migratory fishes exhibit can be distinguished among marine fish species: diadromous and oceanodromous. Diadromous migration is movement of fish between sea and freshwater. While most of the fishes are restricted to either freshwater or seawater, some fish species regularly migrate between seawater and freshwater and have perfect osmotic balance; they are the true migratory fish and their changes in habitat may cause osmotic imbalance in those fishes. Movement of marine fish that spend most of their life living and feeding in sea and migrate from sea to freshwater for spawning is known as anadromous migration (e.g. *Hilsa* spp. and Lampreys). Movement of freshwater fish from river to sea during breeding season for spawning is known as catadromous migration (e.g. *Anguilla* spp.). Migration of fish within sea in search of suitable feeding and spawning ground is known as oceanodromous migration (e.g. herrings, sardines, and tuna species). In addition to the above fish migrate seasonally between different latitudes (latitudinal migration) due to climatic conditions (e.g. swordfish migrate from North to South during autumn and South to North during spring), and also fish are involved in vertical migration from deep to the surface and vice versa for food, protection, and spawning during spawning seasons (e.g. demersal fish, reef fish) (Dorst, 2022). An understanding of different migratory methods used by fish species found in the area affected by the ship accident is very important to understand the impacts of all incidents that occurred during and after the ship accident on different fish species and also to identify the effects of different pollutants released from the ship on different species occupying the area affected by the ship accident, depending on their biology and diurnal and seasonal behavioral patterns. Out of the carcasses of fish revealed from the shores of Sri Lanka, it is evident that fish which occupy a variety of niches have faced negative impacts from the incidents that occurred during and after the ship accident. Fish eggs, larvae (ichthyoplankton), small fish species with delicate body structures, and those with sensitive life styles may have faced serious losses and disappeared in large numbers without leaving any trace of the damage caused. Affected fish species may hold important positions within the complex food chains and food webs operating within the coastal and marine ecosystems.

11.5 POSSIBLE IMPACTS OF SHIP ACCIDENT ON MARINE FISH AND FISHERIES IN SRI LANKA

When considering ocean productivity, Sri Lankan economy is highly dependent on marine fish production for its economic growth and the nutrient dynamics resulting from monsoon currents are known to directly or indirectly affect the marine fish production around Sri Lanka. The South-West monsoon, which is the main monsoon operating during May/June, originates from the Indian Ocean and brings most rain to the Southern and Western coasts. MV X-Press Pearl ship caught fire at a time when the strong South-West monsoons were in operation, resulting in strong winds and long spells of rain. Due to the possible negative impacts on fishing vessels, gear, and those who were involved in fishing, government had to declare a no-fishing area around the location of the shipwreck. Shipwreck has been lying in the accidental zone for more than 6 months and fishing was prohibited in the area around the wreck and in the area scattered with containers and other debris that fell from the ship. Fishing prohibition zone is shown in Figure 11.9.

In Sri Lanka, from a total of 1192 fish species, 972 are recorded as saltwater or marine, out of which 573 are reef associated (benthic or demersal), 112 are pelagic, and 39 are deep sea, while the rest are not clearly classified. Fish are an economically important living resource due to their importance as a source of food and/or ornamental value. Fisheries sector in Sri Lanka accounts for 1.3% of GDP at the current market prices, and contribution from marine fisheries is 1.1%. In 2019 total production of fish in Sri Lanka was around 505,830 metric tons (MT), out of which 415,480 MT came from the marine fisheries sector. Income from export of marine fish was Rs. 53,483 million (US$299.2 million) in 2019 and its contribution to national export earnings was 1.5% (Ministry of Fisheries Statistics, 2020). Prohibition of fishing in the area affected by the ship accident severely affected the fishery industry in the area. It affected the socioeconomics of fisher communities and also the marine fish consumption among local people, especially in the Western Province, due to the suspicion about contamination of marine fish by the pollutants released from the ship during and after the ship accident.

11.5.1 Diversity of Fish in the Area Affected by the Ship Accident

Marine fish are the most diverse group of organisms among the chordates, consisting of more than 16,764 species (February 19, 2010) from a total of 31,362 valid species of fish species (Eschmeyer et al., 2010). They are found in shallow neritic zone above the continental shelf and in the deeper oceanic waters beyond continental shelf. Depending on the zones they occupy in marine environment, fish can be broadly divided into three main categories: pelagic, demersal, and reef fish. Demersal fish also can be categorized into two groups, namely semidemersal or benthopelagic fish, which normally live close to the bottom of the sea, and true benthic fish, which normally stay at the bottom buried under the sand and live in burrows or cracks and crevices and only occasionally show limited movements to fulfill their biological needs or to escape from predators or extreme conditions. Demersal fish live on or near the bottom of the water body, while reef fish are associated with coral reefs (Lal & Fortune, 2000). Marine pelagic fish can be divided into coastal (inshore) fish and oceanic (offshore) fish. Coastal fish inhabit the relatively shallow and sunlit highly productive neritic waters above the continental shelf, while oceanic fish (which may also swim inshore for biological needs) inhabit the vast and deep waters, from surface to deepest trenches beyond the continental shelf. Fish are an important resource worldwide, especially as a source of protein. Commercial and subsistence fishery depend on wild stocks or more intensive cultures of selected fish species under captive conditions. Marine fish are also caught by recreational fishers, kept as pets, raised by fish keepers, and exhibited in public aquaria. Therefore, depending on the use, fish can be divided mainly into major categories: food fish, ornamental fish, and recreational fish. Sri Lanka enjoys an eight times larger maritime zone compared to its land area. A variety of habitats such as sea grass beds, lagoons, estuaries, coral reefs, and fertile coastal waters are distributed around the country. This diverse array of habitats supports a rich marine fish fauna (Kumara & Dalpathadu, 2012).

Impact on Coastal and Marine Fish and Fisheries

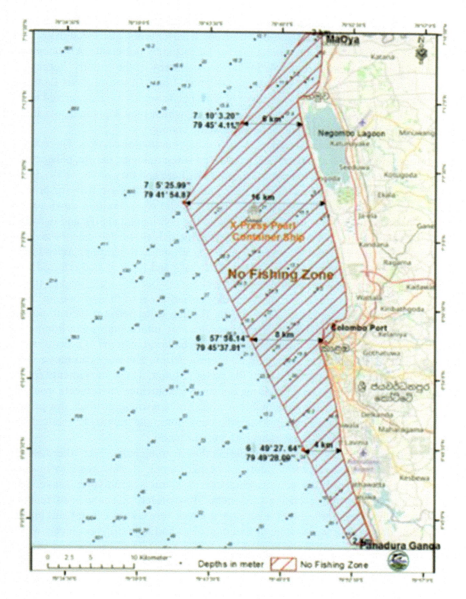

FIGURE 11.9 Fishing prohibition area declared by the Government of Sri Lanka after the X-Press Pearl ship accident.

Marine fish species which can be subjected to direct detrimental effects of the MV X-Press Pearl ship disaster are mostly the fish inhabiting the neritic zone (above the continental shelf), which includes all categories of small pelagic fish, some species of large pelagic fish and demersal fish that also include reef fish. Fish inhabiting other areas also will face indirect impacts of this marine disaster, due to horizontal and vertical dispersal of pollutants to a vast volume of oceanic waters covering all zones of the marine environment as a result of monsoonal currents and also through grazing/predation on marine organisms (including fish) affected by pollutants. All categories of marine fish play an important role in the socioeconomy of the country, because of their importance as a major or only source of animal protein for most Sri Lankans; also among them there are a large number of fish species which are categorized under ornamental fish and few species among food fish, which play a pivoted role in the export market that attracts foreign exchange to the country.

Marine fish can be categorized depending on their habitats and habits. Three major categories of marine fish are identified according to their habitats: neritic, pelagic, and benthic. Neritic fish are also known as coastal or inshore and they are normally found above the continental shelf including the intertidal zone. Pelagic fish are broadly categorized as epipelagic (living at upper levels of the water column down to 200 m), oceanic (living in the water column beyond the continental shelf), and demersal (living close to the sea floor) (NOAA,2021). Benthic fish are further categorized into two types, i.e. true benthics (live on or in the seafloor) and benthopelagics (swim freely in the water column as part of the reef habitat), and latter is also referred to as demersal. True benthic fish are those which either stay buried under sand or live in burrows or cracks and crevices found in rocky or reef areas of the sea bottom. They occasionally move along the sea floor during feeding or to perform their important biological functions or to escape from predators or unacceptable environmental conditions. Fish may display diurnal or seasonal behavioral patterns in order to fulfill their biological needs, such as feeding and reproduction, and they can move vertically or horizontally or migrate within the water column. These movements can be localized movements or extended to long distances that include movements from one ocean to another in order to find suitable feeding and spawning grounds, for protection from predators, survive extreme climatic conditions, and maintain genetic diversity and as adaptive characters for survival and existence.

Fish can be further categorized considering their feeding habits, viz. filter feeders, planktivores (phytoplanktivores and zooplanktivores), herbivores, carnivores, coralivores, piscivores, molluscivores, parasites, omnivores, benthivores, scavengers, detritivores, etc. depending on the food items they prefer, mechanism of feeding, and place of feeding (Table 11.1). Fish play a major role within coastal and marine ecosystems due to their wide range of feeding habits and hold different levels within important food chains and webs that operate within coastal and marine ecosystems. Table 11.1 provides the classification of fish according to their feeding habits.

11.5.2 Experimental Fishing After the Ship Accident

In order to obtain samples of pelagic fish for further monitoring experimental fishing was carried out by scientists from National Aquatic Resources Research and Development Agency (NARA) close to the ship accident zone (Figure 11.10). Figure 11.11 indicates the locations selected for experimental fishing in the open sea.

Table 11.2 provides the details of the total fishing effort applied for experimental fishing until October 15, 2021 using different fishing gear in order to capture fish in the area affected by the ship accident. Wherever fishing nets could not be operated,

TABLE 11.1
Classification of Fish According to their Feeding Habits

Category of Marine Fish According to Their Feeding Habits		Description of Food and Feeding Habit
Herbivores	Phytoplanktivores	Feeding on Phytoplankton, which are also known as microalgae, that contain chlorophyll and require sunlight in order to live and grow, and are buoyant and float in the upper part of the ocean, where sunlight penetrates the water (e.g. cyanobacteria, silica-encased diatoms, dinoflagellates, green algae, and chalk-coated coccolithophores)
	Macroplant feeders	Feeding on submerged or rooted plants such as macroalgae and seagrasses
Carnivores	Zooplanktivores	Feeding on zooplankton (tiny animals found near the surface in aquatic environments) which consists of copepods, jellyfish, etc. which are known as holoplankton that spend their whole life as plankton, and larvae of marine invertebrates (crustaceans, mollusks, echinoderms, etc.) and of fish (ichthyoplankton) which are known as meroplankton that spend only part of their life as plankton
	Coralivores	Feeding (grazing) on corals
	Piscivores	Feeding on fish
	Molluscivores	Feeding on Mollusks (bivalves and gastropods)
Omnivores (feeding on any living or nonliving material)	Filter feeders	Feeding on floating organisms or suspended or food particles by filtering them from large volumes of water taken through the mouth using different biological devices within the pharyngeal region and gills with gill-rakers
	Planktivores	Feeding on plankton, which includes small organisms drifting within the water column and moving passively with the water current, which includes phytoplankton (microscopic plants) and zooplankton (microscopic animals and larvae of large marine vertebrates, which includes ichthyoplankton and invertebrates)
	Benthivores	Feeding on benthos (organisms living on the sea floor)
Parasites		Fish that live in (internal parasites) or on an organism (external parasite) of another species (its host) and benefit by deriving nutrients at the other's (host's) expense
Scavengers		Feeding on dead bodies of organisms
Detritivores		Feeding on decaying dead organisms and on living matter suspended within the water column or deposited among sediments at the sea bottom

FIGURE 11.10 The experimental fishing conducted by scientists from NARA in selected sites in the vicinity of area affected by the ship accident.

especially where reefs are located, identification of different fish species inhabiting the area scuba diving was carried out.

Figure 11.12 indicates the taxonomic information and species composition considering the major fish species captured during experimental fishing and Table 11.3 provides the taxonomic information, feeding habits, and habitats of 107 fish species captured or observed during scuba diving within the area affected by the ship accident. Table 11.4 provides the numbers of fish species having different feeding habits among the fish species inhabiting the ship accident–affected area. All species of fish were subjected to morphological and anatomical investigations and tissues were preserved for further chemical analyses and histopathological studies.

According to the species composition of experimental fishing, highest numbers were represented by sardines, which is an important group of commercial fish. They are planktivorous fish and have a higher risk of exposure to harmful algal blooms which may occur due to high nutrient loading that may result from release of urea and other fertilizers, food items, and other biological material in the cargo. Thick algal blooms were observed in the area (Figure 11.13a), and although dinoflagellate blooms dominated by *Noctiluca* sp. (Figure 11.13c and d) were observed, highly toxic dinoflagellates were also observed in smaller numbers (Figure 11.13e, f, and g). Toxic algae density was not sufficient at the time of sampling to create a significant toxic algal bloom. However, toward the end of the year, with low ambient temperatures toxic algal blooms may appear if the nutrient concentration around the shipwreck remains high. Due to bioaccumulation of biotoxins in planktophagus fish, there is a possibility of further bioaccumulation of these toxins along food chains. Next highest composition in the catch was from narrow-barred Spanish mackerels, a highly consumer-demanded predatory fish which seeks high prices.

Overall observations on the species diversity during experimental fishing indicate that fish with different feeding habits (Table 11.3) and commercially important fish species are inhabiting the area affected by the X-Press Pearl ship. Details of their food, feeding habits, and habitats are also listed (Table 11.4) in order to understand the possible paths of pollutants released by the shipwreck that would affect those fish species. These details are also important to identify their niche within the marine environment, which would be very useful to understand how the lineup of incidents related to the ship accident (e.g. fire, smoke, explosions, acid leakages, release of hazardous chemicals, plastic nurdles, food items, fertilizer, lithium batteries, heavy metals, etc.) and the pollution resulted from the ship accident would affect the fish inhabiting and visiting the area affected by the ship accident.

Impact on Coastal and Marine Fish and Fisheries 229

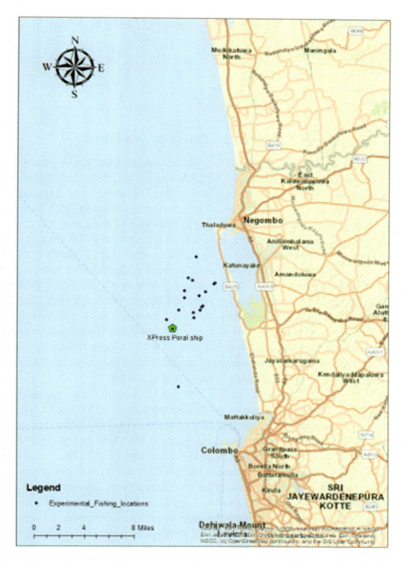

FIGURE 11.11 Maps showing the locations of experimental fishing at open sea by scientists from National Aquatic Resources and Research Agency (NARA).

Figure 11.14 provides a diagrammatic representation of a food web that can exist in the coastal marine environment affected by the MV-X-Press Pearl ship accident, which involves fish and marine mammals considering the food and feeding habits of fish listed in Table 11.4 and also the feeding habits of mammals which involves adult and larval stages of fish.

TABLE 11.2
Total Fishing Effort Applied During Experimental Fishing by NARA

Gear Used for Sample Collection	No. of Operations
Pelagic gill net + bottom set gillnet	13
Trawling	4
Scuba diving	2

11.5.3 Impact of Ship Accident on Fish Eggs and Larvae

Fish eggs and larvae are highly sensitive organisms to pollutants in the aquatic environment (Davydova & Cherkashin, 2007; Cherkashin, et al, 2004; Vashchenko, 2000). Early ontogenesis is a short time interval in fish life cycle during which organisms are most vulnerable to environmental effects (Davydova & Cherkashin 2007; Stepanenko, 1988), including impacts from anthropogenic activities (Tyurin and Khristoforova, 1995; Marty et al., 1997; Cherkashin et al., 2004). Chemical pollution can severely affect entire process of development of eggs or juvenile stages (Courrat et al., 2009; Davydova & Cherkashin 2007). The essential nursery function of habitat may be reduced by these anthropogenic disturbances (Coates et al., 2007; Le Pape et al., 2007). As described in Figure 11.8, fish eggs and larvae which are very often planktonic (ichthyoplankton) are involved in denatant movement (drifting passively) when reaching the nursery grounds. If this phase of the life cycle is exposed to pollutants and environmental disturbances (as in the case of MV X-Press Pearl ship accident), recruitment level, and population size and abundance of the concerned marine fish species may then be severely affected (Peterson et al., 2000).

Understanding the seasonal availability of fish eggs and assemblage of ichthyoplankton in the marine waters off the Western coast of Sri Lanka would provide very important clues for understanding how the lineup of events that disturbed the marine ecosystems and pollutants released from the shipwreck would affect the survival of fish larvae, their distribution, and recruitment to adult stock. Information regarding the distribution and abundance of fish eggs and larvae can provide clues to spawning locations and environmental requirements of important fish species (Gratwicke and Speight, 2005). Moreover, the knowledge of seasonal availability, abundance, and movement of ichthyoplankton is extremely important, because they serve as one of the components of the pelagic food webs; it can represent an important link between smaller planktonic and larger nektonic organisms (Angel and Ojeda, 2001). Further, most fishes inhabit the upper water column during their early life history and have limited avoidance capability of marine disasters; ichthyoplankton surveys provide a relatively low-cost, efficient means to monitor marine fish populations and communities (Rodriguez et al. 2017). At the same time, the mortality of fish larvae due to marine disasters and ability to survive harsh conditions may directly influence future abundances of adult fish stocks. However, the impacts of marine disasters could be clearly understood only by continuous monitoring of the seasonal availability and abundance of fish larvae, their survival up to adolescent stages, and recruitment to adult stocks. Usage of molecular biological techniques and identification keys are very important tools to identify fish eggs and larvae up to

Impact on Coastal and Marine Fish and Fisheries

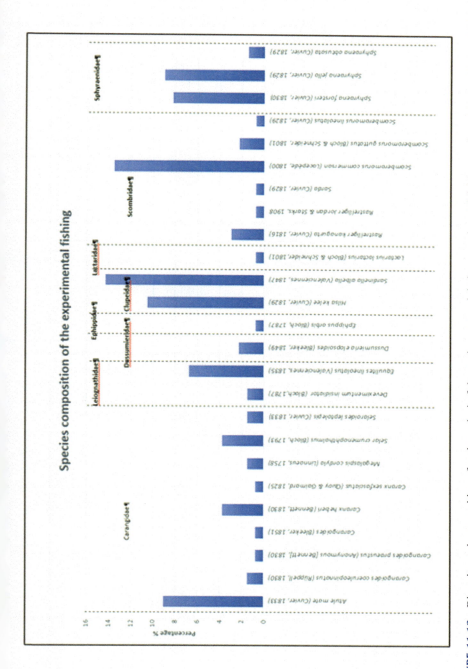

FIGURE 11.12 Diversity and composition of major pelagic fish species captured during experimental fishing conducted by the Marine Biological Division of the National Aquatic Resources Research and Development Agency (NARA).

TABLE 11.3
Number of Fish Species Captured from the Western Coastal Waters of Sri Lanka in the Vicinity of Ship Accident having Different Feeding Habits

Feeding Habit	No. of Fish Species
Planktonic	13
Planktonic/Benthic	01
Pelagic	10
Pelagic/Benthopelagic	03
Benthic	46
Benthopelagic	10
Benthic/Benthopelagic	14
Benthic/Planktonic	01
No information	09
Total	107

species level. To monitor the quality of aquatic bodies, the use of biological indicators that take into account their ecological function is becoming a widespread method (Coates et al., 2007). Indicators based on fish communities are recognized as useful tools to assess anthropogenic impacts (Harrison and Whitfield, 2004; Harrison and Whitfield, 2006; Breine et al., 2007).

Previous studies on ichthyoplankton diversity and density in the same area at the same time period of the year can be used to predict and compare the level of impact due to sudden anthropogenic disasters like MV X-Press Pearl ship accident. However, lack of baseline data in previous years in the same location at the same time is one of the challenging factors in comparing the level of damage due to this particular incident. However, long-term monitoring of the ichthyoplankton density in this area is important to predict the recovery of the biota in the region and the spectrum of the recovery of species after the incident. Accordingly, a survey was conducted by NARA scientists to find out the abundance of fish egg and larval stages available around the areas affected by the MV X-Press Pearl ship accident, shortly after the accident. Due to the strong weather conditions prevailing due to monsoons, during the accident and several weeks after the accident, continuous monitoring of fish egg and larval distribution was not possible considering the facilities available. Therefore, all information on eggs and larvae of fish affected by the ship accident could not be collected.

Initial ichthyoplankton survey was carried out in 16 sites around the MV X-Press Pearl shipwreck on June 10, 2021 at sixteen (16) different locations (Table 11.5) by the scientists in the Marine Biology Research Division of the National Aquatic Resources and Research Agency (NARA). Sampling for fish eggs and larvae was carried out using a 56 cm diameter plankton net with 180 μm mesh size and samples were immediately preserved in 10% formalin. In the laboratory, the total number of ichthyoplankton (eggs and larvae) in each of the locations was counted. Numbers of fish eggs and larvae recorded from 16 locations within the area in the vicinity of the ship accident are given in Figure 11.15. Laboratory examinations of the fish eggs collected during the survey indicated the existence of eggs belonging to families Clupeidae and Engraulidae (Figure 11.16) and fish larvae (Figure 11.17). Large number of fish eggs and larvae are yet to be identified at species level. Compared

Impact on Coastal and Marine Fish and Fisheries

TABLE 11.4
Details of Fish Species Captured from the Vicinity of the MV X-Press Pearl Ship Accident and Their Food, Feeding Habits, and Importance Economic Importance to Fisheries Industry

Family	Species	Common Name	Feeding Habit	Feeding Habit Related to Habitat	Food Items	Importance for Fisheries
Acanthuridae	*Acanthurus* sp.	Surgeonfish	Omnivore	Benthic	Macroalgae/fish carcasses	O ***
Ariidae	*Arius jella*	Black-fin sea catfish	Carnivore	Benthic	Invertebrates	F **
Bothidae	*Grammatobothus polyophthalmus*	Three spot flounder	Carnivore	Benthic	No details	F *
Carangidae	*Trachinotus baillonii*	Small spotted dartfish	Carnivore	Pelagic	Small fish	F **
	Trachinotus blochii	Snubnose Pompano	Carnivore	Benthic	Invertebrates/mollusks and other hard-shelled invertebrates	F **
	Caranx ignobilis	Giant trevally	Carnivore	Benthic	Invertebrates (crabs, spiny lobsters)	F ***
	Selar crumenophthalmus	Bigeye scad	Carnivore	Benthic/Planktonic	Benthic invertebrates (small shrimp, forams); Zooplankton and fish larvae	F ***
	Caranx papuensis	Bassy trevally	Carnivore	Pelagic	Fish	F ***
	Atule mate	Yellowfin scad	Carnivore	Planktonic	Crustaceans, planktonic invertebrates, cephalopods	F **
	Carangoides coeruleopinnatus	Coastal trevally	Carnivore	Benthic	Organisms on sandy bottoms	F ***
	Carangoides ferdau	Blue trevally	Carnivore	Benthic/Benthoplagic	Mollusk, benthic crustaceans, occasionally on small fish	F **

(Continued)

TABLE 11.4 (Continued)
Details of Fish Species Captured from the Vicinity of the MV X-Press Pearl Ship Accident and Their Food, Feeding Habits, and Importance Economic Importance to Fisheries Industry

Family	Species	Common Name	Feeding Habit	Feeding Habit Related to Habitat	Food Items	Importance for Fisheries
	Megalaspis cordyla	Torpedo scad	Carnivore	Pelagic	Fish	F**
	Caranx heberi	Blacktip trevally	Carnivore	Pelagic/Benthopelagic	Fish and Crustaceans	F***
	Selaroides leptolepis	Yellow-striped scad	Carnivore	Pelagic/Benthopelagic	Ostracods, gastropods, euphausiids and small fish	F**
	Caranx sexfasciatus	Bigeye trevally	Carnivore	Pelagic/Benthopelagic	Variety of fishes, cephalopods and crustaceans	F***
Clupeidae	Opisthopterus tardoore	Tardoore	Carnivore	Planktonic	Mysids, Pseudodiaptomus, copepod eggs, prawns, small crustaceans, bivalve eggs and larvae, amphipods, small fishes	F**
	Sardinella fimbricata	Frigescale sardinella	Carnivore	Planktonic	No details	F***
	Sardinella albella	White sardinella	Carnivore	Planktonic	Phytoplankton and Zooplankton	F***
	Hilsa kelee	Kelee shad	Carnivore	Planktonic	Phytoplankton and Zooplankton	F***
Cynoglossidae	Cynoglossus semifasciatus	Bengal tonguesole	Carnivore	Benthic	Benthos	F*
	Cynoglossus bilineatus	Four-lined tonguesole	Carnivore	Benthic	Benthic invertebrates	F*
	Cynoglossus kopsii	Short-headed tonguesole	Carnivore	Benthic	Benthos	F*
Dasyatidae	Neotrygon kuhlii	Blue-spotted sting ray	Carnivore	Benthic	Shrimps and Crabs	F***

Impact on Coastal and Marine Fish and Fisheries 235

Family	Species	Common name	Feeding	Habitat	Food items	
Diodontidae	*Diodon hystrix*	Spotfin porcupine fish	Carnivore	Benthic	Hard-shelled invertebrates like sea urchins, gastropods, and hermit crabs	O **
Drepaneidae	*Drepane punctate*	Spotted sickle fish	Carnivore	Benthic/Benthopelagic	Invertebrates and benthic fish	F **
Dussumieriidae	*Dussumieria elopsoides*	Rainbow sardine	Carnivore	Planktonic	Zooplankton	F **
Engraulidae	*Thryssa malabarica*	Malabar thryssa	Carnivore	Benthic	No details	F *
	Thryssa vitrirostris	Orange mouth anchovy	Carnivore	Planktonic	No details	F *
Ephippidae	*Ephippus orbis*	Orbfish	Carnivore	Benthic/Benthopelagic	Benthic invertebrates and fishes	F **
Gerreidae	*Gerres filamentosus*	Whipfin silver-biddy	Carnivore	Benthic	Small crustaceans, polychaetes and forams on sand or muddy sand bottoms	F **
Haemulidae	*Pomadasys guoraca*	Silver grunter	Carnivore	Benthic	No details	F **
	Pomadasys maculata	Spotted grunter	Carnivore	Benthic/Benthopelagic	Crustaceans and fish	F **
	Pomadasys argenteus	Silver grunt	Carnivore	Benthic	Zoobenthos	F **
	Pomadasys furcatus	Banded grunter	Carnivore	Benthic	No details	F **
	Plectorhinchus sp.	Sweetlip	Carnivore	Benthic	Zoobenthos	F
	Pomadasys kaakan	Javelin grunter	Carnivore	Benthic	Crustaceans and fish	F ***
	Otolithus sp.	Jewfish	Carnivore	Benthic/Benthopelagic	Fish, prawns and other invertebrates	F ***
Holocentridae	*Sargocentron sp.*	Squirrel fish	Carnivore	Benthic	Polychaetes, small clams, solenogastrid, isopods, brachyuran crabs, shrimps and other crustaceans	F **
						O ***
Lactariidae	*Lactarius lactarius*	False trevally	Carnivore	Benthic	Sand-dwelling animals	F ***

(Continued)

TABLE 11.4 (Continued)
Details of Fish Species Captured from the Vicinity of the MV X-Press Pearl Ship Accident and Their Food, Feeding Habits, and Importance Economic Importance to Fisheries Industry

Family	Species	Common Name	Feeding Habit	Feeding Habit Related to Habitat	Food Items	Importance for Fisheries
Latidae	*Psammoperca waigiensis*	Waigieu Seaperch	Carnivore	Benthic/Benthopelagic	Fish and Crustaceans	F ***
Leiognathidae	*Leiognathus equulus*	Common Ponyfish	Carnivore	Benthic/Benthopelagic	Polychaetes, small crustaceans, small fishes and worms	F **
	Gazza achlamys	Small-toothed Ponyfish	Carnivore	Benthic/Benthopelagic	Mainly small fish, crustaceans and polychaetes	F **
	Eubleekeria splendens	Splendid Ponyfish	Carnivore	Benthic/Benthopelagic	Fish, crustaceans, foraminiferans, and bivalves	F **
	Karalla dussumieri	Dussumier's ponyfish	Carnivore	Benthic	Small crustaceans, polychaetes, bivalves, foraminiferans, gastropods and nematodes	F **
	Equulites lineolatus	Ornate Ponyfish	Carnivore	Benthic/Benthopelagic	Fish, crustaceans, chaetognaths, nematodes, bivalves, and gastropods	F **
	Deveximentum insidiator	Pugnose Ponyfish	Carnivore	Planktonic	Zooplankton including copepods, mysids, and fish larvae and crustaceans	F, **
Lethrinidae	*Gymnocranius elongatus*	Forktail large-eye bream	Carnivore	Benthic	Bottom living invertebrates	F **
	Lethrinus lentjan	Pink ear emperor	Carnivore	Benthic/Benthopelagic	Crustaceans and mollusks, echinoderms, polychaetes and fishes	F ***
	Lethrinus obsoletus	Orange-striped Emperor	Carnivore	Benthic	Mollusks, crustaceans, and echinoderms	F ***

Impact on Coastal and Marine Fish and Fisheries

Family	Species	Common name	Feeding	Habitat	Food	
Lutjanidae	*Lutjanus argentimaculatus*	Mangrove Red snapper	Carnivore	Benthic/Benthopelagic	Fish and crustacean	F***
	Lutjanus fulviflamma	Dory Snapper	Carnivore	Benthic/Benthopelagic	Fish, shrimps, crabs and other crustaceans	O** F*** O**
	Lutjanus russellii	Russel's snapper	Carnivore	Benthic/Benthopelagic	Benthic invertebrates and fish	F** O**
Menidae	*Mene maculata*	Moon fish	Carnivore	Benthic	Benthic invertebrates	F**
Monodactylidae	*Monodactylus falciformis*	Full Moony	Carnivore	Planktonic/Benthic	Zooplankton, nekton, zoobenthos	F* O**
Mugilidae	*Crenimugil seheli*	Bluespot Mullet	Carnivore	Benthic/Planktonic	Microalgae, filamentous algae, forams, diatoms, and detritus associated with sand and mud	F***
	Planiliza melinoptera	Greenback Mullet	Omnivore	Benthic	Plant matter, Zoobenthos	F***
Mullidae	*Parupeneus indicus*	Indian goatfish	Carnivore	Benthic	Feed on benthic invertebrates; the diet including small crabs, amphipods, shrimps, small octopuses, polychaete worms, and small fishes	F** O**
Nemipteridae	*Scolopsis bimaculata*	Thumbprint monocle bream	Carnivore	Benthic	Crustaceans, mollusks, echinoderms and fishes	F**
	Scolopsis vosmeri	White cheek monocle bream	Carnivore	Benthic	Benthos	F** O**
	Nemipterus peronii	Threadfin Bream	Carnivore	Benthic	Fish, crustaceans, mollusks and polychaetes	F***
Ostraciidae	*Lactophrys sp.*	Box fish	Carnivore	Benthic	Mostly benthic animals	O**
	Tetrosomus sp.	Turretfish	Carnivore	Benthic	Benthic animals	O**
Platycephalidae	*Grammoplites scaber*	Rough flathead	Carnivore	Benthic	Small Crustaceans	F*
Psettodidae	*Psettodes erumei*	Indian halibut	Carnivore	Benthic	Fish	F**

(Continued)

TABLE 11.4 (Continued)
Details of Fish Species Captured from the Vicinity of the MV X-Press Pearl Ship Accident and Their Food, Feeding Habits, and Importance Economic Importance to Fisheries Industry

Family	Species	Common Name	Feeding Habit	Feeding Habit Related to Habitat	Food Items	Importance for Fisheries
Polynemidae	*Polydactylus sexfilis*	Six-finger threadfin	Carnivore	Benthic	Crustaceans (shrimps and crabs), polychaete worms, other benthic invertebrates; also on teleosts	F *
	Polydactylus plebeius	Striped threadfish	Carnivore	Benthic	Small crustaceans, fishes and other Benthic organisms	F *
	Filimanus similis	Indian Sevenfinger threadfish	No information	No information	No information	F *
Portunidae	*Portunus sanguinolentus*	Threespot Swimming crab	Carnivore	Benthic	Sessile and slow-moving benthic macroinvertebrates	F **
	Portunus pelagicus	Blue swimming crab	Carnivore	Benthic	Bivalves, fish and, to a lesser extent, macroalgae	F ***
Priacanthidae	*Heteropriacanthus cruentatus*	Glass Eye	Carnivore	Benthic	Octopi, pelagic shrimp, stomatopods, crabs, small fish, and polychaetes	F * O *
Pristigasteridae	*Ilisha striatula*	Banded ilisha	Carnivore	Planktonic	Plankton	F *
	Ilisha melastoma	Indian ilisha	Carnivore	Planktonic	Plankton (probably small crustaceans, etc.)	F *
	Opisthopterus tardoore	Tardoore	Carnivore	Pelagic	Mysids, Pseudodiaptomus and copepod eggs, also prawns and other small crustaceans, bivalve eggs and larvae, amphipods and small fishes	F *
	Ilisha striatula		No information	No information	No information	

Impact on Coastal and Marine Fish and Fisheries

Family	Species	Common name	Diet type	Habitat	Food items	Impact
Scaridae	*Scarus ghobban*	Blurbarred Parrotfish	Omnivore	Benthic	Scrape algae from rocks and corals sand facultative scavenger	F*** O**
Scatophagidae	*Scatophagus argus*	Spotted scat	Omnivore	Benthic	Worms, crustaceans, insects and plant matter	F**
Sciaenidae	*Argyrosomus* sp.	Drum	Carnivore	Planktonic	Plankton and Nekton	F***
	Chrysochir aureus	Reeve's croaker	Carnivore	Benthic	Small crustaceans	F***
	Daysciaena albida	Bengal corvina	Carnivore	Benthopelagic	Prawns, teleosts, juvenile crabs, amphipods and isopods	F***
	Johnius amblycephalus	Bearded croaker	No information	No information	No information available	F**
	Johnius carouna	Caroun croaker	No information	No information	No information available	F**
	Otolithes ruber	Tiger tooth croaker	Carnivore	Benthopelagic	Fish, prawns and other invertebrates	F***
	Otolithes cuvieri	Lesser tiger tooth croaker	Carnivore	Benthopelagic	*Acetes* spp., penaeid prawns, deep-sea prawns, fishes, stomatopods, mollusks, isopods, copepods and fish larvae	F***
	Kathala axillaris	Kathala croaker	Carnivore	No information	No information available	F**
	Nibea maculata	Blotched croaker	Carnivore	No information	No information available	F**
Scombridae	*Auxis thazard*	Frigate tuna	Carnivore	Pelagic	Small fish, squids, planktonic crustaceans (megalops), and stomatopod larvae	F***
	Sarda sp.	Bonito	Carnivore	Pelagic	No information available	F**
	Rastrelliger kanagurta	Indian Mackerel	Carnivore	Planktonic	Macroplankton such as larval shrimps and fish	F***
	Scomberomorus commerson	Narrow-barred Spanish mackerel	Carnivore	Benthopelagic	Primarily small fish such as anchovies, clupeids, carangids, also squids and penaeid shrimps	F***

(Continued)

TABLE 11.4 (Continued)
Details of Fish Species Captured from the Vicinity of the MV X-Press Pearl Ship Accident and Their Food, Feeding Habits, and Importance Economic Importance to Fisheries Industry

Family	Species	Common Name	Feeding Habit	Feeding Habit Related to Habitat	Food Items	Importance for Fisheries
	Scomberomorus guttatus	Indo-Pacific king mackerel	Carnivore	Pelagic	Small schooling fishes (especially sardines and anchovies), squids and crustaceans	F ***
Scorpaenidae	*Pterois volitans*	Red Lionfish	Carnivore	Benthopelagic	Carnivore (Hunt small fishes, shrimps, and crabs at night, using its widespread pectorals trapping prey into a corner, stunning it and then swallowing it in one sweep)	O ***
Serranidae	*Cephalopholis sonnerati*	Tomato hind	Carnivore	Benthopelagic	Demersal, carnivore (feed on small fishes and crustaceans including shrimps, crabs and stomatopods)	F ***
	Epinephelus longispinis	Long spine grouper	Carnivore	Benthopelagic	Demersal, carnivore (Feed mainly on crustaceans, especially crabs and stomatopods, rarely on small fishes, squids, and pelecypod flesh)	O **
	Epinephelus faveatus	Barred chest grouper	Carnivore	Benthic	Bottom-dwelling organisms	F ***
Siganidae	*Siganus canaliculatus*	White-spotted spinefoot	Herbivore	Benthic	Benthic algae and to some extent on seagrass	F **
Sillaginidae	*Sillago* sp.	Silago	Carnivore	Benthic	Bottom swelling animals	F ***
	Sillago lutea	Mud silago	Carnivore	Benthic	Small crustaceans, polychaetes and bivalves	F ***

Impact on Coastal and Marine Fish and Fisheries

	Sillago sihama	Silver silago	Carnivore	Benthic	Mainly polychaete worms, small prawns (Penaeus), shrimps and amphipods	F ***
Sparidae	*Acanthopagrus berda*	Gold silk seabream	Carnivore	Benthopelagic	Invertebrates, including worms, mollusks, crustaceans and echinoderms, and small fish	F ***
Sphyraenidae	*Sphyraena obtusata*	Obtuse Barracuda	Carnivore	Pelagic	Mainly fish	F ***
	Sphyraena jello	Pink handle Barracuda	Carnivore	Pelagic	Mainly fish and also squid	F ***
	Sphyraena forsteri	Bigeye Barracuda	Carnivore	Benthopelagic	Fish, penaeid shrimps and squids	F ***
Tromateidae	*Pampus chinensis*	Chinese silver pomfret	Carnivore	Planktonic/Benthic	Ctenophores, salps, medusae, and other zooplankton groups and also prey on small benthic animals	F ***
Synodontidae	*Trachinocephalus myops*	Snake fish	Carnivore	No information	No information available	F *
Terapontidae	*Terapon therps*	Largescaled terapon	Omnivore	No information	Mainly Ascidians	F *
	Terapon puta	Smallscaled terapon	Carnivore	Pelagic	Fish and invertebrates	F *
	Terapon jarbua	Jarbua terapon	Omnivore	Benthopelagic	Fish, insects, algae, and sand-dwelling invertebrates	F *

(F – Food fish; O – Ornamental fish; *** – High economic importance; ** – Moderate economic importance; * – Low economic importance) (NB-Information included herein are gathered from Fish Base (2022) and Munro (1955)).

Maritime Accidents and Environmental Pollution

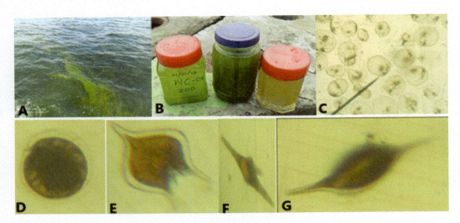

FIGURE 11.13 a) Algal bloom containing dinoflagellates observed near the shipwreck; b) Samples of the algal bloom collected for laboratory analyses. (Photographs provided by Dr. Nilanthi Proiyadharshani of Oceanography Division, NARA) Dinoflagellate species observed in the sample: c) and d) *Noctiluca* sp.; e) *Protoperidinium* sp.; f) *Ceratium* sp.; and g) *Dinophysis* sp.

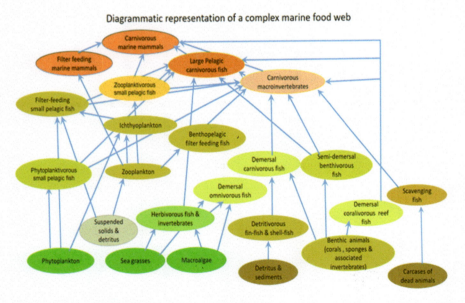

FIGURE 11.14 A complex marine food web that involves fish and marine mammals, which indicates the possible paths for bioaccumulation of pollutants along food chains that may exist in the coastal marine environment affected by the MV-X-Press Pearl ship accident. Prepared by P.R.T. Cumaranatunga using the information in Table 3.

TABLE 11.5
Map Showing the Locations of 16 Sites Selected for Sampling of Fish Eggs and Larvae and Numbers of Fish Eggs and Larvae Found in Each Location

Sampling Site	Latitude	Longitude	No. of Fish Eggs/m^3	No. of Fish Larvae/100 m^3
1	7.076239	79.77284	65	1.22
2	7.080522	79.7674	81	0.407
3	7.085044	79.77092	41	2.033
4	7.081594	79.77626	24	0.407
5	7.036522	79.78099	37	2.033
6	7.080108	79.72668	45	0.813
7	7.125769	79.76378	75	1.22
8	7.169364	79.75548	106	0
9	7.078497	79.68149	74	1.626
10	6.991758	79.78993	92	0.43
11	6.996	79.834	71	0
12	7.035	79.704	101	1.42
13	7.155	79.795	62	1.22
14	7.306	79.763	20	0
15	7.278	79.697	357	2.033
16	7.23143	79.7973	195	0

Map Showing the Sampling Sites

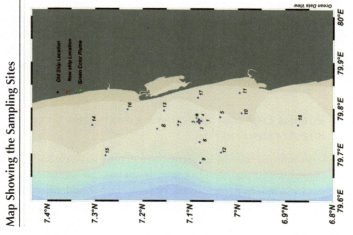

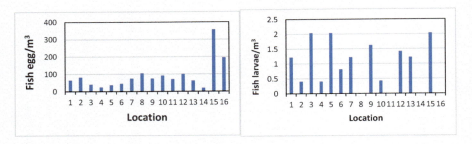

FIGURE 11.15 Number of fish eggs/m^3 (left) and fish larvae/100 m^3 in each of the 16 locations on 10/06/2021.

to the large number of fish eggs which were recorded during this period, number of larvae recorded was very low (Table 11.5, Figure 11.15).

During future years a reduction in juvenile anchovy and other fish species, whose eggs and larvae were available in the area affected by ship disaster, could result. In order to evaluate the total damage to stocks of different species of fish, continuous monitoring of diversity and density of fish eggs, larvae, and adults is essential. In order to identify the species of fish whose eggs and larvae were affected by the ship disaster, monitoring of diversity and density of fish eggs and larvae (ichthyoplankton) during the same months of future years should be carried out. To identify the eggs and larvae at species level they should be subjected to molecular biological investigations.

In addition to monitoring the diversity of fish species it is important to carry out histopathological investigations and molecular biological investigations to identify the pathological and mutagenic impacts on fish caused by the persistent pollutants released from the ship cargo during and after the accident and during and after the removal of the shipwreck and other debris scattered in the ship disaster–affected coastal and marine environment. More than 75% of the eggs found during June 2021 in the area affected by the accident belong to different species of anchovy. Therefore, possible negative impacts on anchovy species cannot be ruled out. Similarly species of fish whose eggs, larvae, and postspawning (spent) populations of fish which were passively drifting (involved in denatant migration) in the area affected by the ship disaster should be monitored and damages to their stocks should be accurately evaluated to understand the total damage caused to fish species inhabiting and visit the territorial waters off the Western coast of Sri Lanka.

11.5.4 Possible Effects of Different Incidents That Occurred During Ship Accident on Fish

Ship disasters included several incidents, which may have caused different effects on marine ecosystems and their biodiversity. Incidents that occurred and their possible impacts on marine fish fauna are listed in Table 11.6. Incidents connected to MV X-Press disaster include base leakage; fire; release of noxious gasses to the atmosphere;

Impact on Coastal and Marine Fish and Fisheries 245

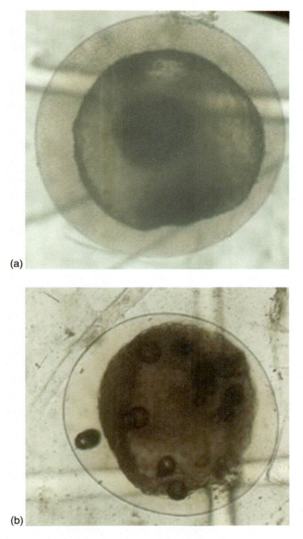

FIGURE 11.16 Eggs of fish belong to families Clupeidae (a, b, c) and Engraulidae (d, e).

explosions; falling of containers and debris into marine waters; and release of thousands of tons of plastic nurdles (unburned and burned), lithium batteries, food items, fertilizer, hazardous chemicals, fuel and crude oil, and other unidentified cargo, which can occur and be distributed in marine waters in different forms (dissolved, floating, suspended and deposited among sediments), causing possible acute and chronic impacts on fish and their sensitive coastal and marine ecosystems, depending on how they come in contact with fish and enter their biological systems (e.g. bioaccumulation along food chains).

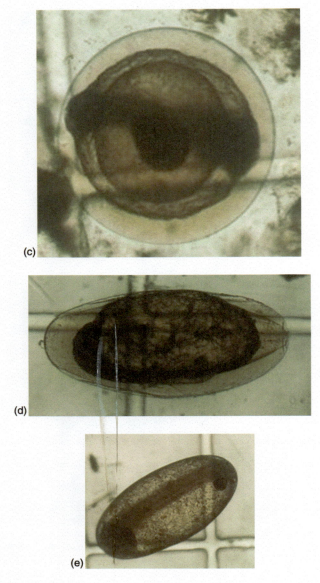

FIGURE 11.16 (Continued)

11.5.5 Observations Made on Carcasses of Fish Encountered on Beaches Immediately After the Accident

Immediate impacts from plastic nurdles on fish were studied through closer investigations of fish carcasses found on the beaches during the accident. Medium-term and long-term impacts on fish, fisheries, and fish consumers are yet to be revealed through careful monitoring and scientific investigations.

FIGURE 11.17 Fish larvae collected from the accident-affected site.

Out of the 39 specimens analyzed, death of 2 fish was concluded to be due to ingestion of plastic pellets/nurdles or due to obstruction to digestive tract or to respiratory process or damages caused by the plastic pellets/nurdles. *Acanthopagrus berda* (Forsskål, 1775) of family Sparidae (Plate 8-A), found on the Kapuhengoda beach (Western Province) on May 28, 2021, and *Caranx ignobilis* (Forsskål, 1775) of family Carangidae (Plate 9-A), found on the beach of Wellawatte (Western Coast of Sri Lanka) (Figure 11.21) on May 31, 2021, were the two fish specimens which indicated a possible death due to plastic pellets/nurdles ingestion. All postmortem information collected from those two fish is given in Table 11.8 and Figures 11.20 and 11.21. The plastic pellets found in both fishes were less than 5 mm in diameter and were in unburned condition. Those plastic pellets found inside both fish species were similar in size and shape. Further, those pellets are superficially similar to those recovered from the bags of plastic pellets that fell from the containers in the cargo, which got damaged during fire and explosions.

Plastic contaminants in the marine environment can cause adverse impacts on marine fish fauna (Mascaró, 2020; Savoca, et al., 2021). Several studies have proven that plastic ingestion by marine organisms commonly occurs due to misidentification of plastic pellets as natural food items (Miranda, et al., 2016). Furthermore, the behavior and feeding strategies of the fish also influence the rate and amount of plastic ingestion in the marine environment and it is reported that active predators and benthic foraging species most commonly ingest plastic during their feeding and respiratory processes (Savoca, et al., 2021).

A. berda is a demersal species that feeds on worms, mollusks, small fish, and plant material (Fishbase, 2022b). The specimen collected from Kepungoda Beach was in moribund state. Death of *A. berda* could have been due to the obstruction of digestive tract by plastic pellets swallowed. Those plastic/nurdles may have been those which were settled at the bottom of the sea or reef environment, and they may have entered the mouth while feeding on benthic organisms such as shrimps and sea urchins (found inside the stomach and rest of the gut) or due to unselectively swallowing floating plastic nurdles while taking water through the mouth during the respiratory process. Presence of plastic pellets among the gills provides evidence for the latter. Absence of plastic pellets inside the stomach and rest of the gut indicates that plastic pellets have not entered the stomach, but they may have blocked the esophageal opening. Absence of plastic pellets and presence of food items inside the stomach and rest of the gut indicate that *A. berda* has ingested plastic pellets during feeding or during the respiratory process while swimming in the water column after feeding. Therefore, it is possible to conclude that obstruction of respiratory process by plastic pellets that

TABLE 11.6
Possible Effects of Different Incidents that Occurred During and after the Outbreak of Fire in the MV X-Press Pearl Ship on Marine Fish Fauna

Incidents That Occurred During and after the Ship Accident	Possible Impact on Fish and Fisheries
Fire and increase in temperature in surface waters in the vicinity	• Even a slight increase in water temperature can be lethal to fish eggs, planktonic larvae (ichthyoplankton) that are passively drifting with the strong surface monsoonal currents, and also adult fish that are exhausted after spawning (too weak to actively swim) and drifting with the strong currents passing through the area of the accident. • Pelagic fish which are swimming against the monsoonal currents toward coastal areas for spawning (breeding) will most probably avoid the location of the ship accident if they sense the danger from a considerable distance and will have to follow a longer route which wastes more energy and causes a delay in reaching the coastal spawning grounds, which may indirectly affect the chance of meeting individuals of opposite sex for courtship behavior and also reduce the quality of eggs produced, hatching rate of eggs, and survival of fish larvae.
Thick smoke containing and soot/ash	• Smoke containing noxious gasses and soot that may fall back into the sea through dissolution in rainwater, and if they contain any toxic substances, planktonic stages of fish (ichthyoplankton), small and large pelagic fish which move in schools (shoals) at the upper layers of water will be affected through contact poisoning or due to uptake of dissolved toxic substances or suspended soot/ash particles through the respiratory and feeding water currents.
Explosions	• Explosions can damage the air bladder of pelagic fish which are moving in shoals in the same manner when explosives are used for illegal fishing activities. Such explosions can cause mass mortalities due to heavy internal bleeding. Observations made on fish killed by blast fishing are shown in Figure 11.18, and if fish were affected by several explosions that occurred in the ship under fire, they also may have been subjected to a similar situation. Explosives can damage the air bladder of fish which is used as a hydrostatic organ to maintain buoyancy when fish are moving at different depths of a water body. To maintain air pressure within the air bladder there is a good blood supply to secrete gasses into the air bladder and to absorb gasses from the air bladder. Damage to air bladder during explosions cannot maintain buoyancy, and they will die also due to internal bleeding and carcasses will sink to the bottom of the sea. Fish of all sizes will be killed due to explosions and all carcasses will disappear from the site; while sinking they will be scavenged and subjected to autolysis due to release of lytic enzymes from damaged tissues.

Impact on Coastal and Marine Fish and Fisheries

TABLE 11.6 (Continued)
Possible Effects of Different Incidents that Occurred During and after the Outbreak of Fire in the MV X-Press Pearl Ship on Marine Fish Fauna

Incidents That Occurred During and after the Ship Accident	Possible Impact on Fish and Fisheries
	• Anterior projections of the air bladder of fish are also connected to the wall of the auditory capsule, and hence the damage caused to air bladder will affect the sound reception and sound generation in certain species of fish which communicate using sound. During explosions communication of fish which are moving in schools (shoals), following the signals given by the leader of the shoal, will also be negatively affected.
Release of nitric acid	• Although concentrated nitric acid can get diluted due to large masses of water moving with monsoonal currents, if concentrated nitric acid directly falls on fish eggs or delicate ichthyoplankton or on a shoal of fish, which are passively drifting with the monsoonal currents, their soft body tissues will be damaged or totally dissolved, without leaving any trace. Such an event will cause mass mortalities of fish and ichthyoplankton which will result in a depletion of marine fish populations.
Release of caustic soda	• Soft tissues and bones of fish can completely be dissolved in strong solutions of caustic soda. Fish eggs, fish larvae, and adult fish which drift with the currents passing the area of the ship accident can be totally dissolved or partially damaged due to dissolution of a large tonnage of caustic soda which entered the marine waters.
Virgin Plastic pellets	• Large predatory and/or filter-feeding fish may misidentify the floating and submerged plastic pellets as fish eggs and feed on them, causing clogging of gills and the pharyngeal cavity, disturbing the respiratory system, and blocking the alimentary canal.
	• If plastic pellets are ingested, fish which are having different structural adaptations within the alimentary canal (Pharyngeal teeth and strong musculature) to grind hard food particles will grind the plastic pellets into smaller micro- and nanoparticles, which may later be subjected to chemical digestion and absorption causing medium-term and long-term detrimental effects to fish who ingest plastic pellets and to marine organisms that feed on affected fish.
	• During grinding of plastic pellets internal tissues of the alimentary canal may get damaged, causing internal bleeding.
Burned plastic	• Large volumes of burned plastic were found floating or suspended within the water column or deposited among sediments. They may mechanically break into micro- and nanoparticles due to strong current and wave action and may stay floating or submerged and may get carried to long distances with currents. Filter feeding or

(Continued)

TABLE 11.6 (Continued)
Possible Effects of Different Incidents that Occurred During and after the Outbreak of Fire in the MV X-Press Pearl Ship on Marine Fish Fauna

Incidents That Occurred During and after the Ship Accident	Possible Impact on Fish and Fisheries
	other foraging pelagic fish may ingest them during filter feeding within the water column and/or during the intake of respiratory water currents through the mouth. Those which are deposited among sediments will be grazed by demersal fish.
	• These suspended particles of plastic may clog and damage the gills or will be chemically and mechanically digested and absorbed while passing through the alimentary canal. Possible effects of different types of plastic and their derivatives on fish are provided in Table 11.7.
Fertilizer, food items, fresh fish, and other materials containing nutrients released from cargo	• These items release nutrients that can cause eutrophication and hence result in algal blooms. Among these blooms there can be toxic algae (Figure 11.7) that would cause harmful effects on fish which are filter-feeding on them. There can be lethal toxins that may cause mass mortality in fish or they can accumulate within large predatory fish along food chains and food webs. If these toxins are accumulated in economically important food fish species, their consumption can cause harmful effects to humans.
Any type of fuel oil	• Oil will form a slick at the sea surface, preventing mixing of atmospheric air which will reduce the availability of oxygen for respiration of fish and other marine organisms. Furthermore, due to rough sea conditions oil will make small globules that will sink (Figure 11.19). Oil in the slick and sinking oil droplets will be taken up by fish with respiratory water flow and also by filter feeders during feeding, resulting in detrimental effects on gas exchange through gills, which may ultimately result in death.

blocked the opening to the pharyngeal cavity and/or the gills may have been one of the reasons, among many other environmental changes that occurred during the ship accident, which caused the death of *A. berda*. Structure of teeth on the upper and lower jaws of *A. berda* indicates that it is capable of crushing hard food items (Plate 1C). It is known to possess six peg-like curved incisors in front of both upper and lower jaws and rounded molars: 4–5 series in upper jaw and 3–4 series in lower jaw (Munro, 1955). *A. berda* is known to feed mostly on barnacles, crabs, *Modiolus* spp., oysters, shrimps, etc. (Shilta *et al.,* 2018), which are benthic organisms. In the gut contents of *A. berda* found on the Kepungoda beach, stomach contained shrimps and shell parts of sea urchins, and it provides evidence for this fish to be feeding on benthic organisms prior to ingestion of plastic pellets. Therefore, it is possible to conclude that above fish could have possibly ingested plastic pellets settled at the bottom

Impact on Coastal and Marine Fish and Fisheries

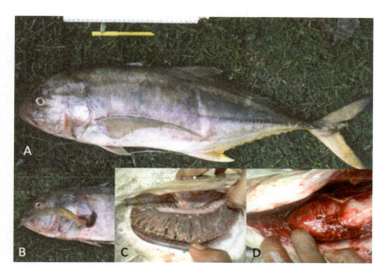

FIGURE 11.18 Damages caused to giant trevally due to blast fishing using hand grenades. a) Red-colored patches on the body surface, b) external bleeding under the pectoral fin, c) pale-colored gills due to loss of blood as a result of internal bleeding, and d) internal bleeding due to the rupture of the air bladder as a result of blast fishing.

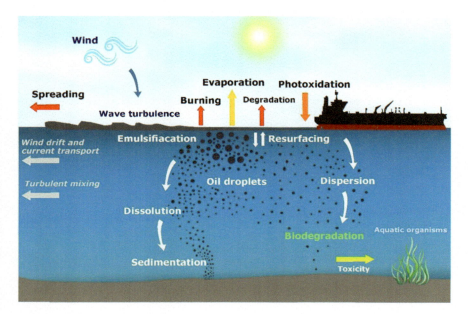

FIGURE 11.19 Main transport and weathering processes during an oil spill (Keramea et al., 2021). Oil may come in contact with body surfaces of fish and also during emulsification, dispersion, and resurfacing, fish can ingest the oil droplets with respiratory water currents, and they may deposit on gills, affecting gas exchange.

TABLE 11.7
Pollutants Released from the Ship and Possible Effects on Fish

Type of Pollutant	Possible Effects on Fish	Possible Uptake of the Pollutant with Respect to Feeding Habit	Human Health Hazard due to Consumption of Contaminated Fish
\multicolumn{4}{l}{Heavy Metals}
• Lead (Pb)	Fish can normally accumulate heavy metals from water, food, and sediment (Sarah et al., 2019). Pb accumulation in fish tissues could cause oxidative stress due to excessive reactive oxygen species (ROS) production and induce synaptic damage and neurotransmitter malfunction in fish as neurotoxicity. Exposure to Pb influences immune responses in fish as an immune-toxicant (Lee et al., 2019).	Fish positioned at the top of the aquatic food chain accumulate pollutants from the consumption of other aquatic organisms and, along with essential metals, toxic metals are also accumulated in fish tissues (Saleem et al., 2022). Fish are the most susceptible to the toxic effects of Pb exposure. Pb-induced toxicity in fish is primarily induced by bioaccumulation in specific tissues, and the accumulation mechanisms vary depending on water habitat (freshwater or seawater) and pathway (waterborne or dietary exposure) (Lee et al., 2019).	Lead (Pb) is a potent environmental pollutant and toxic to human health. It can accumulate in the muscles, bones, blood, and fats (Rossi and Jamet, 2008). Can cause severe damage to liver, kidney, brain, nervous, and reproductive disorders; heart diseases like high blood pressure especially in men and anemia. Extensive exposure to lead can cause memory problems, behavioral disorders, and mental retardation. Lesser levels of Pb damage the nerve and brain in fetuses and young children, resulting in lowered IQ and learning deficits (Sehar et al., 2014). Can cause renal failure, liver damage, cardiovascular diseases. Neurotoxic, carcinogenic effects, and reduced immune systems (Oguguah et al., 2017; Rahman et al., 2017).

• Copper (Cu)	In fish and other vertebrates, although Cu is the key constituent of many metabolic enzymes and glycoprotein, at high concentrations, it causes toxic effects (Richard Bull, 2000). Chronic toxicity of Cu in fish causes poor growth, shortening of life span, decreased immune response, and fertility problems (Yacoub and Gad, 2012). Cu is neurotoxic to fish and interferes with the function of olfactory neurons (Mcintyre et al., 2008). Depending on the fish species Cu may cause induction in apoptosis (Monteiro et al. 2009), biochemical and morphological changes in the liver tissue (Varanka et al., 2001), impaired complex fish behaviors such as social interaction, avoidance of predators, and reproductive behavior (Garari et al., 2021), lateral line dysfunction (Johnson et al., 2007), and high rate of body deformities and mortality (Kong et al., 2013). Bioaccumulation of this trace element influenced oxidative metabolism, lipid peroxidation, and protein content in fish (e.g. carp) tissue (Radi and Matkovics., 1988).	Fish uptake copper mainly through the dietary route or ambient exposure (Dang et al. 2009).	Trace metals in fish may be transferred and pose serious health problems to fish consumers (Esilaba et al., 2020). Cu levels above the recommended limit may lead to brain, liver, and kidney disorders. Short-term exposure to high doses of Cu can cause diarrhea, stomach pains, vomiting, and death due to liver and kidney failure or due to the depression of the central nervous system (Dorsey et al., 2014).
• Aluminum (Al)	Al may be associated with gill damage in fish, due to its deposition and changes in osmoregulation, as well as with oxidative stress in lymphocytes (Galar-Martinez et al., 2010; Garcia-Medina et al., 2010). It can cause direct damage to mitochondrion and affect electron transport in the respiratory chain, increasing Lipid peroxidation (LPO) and, subsequently, Reactive Oxygen species	Al enters fish through surface water/sediment uptake through gills or integument and through ingestion of aluminum-accumulated aquatic plants (www.epa.gov/sites/product ion/files/2018-12/documents/alumi num-final-national-recommended-awqc.pdf ↩).	Al accumulation in the brain has been suggested to be involved in the development of neurodegenerative disorders, amyotrophic lateral sclerosis, and dialysis encephalopathy (Flora et al., 2003; Bondy 2010). Al can evoke oxidative stress through stimulation

(*Continued*) |

TABLE 11.7 (Continued)
Pollutants Released from the Ship and Possible Effects on Fish

Type of Pollutant	Possible Effects on Fish	Possible Uptake of the Pollutant with Respect to Feeding Habit	Human Health Hazard due to Consumption of Contaminated Fish
	(ROS) production (Stohs and Bagchi, 1995; Bondy and Cambell, 2001; Fernandez-Davila et al., 2012). Al accumulation causes injury to the gill epithelium, apoptosis, and necrosis of gill ion-transporting cells (Eeckhaoudt, 1994). It causes dysfunctioning of ion-regulation and osmoregulation (Witters et al., 1996).		of reactive oxygen species (ROS) production in cells (Li et al., 2006; Sinha et al., 2007). It can also induce Lipid peroxidation (LPO) and induce and disrupt activity of antioxidant enzymes including superoxide dismutase (SOD), catalase (CAT) and glutathione peroxidase (GPx) and facilitation of protein oxidation (Almroth et al. 2005; Parvez and Raisuddin, 2005; Vlahogianni et al., 2007).
• Lithium (Li)	Li has a long oceanic residence time (~1.2 million years) and it has a weak capacity to adsorb onto marine particles (Decarreau et al., 2012). Li is reported to be homogeneously distributed throughout the water column (Misra & Froelich, 2012). Li is toxic to aquatic organisms, affecting some of their metabolic functions. Li-enriched seawater disrupts embryogenesis in urchins, zebrafish, and amphibians (Hall, 1942; Kao et al., 1986; Kiyomoto et al., 2010; Stachel et al., 1993). Li might secondarily contribute to ambient toxicity by encouraging the activity of naturally occurring toxic organisms (Stewart et al., 2003).	Thibon et al. (2021) have found that highest Li concentrations are found in filter feeders and lowest in predatory fish and strong variations have been found among organs. Biochemical similarity has been observed between Na and Li during transport in the brain and in osmoregulatory organs. Relatively high Li concentrations have been observed in fish gills and kidneys (0.26 and 0.15 µg/g, respectively), and a large range of Li contents	It causes neurotoxicity and several other adverse effects (Diserens et al., 2021; Jacob et al., 2020; Verdoux et al., 2021). High concentrations of Li could cause severe damage to human nervous system (coarse tremor and hyperreflexia), kidney (sodium-losing nephritis and nephrotic syndrome), and endocrine system (hypothyroidism) (Davis et al., 2018; Kibirige et al., 2013).

Impact on Coastal and Marine Fish and Fisheries 255

	Adverse effects in animals exposed to Li are neurotoxicity (Viana et al., 2020; Oliveira et al., 2011), Hepatotoxicity (Pinto-Vidal et al., 2021), Nephrotoxicity (Jing et al., 2021) and Reproductive toxicity (Kszos et al., 2003) (up to 0.34 μg/g) in fish brains, while depleted levels have been observed in fish liver and muscles (0.07 ± 0.03 and 0.06 ± 0.08 μg/g, respectively). Li is uptaken by living organisms and accumulates in several species (Tkaitcheva et al., 2015; Viana et al., 2020). It is present in terrestrial and aquatic trophic webs (Aral and Vecchio-Sadus, 2008; Bolan et al., 2021; Thibon et al., 2021).	It inhibits the activity of cardiomyocytes and promotes cardiomyocyte apoptosis (Shen et al., 2020). The symptoms of lithium toxicity depend on the level of lithium in blood. Symptoms of mild to moderate lithium toxicity are diarrhea, vomiting, stomach pains, fatigue, tremors, uncontrollable movements, muscle weakness, drowsiness, and weakness. Severe lithium toxicity can occur when serum lithium levels exceed 2.0 mEq/L., it can show additional symptoms, including heightened reflexes, seizures, agitation, slurred speech, kidney failure, rapid heartbeat, hyperthermia, uncontrollable eye movements, low blood pressure, confusion, coma, delirium, and death (www.healthline.com/health/lithium-toxicity). Related Literature not available.
• Conc. Nitric Acid (HNO_3)	Fish could die at low pH (3.0–4.0) by exhibiting classic symptoms of acid toxicity (Swift & Mogan., 1983). Acid deposition has many harmful ecological effects when the pH of most aquatic systems falls below 6 and especially below 5. As the pH approaches 5, nondesirable species of plankton and mosses may begin to invade, and populations of fish such as smallmouth bass disappear (www.lenntech.com/aquatic/acids-alkalis.htm)	Related Literature not available.

(Continued)

TABLE 11.7 (Continued)
Pollutants Released from the Ship and Possible Effects on Fish

Type of Pollutant	Possible Effects on Fish	Possible Uptake of the Pollutant with Respect to Feeding Habit	Human Health Hazard due to Consumption of Contaminated Fish
spills cause mass mortality of fish. Fish eggs Caustic Soda (NaOH)	Extreme pH can kill adult fish and invertebrate life directly and can also damage developing juvenile fish. It will strip a fish of its slime coat, and high pH level 'chaps' the skin of fish because of its alkalinity (www.lenntech.com/aquatic/acids-alkalis.htm)		
Heavy fuel oil and its derivatives			
• Petroleum (Crude Oil)		Intake of oil can occur with inflow of respiratory and filter-feeding water, dermal Contact, and contact with bottom sediments and coastal sands polluted with crude oil.	Chronic exposure affects physiological functions such as Hematologic, hepatic, respiratory, and Renal and neurological functions.
• Petroleum hydrocarbon	Oil spills cause mass mortality of fish. Fish eggs and larvae are regularly powerless against poisonous oil mixes because of their little size, ineffectively created films, and detoxification frameworks, just as their situation in the water segment (Khan, 1990; Hjermann et al., 2007; Langangen et al., 2018). Oil mixes (for the most part, polycyclic aromatic hydrocarbons, PAHs) at low fixations can execute or cause subdeadly harm to eggs and larvae. Subdeadly impacts incorporate, for example, morphological disfigurements, decreased sustaining, and development rates and are probably going to build helplessness to predators and starvation. Species that feed vigorously on	Consumption of prey items with petroleum hydrocarbon. Ingestion of tainted soils, residue, and contaminated food (Alzahrani and Rajendran, 2019).	Can harm any organ system in the human body such as sensory system, respiratory system, circulatory system, immune system, regenerative system, tactile system, endocrine system, liver, kidney, and so on, and subsequently can cause a wide scope of ailments and disarranges. Long-term exposure to low levels of petroleum hydrocarbons may impair behavior and memory. A single exposure to a moderately high concentration of virtually any hydrocarbon

	sediment-related invertebrates will generally be in more serious danger of PAH exposure with respect to higher-order consumers (Alzahrani and Rajendran, 2019)	solvent vapor will cause a general depression of CNS which, at high doses, will lead to unconsciousness (Alzahrani and Rajendran, 2019)
Polycyclic Aromatic Hydrocarbons (PAHs)	Crude oil containing PAHs could cause toxic effects, such as immunotoxicity, embryonic abnormalities, and cardiotoxicity, for wildlife including fish, benthic organisms, and marine vertebrates (Barron, 2012; Romero et al., 2018; Snyder et al., 2015). In aquatic animals, such as fish, epizootic neoplasia may occur (Bunton, 1996). PAHs can affect bone metabolism, liver metabolism, and reproduction in fish. It can cause disruption of cardiac function, dysregulation of genes important in eye development and function, and morphological abnormalities of the eye (Magnuson, 2018). May possess a strong toxic effect on the endocrine system of vertebrates (Hayakawa et al., 2006).	Cause mutagenic and carcinogenic effects, developmental toxicity, genotoxicity, immunotoxicity, oxidative stress, and endocrine disruption (Bekki et al., 2009; Cherr et al., 2017; Hannam et al., 2010; Lee et al., 2011; MacDonald et al., 2013). Bioaccumulation can occur via water, sediments, and microplastics.

Plastic types and their derivatives due to burning and mechanical and chemical degradation

Microplastics (MPs)	MPs exposed to biota pose risk for bioaccumulation and biomagnification at the trophic level and cause multiple ecological repercussions (Barnes et al., 2009; Phuong et al., 2018). They could be attached and collected in the epidermis of marine animals, serve as bacterial transport channels, and absorb chemical compounds on their surface (Auta et al., 2017; Kumar et al., 2022; Roch et al., 2020; Rubin et al., 2021). MPs have been shown to accumulate in the gastrointestinal system of aquatic animals (Li et al., 2022; Ugwu et al., 2021). MPs increase	Ingestion or intake of contaminated water or food. Foods from fish farms and marine culture zones (Feng et al., 2019). Inhalation of contaminated air. The chemical composition of these particles may cause acute and chronic respiratory problems in the short and long term (Sangkham et al., 2022). Atmospheric MPs and NPs can be directly inhaled due to	MPs can cause oxidative stress, cytotoxicity, DNA damage, inflammation, immune response, neurotoxicity, and metabolic disruption, ultimately affecting digestive system, immunology, respiratory system, reproductive system, and nervous system. Recent study reported the first evidence of MPs (polypropylene) in the human placenta (Ragusa et al., 2021).

(*Continued*)

TABLE 11.7 (Continued)
Pollutants Released from the Ship and Possible Effects on Fish

Type of Pollutant	Possible Effects on Fish	Possible Uptake of the Pollutant with Respect to Feeding Habit	Human Health Hazard due to Consumption of Contaminated Fish
	catalase, glutathione reductase, glutathione-s-transferase, and induced antioxidant defenses in the liver of fish (*Sparus aurata*) (Cap'o et al., 2021). It was reported that MPs were ingested by planktivorous fish (*Acanthochromis polyacanthus*), a common and abundant species on Indo-Pacific coral reefs. It has been observed that the number of plastics in the GIs vastly increases when the size of plastic particles is reduced to approximately one-quarter of the size of the food particles, with a maximum of 2102 (300 μm) particles present in the gut of an individual fish after 1 week of plastic exposure under 5 different plastic concentration treatments, with the plastics being the same size as the natural food particles (mean of 2 mm diameter) (Critchell and Hoogenboom, 2018). Particles smaller than 5 μm might pass through the gastrointestinal tract wall, resulting in bioaccumulation when absorption exceeds release or particles are digested in tissues or organs (Roch et al., 2020). Oxidative stress has been seen in zebrafish (*Danio rerio*) following exposure to MPs (Lu et al., 2016). At aquatic exposure, the exposure route impacts the distribution and toxicity of polystyrene nanoplastics in zebrafish and antioxidant gene expression and	their small size and pose human health risks by accumulating in the respiratory tracts and potentially crossing the blood-brain barrier (BBB) (Chen et al., 2020; Prata, 2018).	

Impact on Coastal and Marine Fish and Fisheries 259

hypoactivity (Zhang et al., 2020). PS particles of different sizes have been in zebrafish's livers, gills, and intestines, causing inflammation (Lu et al., 2016). MP particles may also reach the circulatory system and subsequently the brain by absorption through the gills, intestines, and lungs or directly through the nasal cavity (Sangkham et al., 2022).

Resins

- Epoxy resin

Harmful to aquatic life with long-lasting effects (SAFETY DATA SHEET Epoxy Resin ER2224, part A). According to Regulation (EC) No 1907/2006, Annex II, as amended). Coatings based on epoxy resins can contain a wide range of organic solvents, metallic pigments, UV stabilizers, biocides, and other potentially hazardous substances such as bisphenol A/F, alkyl phenols, and other phenolic substances and polyamines (Jin et al., 2015; Verma et al., 2020). Estrogenic activity and toxic effects on water fleas, luminescent bacteria, and cypris larvae of barnacles were detected in previous studies (Bell et al., 2020; Vermeirssen et al., 2017; Watermann et al., 2005). Potential release of hazardous substances into the environment along with elevated toxicity to luminescent bacteria and estrogen-like effects (Vermeirssen et al., 2017; Bell et al., 2021)

Uptake can be oral and dermal, from sediment and water (Rudawska et al., 2021).

Causes skin irritation and may cause an allergic skin reaction and causes serious eye irritation. (Alzahrani and Rajendran, 2019) (SAFETY DATA SHEET Epoxy Resin ER2224, part A). According to Regulation (EC) No 1907/2006, Annex II, as amended). It can cause cytotoxicity (Rudawska et al., 2021). BPA is able to interact with human estrogen receptors (ER). This is expressed very strongly in the mammalian fetal brain and placenta at sites that could have important outcomes for newborns (Ikhlas et al., 2019). Bisphenols are capable of inducing cytotoxicity through oxidative stress and genotoxicity, causing harmful effects on human health and the environment (O'Boyle et al., 2012).

TABLE 11.8
Postmortem Observations on Carcasses of *Acanthopagrus berda* Found on the Beach of Kepungoda and *Caranx ignobilis* (Forsskål, 1775) on the Beach of Wellawatte of Western Province of Sri Lanka

Taxonomy of fish	*Species: Acanthopagrus berda* (Forsskål, 1775) Family: Sparidae	Species: *Caranx ignobilis* (Forsskål, 1775) Family: Carangidae
Common name English (E) and Sinhala (S)	Black bream (E) Thiraliya (S)	Giant trevally (E) Atanagul parava (S)
Sampling date	May 28, 2021	May 31, 2021
Place of sampling	Kepungoda Beach	Wellawatte Beach
State of fish at the time of sampling	Moribund	Fresh
Total length	38.5 cm	22.0 cm
Total body weight	1063.38 g	366.12 g
Postmortem observations		
External characters	Mouth closed; gill operculum closed. No external injuries or peculiarities	Mouth open, closed gill operculum, No external injuries or peculiarities; No external damages or contamination by oil or other pollutants
Eyes	Normal	Bloodstain around the pupil
Mouth cavity	Plastic pellets observed	0.1 g plastic pellets found inside the mouth cavity.
Gills	Plastic pellets found among the gills	Pale in color; 0.44 g plastic pellets found outside and among the gills.
Pharyngeal cavity	Empty	0.41 g of plastic pellets were found inside and they were obstructing the opening to the esophagus; Internal bleeding was observed.
Stomach	Filled with small shrimps and urchin shell parts; No plastic pellets	0.02 g of broken plastic pellets were found; No food or digested matter found.
Gut	Digested food items; No plastic pellets:	Empty and no plastic particles were observed.
Possible cause of death	Possibly due to blocking of the digestive tract by plastic pellets and obstruction of respiratory process	Blocking of digestive tract and mechanical damage to internal epithelial tissues and other tissues of the digestive tract causing internal bleeding

Impact on Coastal and Marine Fish and Fisheries 261

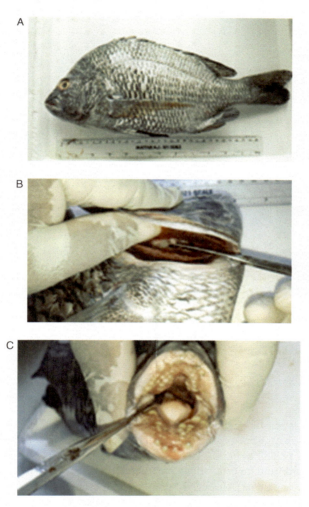

FIGURE 11.20 A) Dead specimen of *Acanthopagrus berda* found on the beach of Kepungoda; B) plastic pellets/nurdles trapped among gills; C) plastic pellets/nurdles inside the mouth obstructing the digestive tract.

of the sea when it was feeding at the bottom of the sea on benthic organisms or it has ingested plastic pellets floating/suspended in the water column during the intake of respiratory water current.

C. ignobilis (giant trevally), found on Wellawatte Beach, is a pelagic predator that actively feeds on crustaceans (like crabs and spiny lobsters) and fish at night. Postmortem observations of *C. ignobilis* indicated internal bleeding. Pale gills and bleeding inside the pharyngeal cavity are evidence of internal bleeding. Wide open mouth at the time of death also indicates an impact on the respiratory process. Plastic pellets have blocked the pharyngeal cavity and also may have obstructed

262 Maritime Accidents and Environmental Pollution

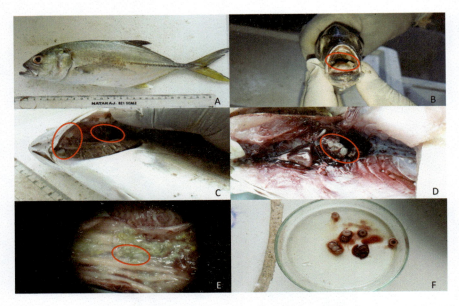

FIGURE 11.21 a) Giant trevally, *Caranx ignobilis,* found on the Wellawatte Beach; b) mouth cavity of the fish with plastic pellets/nurdles; c) gills of the fish pale in color with few plastic pellets/nurdles; d) plastic pellets clogged inside the pharyngeal cavity causing a hemorrhagic condition; e) small broken plastic pellets inside the otherwise empty stomach of the fish; and f) plastic pellets collected from the pharyngeal cavity of the fish.

the respiratory water flow, and hence the respiratory process. Considering the above observations it is also reasonable to state that the fish had mistakenly identified the floating plastic pellets as food items (school of small fish or eggs of fish or other marine organisms) and ingested them, obstructing the gills, pharyngeal cavity and the opening to the esophagus. Absence of undigested food items inside the stomach and digested food inside the rest of the gut could be considered evidence of obstruction of the digestive tract by the plastic pellets.

Small plastic pieces (mechanically ground into smaller pieces) observed inside the stomach indicate that the plastic pellets have been subjected to grinding or crushing process inside the pharyngeal cavity, which may have damaged the epithelium of the pharyngeal cavity, resulting in internal bleeding. Blocking of respiratory water flow and the internal bleeding during the process of grinding/crushing of plastic pellets could be the cause of death of *C. ignobilis* found on the beach of Wellawatte. As *C. ignobilis* is a fish species with independent predators that thrive solitarily in its habitat (Fishbase, 2022a), and therefore it is not possible to expect mass mortality of this species and therefore occurrence of one affected fish on Wellawatte Beach could have contributed to their solitary behavior at the time of exposure to floating plastic pellets in coastal waters. Presence of small pieces of plastic and absence of food items inside the stomach or absence of digested food inside the rest of the gut also indicate an obstruction of the digestive tract by plastic pellets, preventing ingestion of food.

Impact on Coastal and Marine Fish and Fisheries

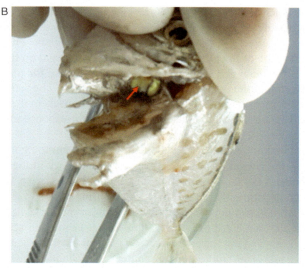

FIGURE 11.22 A) Black blotches on the skin of false trevally, *Lactarius lactarius* and B) yellow-colored liver of pugnose ponyfish, *Deveximentum insidiator* (Bloch, 1787), captured by trawling in the area affected by ship accident.

The diet of *C. ignobilis* is known to include fish (eels), mollusks (squid, octopus, etc.), and crustaceans (shrimps, lobsters, etc.), which indicates that they can feed on large prey items, some of which have hard shells.

Plastic pellets can also cause toxicological effects on fish directly due to active transportation of micro- or nanoplastic particles by the internal epithelia of the digestive tract or indirectly due to various types of chemicals, physical components of the material, and microorganisms adsorbed/attached to the surface of plastic pellets (Teuten, et al., 2009; Rochman, et al., 2013). Thus starvation, internal bleeding caused by the plastic pellets, and possible harmful effects of the adsorbed material on the plastic pellets may have been the cause of death of *C. ignobilis* found on Wellawatte Beach. Several studies have shown that the family Carangidae is more susceptible to plastic ingestion in the marine environment (Savoca, et al., 2021). This may further fortify the statement that the plastic pellets could be the possible cause of the death of the *C. ignobilis* specimen. Esophagus of *C. ignobilis* is short, thick-walled, and many folds present in the esophagus indicate its elasticity and capability of swallowing large prey items and ability to crush food items. They have a wide mouth gape and large J-shaped, thick-walled stomach, which is capable of contraction (Phuong, et al., 2018) and also indicates its ability to hold large volumes of food. Large plastic objects have been detected in the stomach of giant trevally captured off Maldives, which also provide evidence for ability of *C. ignobilis* to swallow large objects.

Most of the studies were concentrated on the effects of microplastic on fish and other marine biota. There were extremely few studies dealing with the effect of macroplastic (such as the plastic pellets originating from the MV X-Press Pearl ship incident) on marine organisms. Thus it is not possible to validate the hypothesis that the fish death occurred via the influence of the plastic pellets, without careful histological examination of the alimentary canal. However, based on the evidence, it was reasonable to conclude that there was significant impact of the plastic pellets on the cause of the death of the two fish specimens we collected. The authors highly recommend the necessity of further studies on the possible impacts of the macroplastic such as plastic pellets/nurdles on the marine biota and their response to the various levels of contamination.

In a considerable number of pelagic fish, stomach and the gut were empty. This can be due to the reason that fishing was carried out early in the morning and they may have digested the food by that time and undigested materials have been expelled with fecal matter. Normally pelagic fish forage during dusk and dawn. This could be the reason for not detecting any undigested macro- or microplastic inside the alimentary canals of fish subjected to investigation. Diurnal sampling at two-hour intervals will have to be carried out to capture fish with full stomach to identify the food items and to obtain sufficient amount of gut contents for chemical analyses to prove the ingestion of virgin and burned plastic by fish.

There were some abnormalities observed in certain fish samples and most prominent were the peculiar dark marks observed on the skin of false trevally (*Lactarius lactarius*) and yellow-colored liver found in pugnose ponyfish, *Deveximentum insidiator* (Froese and Pauly, 2023), captured by trawling in the area affected by ship accident. Peculiar dark marks observed on the skin of false trevally could be due to an acid burn or due to contact with a chemical that reacts with the skin. In addition to the above internal organs of this fish were pale red in color when compared to that of other individuals of the same species. This sample was preserved for further investigations.

Yellow color in the liver of the fish occurs due to various reasons. According to several studies, this can be a reason for poor water quality (Smyrli et al., 2017) and an alteration of the liver lipid metabolism due to exposure to polystyrene nanoplastics (Lai et al., 2021). Although pugnose ponyfish are plankton feeders, which also filter plankton from the gill rakers on their gills, they feed close to the bottom of the sea because they usually inhabit shallow waters. Therefore, there is a strong possibility for them to filter suspended pollutants such as plastic and heavy metals suspended in bottom waters. Further investigations on this matter are strongly recommended for verification of possible reasons.

11.6 RECOMMENDATIONS

- Continuous monitoring of diversity and density of adult fish species and their planktonic eggs and larvae should be carried out, in order to determine the possible medium-term and long-term impacts of pollutants released during and after the ship accident and removal of the wreck.
- Samples of fish, water, and sediments should be periodically subjected to chemical analyses to determine the methods of uptake and bioaccumulation of pollutants, especially on species which are important for human consumption.
- Fish eggs and larvae should be identified using molecular biological techniques to determine the availability of planktonic fish eggs of larvae at different times of the year and to determine the species affected by incidents related to ship accident.
- Histopathological studies should be carried out to determine the harmful effects of pollutants, especially on tissues of economically important fish species.
- To identify the mutagenic effects of pollutants on different species and determine the possible involvement of pollutants released from the MV X-Press Pearl.

REFERENCES

Almroth, B.C., Sturve, J., Berglund, Å. and Förlin, L. (2005) Oxidative damage in eelpout (Zoarces viviparus), measured as protein carbonyls and TBARS, as biomarkers. *Aquatic Toxicology*, 73(2), 171–180.

Alzahrani, Abdullah M. and Rajendran, Peramaiyan (2019). Petroleum Hydrocarbon and Living Organisms. In, "Hydrocarbon Pollution and its Effect on the Environment". Muharrem Ince and Olcay Kaplan Ince (Eds.) DOI: 10.5772/intechopen.86948., https://www.intechopen.com/profiles/295849

Angel, A. and Ojeda, F.P., (20010. Structure and trophic organization of subtidal fish assemblages on the northern Chilean coast: the effect of habitat complexity. *Marine Ecology Progress Series, 217*, pp.81–91.

Aral, H. and A. Vecchio-Sadus, Toxicity of lithium to humans and the environment—A literature review. *Ecotoxicology and Environmental Safety*, 2008. 70(3): p. 349–356.

Auta, H.S., Emenike, C.U., Fauziah, S.H., (2017). Distribution and importance of microplastics in the marine environment: a review of the sources, fate, effects, and potential solutions. Environ. Int. 102, 165–176. https://doi.org/10.1016/j. envint.2017.02.013

Barnes, D.K.A., Galgani, F., Thompson, R.C., Barlaz, M., (2009). Accumulation and fragmentation of plastic debris in global environments. Philos. Trans. R. Soc. Lond. Ser. B Biol. Sci. 364, 1985–1998. https://doi.org/10.1098/rstb.2008.0205

Barron, M.G., (2012). Ecological impacts of the Deepwater Horizon oil spill: Implications for immunotoxicity. Toxicol. Pathol. 40, 315–320. [CrossRef].

Bekki, K.; Takigami, H.; Suzuki, G.; Tang, N.; Hayakawa, K., (2009). Evaluation of toxic activities of polycyclic aromatic hydrocarbon derivatives using in vitro bioassays. J. Health Sci., 55, 601–610. [CrossRef].

Bell, A.M., Baier, R., Kocher, B., Reifferscheid, G., Buchinger, S., Ternes, T., (2020). Ecotoxicological characterization of emissions from steel coatings in contact with water. Water Res. 173, 13. https://doi.org/10.1016/j.watres.2020.115525

Bolan, N., et al., *From mine to mind and mobiles – Lithium contamination and its risk management. Environmental Pollution*, 2021. 290: p. 118067.

Boucher, J. & Friot D. (2017). *Primary Microplastics in the Oceans: A Global Evaluation of Sources*. Gland, Switzerland: IUCN. 43pp.

Breine, J.J., Maes, J., Quataert, P., Van den Bergh, E., Simoens, I., Van Thuyne, G., Belpaire, C., (2007). A fish-based assessment tool for the ecological quality of the brackish Schelde estuary in Flanders (Belgium). Hydrobiologia 575, 141–159.

Bunton, T.E.,(1996). Experimental chemical carcinogenesis in fish. Toxicol. Pathol., 24, 603–618. [CrossRef] [PubMed].

Cap´o, X., Company, J.J., Alomar, C., Compa, M., Sureda, A., Grau, A., Hansjosten, B.,P L´opez-V´azquez, J., Quintana, J.B., Rodil, R., Deudero, S., (2021). Long-term exposure to virgin and seawater exposed microplastic enriched-diet causes liver oxidative stress and inflammation in gilthead seabream *Sparus aurata*, Linnaeus 1758. *Sci. Total Environ.* 767, 144976. https://doi.org/10.1016/j.scitotenv.2021.144976

C Bondy, S. (2012). Can environmentally relevant levels of aluminium promote the onset and progression of neurodegenerative diseases?. *Current Inorganic Chemistry (Discontinued)*, 2(1), 40-45.

Chen, G., Feng, Q., Wang, J., (2020). Mini-review of microplastics in the atmosphere and their risks to humans. *Sci. Total Environ.* 703, 135504 https://doi.org/10.1016/j.scitotenv.2019.135504

Cherkashin, A.S., Nikiforov, M. V. and Shelekhov, V. A., "The Use of the Mortality Rate of Marine Fish Prolarvae for the Estimation of Zinc and Lead Toxicity," *Biol. Morya* 30 (3), 247–252 (2004)

Cherr, G.N.; Fairbairn, E.; Whitehead, A., (2017). Impacts of petroleum-derived pollutants on fish development. *Annu. Rev. Anim. Biosci.,* 5, 185–203. [CrossRef] [PubMed].

Coates, S., Waugh, A., Anwar, A., Robson, M., (2007). Efficacy of a multi-metric fish index as an analysis tool for the transitional fish component of the Water Framework Directive. *Marine Pollution Bulletin* 55, 225–240.

Courrat, A., Lobry, J., Nicolas, D., Laffargue, P., Amara, R., Lepage, M., Girardin, M. and Le Pape, O., (2009). Anthropogenic disturbance on nursery function of estuarine areas for marine species. *Estuarine, Coastal and Shelf Science*, *81*(2), pp.179–190.

Critchell, K., Hoogenboom, M.O., (2018). Effects of microplastic exposure on the body condition and behaviour of planktivorous reef fish (*Acanthochromis polyacanthus*). *PLoS One* 13, e0193308. https://doi.org/10.1371/journal.pone.0193308

Cumaranatunga, P.R.T., Ranawickreme, A.S.K., Wickstrom, H. and Vithanage, K.V.S. (1997). Factors Affecting the Distribution of Anguilla bicolor bicolor McClelland and Anguilla nebulosa nebulosa McClelland (Anguilliformes; Anuillidae) in a River System of Southern Sri Lanka. *ASIAN FISHERIES SCIENCE* 10, 9–22.

Dang F, Zhong H, Wang WX., (2009). Copper uptake kinetics and regulation in a marine fish after waterborne copper acclimation. *Aquat Toxicol.* 94(3):238–44.

Davis, P.J., Mousa, S.A., Schechter, G.P. (2018). New interfaces of thyroid hormone actions with blood coagulation and thrombosis. *Clin Appl Thromb Hemost.* 24(7): 1014–1019.

Davydova, S. V., & Cherkashin, S. A. (2007). Ichthyoplankton of the eastern shelf of Sakhalin Island and its use as an environmental state indicator. *Journal of Ichthyology, 47*(6), 438–448.

Diserens, L., Porretta, A.P., Trana, C., Meier, D., 2021. Lithium-induced ECG modifications: Navigating from acute coronary syndrome to Bugada syndrome. *BMJ Case Rep.* 14, e241555.

De Vos, A., Pattiaratchi, C.B. and Wijeratne, E.M.S. (2014). Surface circulation and upwelling patterns around Sri Lanka. *Biogeosciences, 11*(20), p.5909.

Dorst, J. P. (2022). Migration-Fish/ Encyclopaedia Britannica. https://www.britannica.com/science/migration-animal/Fish (accessed on 01-05-2022)

Eschmeyer, W. N., Ronald F., Fong, Jon D., & Polack, D. A. (2010). Marine fish diversity: history of knowledge and discovery (Pisces) Zootaxa 2525: 19–50 (2010). ISSN 1175-5334 (online edition)

Esilaba, F., Moturi, W.N., Mokua, M. and Mwanyika, T. (2020). Human health risk assessment of trace metals in the commonly consumed fish species in Nakuru Town, Kenya. *Environmental Health Insights*, 14, 1178630220917128.

Feng, Z., Zhang, T., Li, Y., He, X., Wang, R., Xu, J., Gao, G., (2019). The accumulation of microplastics in fish from an important fish farm and mariculture area, Haizhou Bay, China. Sci. Total Environ. 696, 133948. https://doi.org/10.1016/j.scitotenv.2019.133948

Fishbase (2022a) list of marine fish reported in Sri Lanka. www.fishbase.se/Country/CountryChecklist.php?c_code=144&vhabitat=saltwater&csub_code=#:~:text=List%20of%20Marine%20Fishes%20reported%20from%20Sri%20Lanka&text=Table%201%3A%20972%20species%20currently%20present%20in%20the%20country%2Fisl and (Accessed on 01-05-2022)

Fishbase (2022b). http://www.fishbase.org/

Fisheries Statistics, Ministry of Fisheries and Aquatic Resources Development (MFRD), Sri Lanka, 2020. www.fisheriesdept.gov.lk/web/images/Statistics/FISHERIES-STATISTICS--2020-.pdf

Flora, S.J.S., M. Pande, and A. Mehta, (2003). Beneficial effect of combined administration of some naturally occurring antioxidants (vitamins) and thiol chelators in the treatment of chronic lead intoxication. *Chemico-Biological Interactions, 145*(3): 267–280.

Froese, R., Pauly, D., (eds.) (2023). FishBase. *Siganus guttatus* (Bloch, 1787). Accessed through: World Register of Marine Species at: https://www.marinespecies.org/aphia.php?p=taxdetails&id=273913 on 2023-11-05

Garai, P., Banerjee, P., Mondal, P., & Saha, N. C. (2021). Effect of Heavy Metals on Fishes: Toxicity and Bioaccumulation. *Journal of Clinical Toxicology, 11*, 1.

Gratwicke, B. and Speight, M.R. (2005) The Relationship between Fish Species Richness, Abundance and Habitat Complexity in a Range of Shallow Tropical Marine Habitats. *Journal of Fish Biology*, 66, 650–667.http://dx.doi.org/10.1111/j.0022-1112.2005.00629.x

Hannam, M.L., Bamber, S.D., Moody, A.J., Galloway, T.S., Jones, M.B. (2010). Immunotoxicity and oxidative stress in the Arctic scallop *Chlamys islandica*: Effects of acute oil exposure. *Ecotoxicol. Environ. Saf.* 73, 1440–1448. [CrossRef] [PubMed].

Harrison, T.D., Whitfield, A.K. 2004. A multi-metric fish index to assess the environmental condition of estuaries. Journal of Fish Biology 65, 683–710.

Harrison, T.D., Whitfield, A.K. (2006). Application of a multimetric fish index to assess the environmental condition of South African estuaries. Estuaries and Coasts 29, 1108–1120.

Hayakawa, K., Nomura, M., Nakagawa, T., Oguri, S., Kawanishi, T., Toriba, A., Kizu, R., Sakaguchi, T., Tamiya, E. (2006). Damage to and recovery of coastlines polluted with C-heavy oil spilled from the Nakhodka. Water Res.40, 981–989. [CrossRef] [PubMed].

Healthline: The facts about Lithium Toxicity. www.healthline.com/health/lithium-toxicity (visited in November 2002)

Hjermann, D.Ø., Melsom, A., Dingsør, G.E., Durant, J.M., Eikeset, A.M., Roed, L.P., Ottersen, G., Storvik, G., Stenseth, N.C. (2007). Fish and oil in the Lofoten-Barents Sea system: synoptic review of the effect of oil spills on fish populations. *Marine Ecology Progress Series*, 339, 283–299.

Ikhlas, S., Usman, A., Ahmad, M. (2019). In vitro study to evaluate the cytotoxicity of BPA analogues based on their oxidative and genotoxic potential using human peripheral blood cells. Toxicol. Vitr. 60, 229–236. [CrossRef] [PubMed].

Jacob, H., Besson, M., Swarzenski, P.W., Lecchini, D. and Metian, M. 2020. Effects of virgin micro-and nanoplastics on fish: trends, meta-analysis, and perspectives. *Environmental Science & Technology*, 54(8), pp.4733–4745.

Jin, F.L., Li, X., Park, S.J.(2015). Synthesis and application of epoxy resins: A review. J. Ind. Eng. Chem. 29, 1–11. https://doi.org/10.1016/j.jiec.2015.03.026

Johnson, A., Carew, E., Sloman, K.A. (2007). The effects of copper on the morphological and functional development of zebrafish embryos. *Aquatic Toxicology* 84(4):431–8.

Keramea, P., Spanoudaki, K., Gikas, G., & Sylaios, G. (2021). Oil Spill Modeling: A Critical Review on Current Trends, Perspectives, and Challenges. *J. Mar. Sci. Eng.* 2021, 9(2), 181; https://doi.org/10.3390/jmse9020181

Khan, R.A. (1990). Parasitism in marine fish after chronic exposure to petroleum hydrocarbons in the laboratory and to the Exxon Valdez oil spill. *Bulletin of Environmental Contamination and Toxicology*, 44, 59–763.

Kibirige, D., Luzinda, K., Ssekitoleko, R. (2013). Spectrum of lithium induced thyroid abnormalities: A current perspective. *Thyroid Research*, 6(1), 3

Kong, X., Jiang, H., Wang, S., Wu, X., Fei, W., Li, L., et al. (2013). Effects of copper exposure on the hatching status and antioxidant defense at different developmental stages of embryos and larvae of goldfish Carassius auratus. *Chemosphere*. 92(11):1458–64.

Kumar, R., Manna, C., Padha, S., Verma, A., Sharma, P., Dhar, A., Ghosh, A., Bhattacharya, P. (2022). Micro(nano)plastics pollution and human health: how plastics can induce carcinogenesis to humans? *Chemosphere* 298, 134267. https://doi.org/10.1016/j.chemosphere.2022.134267

Kumara, P.B., Terney P. & Kasun R. Dalpathadu. (2012) Provisional checklist of Marine Fish of Sri Lanka. IN: The National Red List 2012 of Sri Lanka; Conservation Status of the Fauna & Flora. Weerakoon, D.K. & S. Wijesundara Eds, Ministry of Environment, Colombo, Sri Lanka. pp. 411–430.

Lai, W., Xu, D., Li, J., Wang, Z., Ding, Y., Wang, X., Li, X., Xu, N., Mai, K. and Ai, Q., 2021. Dietary polystyrene nanoplastics exposure alters liver lipid metabolism and muscle nutritional quality in carnivorous marine fish large yellow croaker (*Larimichthys crocea*). *Journal of Hazardous Materials, 419*, p.126454.

Lai, W., Xu, D., Li, J., Wang, Z., Ding, Y., Wang, X., Li, X., Xu, N., Mai, K. and Ai, Q., 2021. Dietary polystyrene nanoplastics exposure alters liver lipid metabolism and muscle nutritional quality in carnivorous marine fish large yellow croaker (*Larimichthys crocea*). *Journal of Hazardous Materials, 419*, p.126454.

Langangen, Ø., Stige, L.C., Kvile, K.Ø., Yaragina, N.A., Skjæraasen, J.E., Vikebø, F.B. and Ottersen, G. (2018) Multi-decadal variations in spawning ground use in Northeast Arctic haddock (Melanogrammus aeglefinus). Fisheries Oceanography 27(5), 435–444.

Lal, Brij V. & Fortune, Kate. (2000). *The Pacific Islands: an encyclopedia*. Honolulu: University of Hawai'i Press.

Lee, H.J.; Shim, W.J.; Lee, J.; Kim, G.B., (2011).Temporal and geographical trends in the genotoxic effects of marine sediments after accidental oil spill on the blood cells of striped beakperch (*Oplegnathus fasciatus*). Mar. Pollut. Bull. 62, 2264–2268. [CrossRef] [PubMed].

Lee JW, Choi H, Hwang UK, Kang JC, Kang YJ, Kim K Il, et al., (2019). Toxic effects of lead exposure on bioaccumulation, oxidative stress, neurotoxicity, and immune responses in fish: A review. *Environ Toxicol Pharmacol*. 68:101–108.

Le Pape, O., Gilliers, C., Riou, P., Morin, J., Amara, R., Désaunay, Y. (2007). Convergent signs of degradation in both the capacity and the quality of an essential fish habitat: state of the Seine estuary (France) flatfish nurseries. *Hydrobiologia*, 588(1), pp. 225–229.

Li, W., Chen, X., Li, M., Cai, Z., Gong, H., Yan, M., (2022). Microplastics as an aquatic pollutant affect gut microbiota within aquatic animals. *J. Hazard. Mater.* 423, 127094 https://doi.org/10.1016/j.jhazmat.2021.127094

Lu, Y., Zhang, Y., Deng, Y., Jiang, W., Zhao, Y., Geng, J., Ding, L., Ren, H., (2016). Uptake and accumulation of polystyrene microplastics in zebrafish (Danio rerio) and toxic effects in liver. *Environ. Sci. Technol.* 50, 4054–4060. https://doi.org/10.1021/acs.est.6b00183

MacDonald, G.Z.; Hogan, N.S.; Köllner, B.; Thorpe, K.L.; Phalen, L.J.; Wagner, B.D.; Van Den Heuvel, M.R., (2013). Immunotoxic effects of oil sands-derived naphthenic acids to rainbow trout. *Aquat. Toxicol.* 126, 95–103. [CrossRef] [PubMed].

Magnuson, J.T.; Khursigara, A.J.; Allmon, E.B.; Esbaugh, A.J.; Roberts, A.P., (2018). Effects of Deepwater Horizon crude oil on ocular development in two estuarine fish species, red drum (Sciaenops ocellatus) and sheepshead minnow (*Cyprinodon variegatus*). *Ecotoxicol. Environ. Saf.* 166, 186–191. [CrossRef].

Martenstyn, H. 2019. Protected Waters of Sri Lanka. www.slam.lk/protected-waters. (Accessed on 01-05-2022)

Marty, G. D. I. E. Hose, M. D. McGurk, and E. D. Brown, "Histopathology and Cytogenetic Evaluation of Pacific Herring Larvae Exposed to Petroleum Hydrocarbons in the Laboratory or in Prince William Sound, Alaska, after the Exxon Valdez Oil Spill," *Can. J. Fish. Aquat. Sci.* **54**, 1846–1857 (1997)

Mcintyre J.K, Baldwin D.H, Meador J.P, Scholz N.L., (2008). Chemosensory deprivation in juvenile coho salmon exposed to dissolved copper under varying water chemistry conditions. *Environ Sci Technol*. 2008;42(4):1352–8.

Miranda, D.d.A., de Carvalho-Souza, G.F., Are we eating plastic-ingesting fish? *Marine Pollution Bulletin*, 103(1): 109–114.

Monteiro S.M, dos Santos N.M.S, Calejo M, Fontainhas-Fernandes A, Sousa M., (2009). Copper toxicity in gills of the teleost fish, Oreochromis niloticus: Effects in apoptosis induction and cell proliferation. *Aquat Toxicol*. 94(3):219–228.

Ministry of Fisheries. (2020). *Fisheries Statistics 2020*. Ministry of Fisheries: Colombo, Sri Lanka.

Munro, I. S. R., 1955. The marine and freshwater fishes of Ceylon. Dept. of External Affairs.

NOAA, (2021). What are Pelagic Fish? National Ocean Service website, https://oceanservice.noaa.gov/facts/pelagic.html, 02/26/21.

Paul, T.T., Dennis, A., George, G. (2016). A Review of Remote Sensing Techniques for the Visualization of Mangroves, Reefs, Fishing Grounds, and Molluscan Settling Areas in Tropical Waters. In: Finkl, C., Makowski, C. (eds) Seafloor Mapping along Continental Shelves. Coastal Research Library, vol 13. Springer, Cham. https://doi.org/10.1007/978-3-319-25121-9_4

Phuong, N.N., Poirier, L., Pham, Q.T., Lagarde, F., Zalouk-Vergnoux, A., (2018). Factors influencing the microplastic contamination of bivalves from the French Atlantic coast: location, season and/or mode of life? *Mar. Pollut. Bull.* 129, 664–674. https://doi.org/10.1016/j.marpolbul.2017.10.054

Prata, J.C., (2018). Airborne microplastics: consequences to human health? *Environ. Pollut.* 234, 115–126. https://doi.org/10.1016/j.envpol.2017.11.043

O'Boyle, N.M.; Delaine, T.; Luthman, K.; Natsch, A.; Karlberg, A.-T., (2012). Analogues of the epoxy resin monomer diglycidyl ether of bisphenol F: Effects on contact allergenic potency and cytotoxicity. *Chem. Res. Toxicol.* 25, 2469–2478. [CrossRef]

Oguguah, N.M., Onyekachi, M and Ikegwu, J. (2017). Concentration and Human Health implication of Trace metals in fish of economic importance in Lagos lagoon, Nigeria. *Journal of Health and Pollution*, 7(13)68–72.

Peterson, C.H., Summerson, H.C., Thomson, E., Lenihan, H.S., Grabowski, J., Manning, L., Micheli, F., Johnson, G., (2000). Synthesis of linkages between benthic and fish communities as key to protecting essential fish habitat. *Bulletin of Marine Science* 66, 759–774.

Radi AAR, Matkovics B.(1988). Effects of metal ions on the antioxidant enzyme activities, protein contents and lipid peroxidation of carp tissues. *Comp Biochem Physiol Part C, Comp.* 90(1):69–72.

Ragusa, A., Svelato, A., Santacroce, C., Catalano, P., Notarstefano, V., Carnevali, O.,Papa, F., Rongioletti, M.C.A., Baiocco, F., Draghi, S., D'Amore, E., Rinaldo, D., Matta, M., Giorgini, E., (2021). Plasticenta: first evidence of microplastics in human placenta. *Environ. Int.* 146, 106274 https://doi.org/10.1016/j.envint.2020.106274

Rahman, M.S., Molla, A.H., Saha, N. and Rahaman, A. (2012). Study on heavy metals levels and its risk assessment in some edible fishes from Bangshi River, Savar Dhaka, Bangladesh. Food Chemistry, 134(2)1847–1854.

Richard Bull. Copper in Drinking Water. (20000. Paperback: 978-0-309-06939-7.

Rudawska, A., (2021). Mechanical properties of epoxy compounds based on bisphenol A aged in aqueous environments. *Polymers* 2021,13, 952. [CrossRef].

Roch, S., Friedrich, C., Brinker, A., (2020). Uptake routes of microplastics in fishes practical and theoretical approaches to test existing theories. *Sci. Rep.* 10, 3896. https://doi.org/10.1038/s41598-020-60630-1

Romero, I.C.; Sutton, T.; Carr, B.; Quintana-Rizzo, E.; Ross, S.W.; Hollander, D.J.; Torres, J.J.,(2018). Decadal assessment of polycyclic aromatic hydrocarbons in mesopelagic fishes from the Gulf of Mexico reveals exposure to oil-derived sources. *Environ. Sci. Technol.* 52, 10985–10996. [CrossRef] [PubMed].

Rochman, C.M., Hoh, E., Kurobe, T., Teh, S.J., 2013. Ingested plastic transfer contaminants to fish and induces hepatic stress. *Scientific Reports*.

Rodriguez, J.M., Alemany, F. and Garcia, A., 2017. A guide to the eggs and larvae of 100 common Western Mediterranean Sea bony fish species. *Rome, Italy: FAO*.

Rossi, N. R. and Jamet, J.L. (2008). In Situ heavy metals (coppers, lead and cadmium) in different plankton Compartments and suspended particulate matter in two coupled Mediterranean coastal ecosystems (Toulon Bay, France). *Marine Pollution Bulletin*,56:1862–1870.

Rubin, A.E., Sarkar, A.K., Zucker, I., (2021). Questioning the suitability of available microplastics models for risk assessment – a critical review. *Sci. Total Environ.* 788–147670 https://doi.org/10.1016/j.scitotenv.2021.147670

Sangkham, S., (2022). Impact of the COVID-19 outbreak on the generation of plastic waste. In: Shahnawaz, M., Sangale, M.K., Daochen, Z., Ade, A.B. (Eds.), Impact of Plastic Waste on the Marine Biota. Springer Singapore, Singapore, pp. 37–47. https://doi.org/10.1007/978-981-16-5403-9_3

Sarah, R., Tabassum, B., Idrees, N., Hashem, A., & Abd_Allah, E. F. (2019). Bioaccumulation of heavy metals in *Channa punctatus* (Bloch) in river Ramganga (U.P.), India. *Saudi Journal of Biological Sciences*, *26*(5), 979–984. https://doi.org/10.1016/j.sjbs.2019.02.009

Savoca, M.S., McInturf, A.G., & Hazen, E.L. (2021). Plastic ingestion by marine fish is widespread and increasing. *Global Change Biology*, 27(10), 2188–2199.

Sehar, A., Shafaqat, A., Uzma, S.A. Mujahid, F., Salima, A.B., Fakhir, H. and Rehan, A. (2014). Effect of different Heavy Metal Pollution on fish. *Research Journal of chemical and environmental Scal.* 2:74–79.

Shen, C., Wang, Z., Zhao, F., Yang, Y., Li, J., Yuan, J. et al. (2020). Treatment of 5 Critically Ill Patients with COVID-19 with Convalescent Plasma. JAMA, 323, 1582–1589.

Shilta, M.T., Chadha, N.K., Suresh Babu, P.P., Asokan, P.K., Vinod, K., Imelda, J., Sawant, Paramita Banerjee and Abhijith, Ramya (2018) Food and feeding habits of goldsilk seabream, *Acanthopagrus berda* (Forsskal, 1775). *Turkish Journal of Fisheries and Aquatic Sciences*, *19*(7), 605–614.

Sinha, P., Clements, V.K., Bunt, S.K., Albelda, S.M., Ostrand-Rosenberg, S. (2007). Cross-talk between myeloid-derived suppressor cells and macrophages subverts tumor immunity toward a type 2 Response HYPERLINK "javascript:;" 1. *J Immunol* 15 179(2), 977–983. https://doi.org/10.4049/jimmunol.179.2.977

Snyder, S.M.; Pulster, E.L.; Wetzel, D.L.; Murawski, S.A., (2015). PAH exposure in Gulf of Mexico demersal fishes, post-Deepwater Horizon. *Environ. Sci. Technol.* 49, 8786–8795. [CrossRef] [PubMed].

Stepanenko, M. A. (1988). "Interannual Variation in Conditions of Reproduction and Forecasting of the Yield of Generations of Pacific Hake (*Macruronus productus*) and *the Ichthyofauna Composition, Yield of Generations, and Methods of Forecasting Fish Resources in the Northern Part of the Pacific Ocean* (TINRO, Vladivostok, 1988), pp. 28–36.

Stohs, S. J., Bagchi, D. (1995). Oxidative mechanisms in the toxicity of metal ions. *Free Radic Biol Med.* 18:321–336. [PubMed: 7744317].

Teuten, E.L. et al. Transport and release of chemicals from plastics to the environment and to wildlife. *Royal Society*, 364(1526), 2027?2045.

Thibon, F., Weppe, L., Vigier, N., Churlaud, C., Lacoue-Labarthe, T., Metian, M., Cherel, Y. and Bustamante, P. (2021) Large-scale survey of lithium concentrations in marine organisms. *Science of the Total Environment* 751, 141453.

Tkatcheva, V., Poirier, D., Chong-Kit, R., Furdui, V.I., Burr, C., Leger, R., Parmar, J., Switzer, T., Maedler, S., Reiner, E.J., Sherry, J.P. and Simmons, D.B.D. (2015) Lithium an emerging contaminant: Bioavailability, effects on protein expression, and homeostasis disruption in short-term exposure of rainbow trout. Aquatic Toxicology 161, 85–93.

Tyurin, A. N. and Khristoforova, N. K. "Test Selection for the Assessment of Pollution of the Sea Environment," *Biol. Morya* 21 (6), 361–368 (1995)

Ugwu, K., Herrera, A., G´omez, M., (2021). Microplastics in marine biota: a review. *Mar. Pollut. Bull.* 169, 112540 https://doi.org/10.1016/j.marpolbul.2021.112540

Varanka Z, Rojik I, Varanka I, Nemcsók J, Ábrahám M., (2001).Biochemical and morphological changes in carp (Cyprinus carpio L.) liver following exposure to copper sulfate and tannic acid. *Comp Biochem Physiol – C Toxicol Pharmacol.* 128(2):467–77.

Viana, T., Ferreira, N., Henriques, B., Leite, C., De Marchi, L., Amaral, J., Freitas, R. and Pereira, E. (2020) How safe are the new green energy resources for marine wildlife? The case of lithium. Environmental Pollution 267, 115458.

Verma, C., Olasunkanmi, L.O., Akpan, E.D., Quraishi, M.A., Dagdag, O., El Gouri, M.,Sherif, E.S.M., Ebenso, E.E, (2020). Epoxy resins as anticorrosive polymeric materials: A review. *React. Funct. Polym.* 156, 20. https://doi.org/10.1016/j.reactfunctpolym.2020.104741

Vermeirssen, E.L.M., Dietschweiler, C., Werner, I., Burkhardt, M, (2017). Corrosion protection products as a source of bisphenol A and toxicity to the aquatic environment. *Water Res.* 123, 586–593.https://doi.org/10.1016/j.watres.2017.07.006

Verdoux, H., Debruyne, A.-L., Queuille, E. and De Leon, J. (2021) A reappraisal of the role of fever in the occurrence of neurological sequelae following lithium intoxication: a systematic review. *Expert Opinion on Drug Safety 20*(7), 827–838.

Vashchenko, M. A. (2000). "Pollution of Peter the Great Bay, Sea of Japan, and Its Biological Consequences," *Biol. Morya* 26 (3), 149–159 (2000)

Watermann, B., Daehne, B., Sievers, S., Dannenberg, R., Overbeke, J., Klijnstra, J., Heemken, O.J.C., (2005). Bioassays and selected chemical analysis of biocide-free antifouling coatings. *Chemosphere* 60 (11), 1530–1541. https://doi.org/10.1016/j.chemosphere.2005.02.066

Yacoub A.M, Gad N.S. (2012). Accumulation of some heavy metals and biochemical alterations in muscles of Oreochromis niloticus from the River Nile in Upper Egypt *Int. J. Environ. Sci. Engg.* 2012;3:1–10.

Zhang, R., Silic, M.R., Schaber, A., Wasel, O., Freeman, J.L., Sepúlveda, M.S., (2020). Exposure route affects the distribution and toxicity of polystyrene nanoplastics in zebrafish. *Sci. Total Environ.* 724, 138065 https://doi.org/10.1016/j.scitotenv.2020.138065

12 Impact of the X-Press Pearl Disaster on Coastal and Marine Birds

Sampath S. Seneviratne and Jude Janith Niroshan

12.1 SUMMARY

The maritime zone of Sri Lanka faced an environmental catastrophe in May 2021 as the MV X-Press Pearl cargo ship burned and sank with partially burned cargo dispersed into the ocean (Figure 12.1). Here we discussed the immediate and long-term impacts of this accident on coastal and marine birds. We compared similar global incidents to evaluate how such accidents were handled to minimize the environmental impact and the steps taken to prevent similar accidents in the future. The immediate impact of the disaster caused the death of a large number of marine life including marine mammals, sea turtles, and fish. The amount of immediate damage caused to maritime and coastal birds is unclear because there was no direct mortality recorded. By looking at the profile of the discharged material and the debris deposited on the beaches across 1000 km of the shoreline, it is clear that there is much more to the effects of the X-Press Pearl incident than just what we have seen so far. There might be population-level, long-term impacts on marine birds, affecting their foraging, breeding, and resilience to withstand changes in the maritime environment. Therefore, the true scale of the adverse effects of this accident on the coastal and marine birds is yet to be determined.

12.2 EXAMPLES OF MARITIME DISASTERS AFFECTING BIRDS

Hydrocarbons such as plastics and crude oil have contaminated pelagic zones ever since the mass extraction and refining of such hydrocarbons. Even though major spills have captured the attention of the public and policymakers in recent years, small- to medium-scale plastic and oil pollution in the marine environment is widespread (Faksness et al., 2016a).

Marine birds including both coastal seabirds and pelagic birds spend their lives above or on the water and hence are especially vulnerable to marine pollution (Croxall et al., 2012). However, the effect of oil spills on birds cannot be easily predicted or estimated due to the complex nature of these interactions (Faksness et al., 2016b). For example, approximately 300,000 seabirds are being killed each year off the Grand Banks of the North-western Atlantic Ocean as a result of illegal oil discharges from ships (Roberts, 2005; Wright, 2004, Wright et al., 2022). Samples collected from

FIGURE 12.1 X-Press Pearl being caught on fire captured by Landsat 8 (NASA).

Source: https://visibleearth.nasa.gov/images/148423/satellite-observes-ship-fire-off-sri-lanka/148423f

the feathers of beached seabirds in the Atlantic Ocean over a 10-year period showed that more than 90% of oil residues were heavy fuel oil, mixed with lubricants, which is only found in the bilges of large ships (Camphuysen & Heubeck, 2001). On the other hand, large oil spills such as the Amoco Cadiz spill (spilled over 220,000 tons of oil) killed only 5000 birds off the coast of France in 1978 (Whitton, Phillips et al., 1994), but two years later, 600 tons of oil spilled in the strait separating Norway and Denmark killed about 30,000 birds (Dempsey, 1985). Better documented large oil spills, such as Exxon Valdez (37,000 tons of oil off Alaska in 1989 (Prince & Bragg, 1997)) and Deepwater Horizon (780,000 tons of oil off the Gulf Coast in 2010) collectively killed an estimated 0.4–2.0 million marine birds (see below).

12.2.1 Deepwater Horizon Oil Spill (2010)

The Deepwater Horizon oil spill was a maritime disaster that occurred in April 2010 in the Gulf of Mexico and is regarded as the largest marine oil spill in the history of the petroleum industry. The total discharge was estimated by the US federal government to be 210 million US gallons or 780,000 cubic meters. Even though on September 19, 2010, the well was declared sealed after several failed attempts to contain the flow, some reports indicated that the well site was still leaking in early 2012. The spill waters contained 40 times more polycyclic aromatic hydrocarbons (PAHs) which are frequently associated with oil spills and contain carcinogens and chemicals that pose a variety of health risks to marine life. The spill zone was home

Impact on Coastal and Marine Birds

to 8332 species, including over 218 bird species. The damage to birds in this disaster is estimated between 300,000 and 2 million lives (Bursian et al., 2017; Fallon et al., 2018; Haney et al., 2014; Tran et al., 2014).

12.2.2 EXXON VALDEZ OIL SPILL (1989)

On March 24 1989, Exxon Valdez, an oil supertanker bound for Long Beach, California, collided with Prince William Sound's Bligh Reef near Alaska and spilled 10.8 million US gallons (or 37,000 tons) of crude oil, recording the second largest oil spill in US waters. The immediate consequences included the deaths of 100,000 to 250,000 seabirds, 247 bald eagles, and many millions of marine lives. Evidence of negative oil spill effects on marine birds was discovered even after nine years in several species such as brown pelicans ((Piatt et al., 1990)). Although the volume of oil has decreased significantly, with only about 0.14–0.28% of the original spilled volume remaining, studies indicate that the area of the oiled beach has changed little since 1992 (Day et al., 1997; Irons et al., 2000; Murphy et al., 1997; Piatt et al., 1990; Wiens et al., 1996).

12.2.3 RECENT PLASTIC PELLET SPILLS

Over the past decades, thousands of tons of plastics have been released into the ocean as part of garbage dumping and maritime accidents (Figure 12.2; Abelson, 1971; Katsanevakis, 2008; Vikas & Dwarakish, 2015). Unlike widely publicized spills of

FIGURE 12.2 Multiple spills of plastic pellets into oceans over the past decade.

Source: https://ipen.org/sites/default/files/documents/ipen-sri-lanka-ship-fire-v1_2aw-en.pdf (Rubesinghe et al., 2022).

oil, it seems that cases of plastic pollution have easily passed through authorities due to loopholes in international and national maritime laws and regulations (Tunnell et al., 2020). In 2020, a storm caused 13 tons of pellets to spill from a ship in the North Sea. Only about a ton of the spilled plastic had been collected four months later with no substantial prosecution, even though 700 locations had been reported to be affected by the spill (Rubesinghe et al., 2022). A large plastic pellet spill occurred in the Mississippi River (USA) in 2020, with an estimated 743 million nurdles being released into the river. The ship's operator hired a small cleanup crew, but pellets continued to wash up on the river's beaches six months later. Following a storm in 2017, 49 tons of plastic pellets spilled from a container ship outside of South Africa. Similarly, 150 tons of plastic pellets were leaked in 2012 in Hong Kong. Large pellet mounds were still present along the beaches even after six years from the disaster. Following a spill linked to a South Carolina shipping facility in the USA, the relevant organizations reached a $1 million settlement in 2019. These are several examples (Figure 12.2) of relatively simple attention paid to plastic-related accidents (Rubesinghe et al., 2022; Tunnell et al., 2020).

12.3 WHAT MAKES SEABIRDS VULNERABLE TO MARITIME DISASTERS? ASSESSING ROLE FACTORS

Maritime disasters are harmful to birds in different ways, both short-term (death, injury and breeding failure) and long-term (complete to partial reproductive failure and chronic physiological effects), as the resultant chemical and physical effects either are toxic or cause injuries to them and their marine environment (Croxal et al 2012).

12.3.1 Derivatives of Crude Oils or Petroleum Oils

The hydrophobic nature of seabird feathers is critical for survival in the marine environment (Wiese et al., 2001) as birds rest and sleep on water, feed off the surface, or dive into the water to forage. Feathers protect them from getting wet in cold ocean water or prevent salts from contacting the skin. Crude oils and gums, which can adhere to the feathers, destroy the structural integrity of feathers by destroying or deforming the bonds between hooks and barbules in the feather (Figure 12.3), which could damage the ability of the feather to repel water and retain heat.

A wet seabird is a sick bird (except in a few groups such as pelicans), as they could get hypothermia (Clark, 1984; Doerffer, 1992; Helm et al., 2015; Walter et al., 2014). A long list of substances found in crude oil causes injuries to the physiopathology of seabirds. Accidental ingestion of crude oil when birds attempt to clean themselves (Figure 12.4), as well as in the foraging attempts, could damage their digestive epithelium, liver, and internal physiology (Weis, 2014).

Bird eggs are vulnerable to petroleum oil pollution and can be harmed by even trace amounts of oil on eggs. Oils can contact the surface of the egg if the nest is located near water; hence the contaminated water could splash on the nest or on the eggs, and oiled nesting material could be brought into the nest by the parent birds, or by the returning parents from the sea with oil-contaminated feathers. Since birds can't remove surface-floating petroleum oils from their plumage, a contaminated parent

Impact on Coastal and Marine Birds

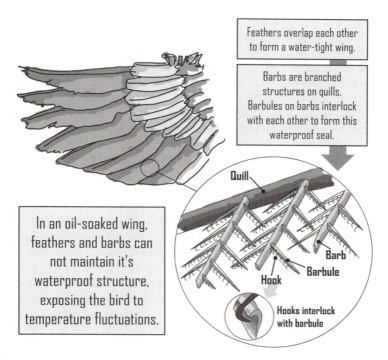

FIGURE 12.3 Oil-soaked wings kill birds.

Source: International Bird Rescue (IBR). www.birdrescue.org/wp-content/uploads/2020/10/image-58.jpeg

FIGURE 12.4 A brown pelican dipped in an oil spill. The oil will destroy the structure of the feather and damage the internal gut epithelium and other vital organs such as the liver. There is no chance of survival of this individual. It will be a slow and painful death.

Source: www.marinedefenders.org/impact-on-mammals-birds-and-fish.html

has no option but to visit the nest for incubation and chick rearing. The oiled eggs hatch at a lower rate and eggs that hatch frequently produce offspring with bone and bill deformities, as well as smaller liver sizes and slower growth rates (Albers, 1980; Goldsworthy et al., 2000; Hoffmann, 1990; Vidal et al., 2011; Weis, 2014).

In the wake of an oil spill in Antarctica near Palmer Station in 1989, a breeding colony of South Polar skuas (*Stercorarius maccormicki*) had 100% chick mortality (Eppley, 1992). In that colony, the adult skuas that became oiled while hunting for food traveled to freshwater ponds to clean themselves rather than returning to their nesting areas. This delay caused a breakdown in pair coordination, leaving chicks vulnerable and starved. Despite never coming into direct contact with oil, all the chicks died within a few weeks of the spill, mainly killed by other skuas (Eppley,1992; Eppley & Rubega, 1990; Smith et al., 1995).

In addition to the effects of hydrocarbons as oils, various other types of hydrocarbons can affect marine birds in a different manner. Plastics laden with DDT (dichlorodiphenyltrichloroethane) and its metabolites cause reproductive failure in brown pelicans (Gross et al., 2003), PCB (polychlorinated biphenyl) causes immunoincompetence and reproductive impairment in seabirds (Focardi et al., 1995; Luke et al., 1989; Stronkhorst, 1992), and PAHs cause lack of weight gain in nestlings and chronic toxicity in seabirds. PAHs enter the circulation, leading to tissue and plasma contamination and causing long-term chronic effects on common guillemots (*Uria aalge*) (Focardi et al., 1995). Heavy metals such as mercury (Hg) have adverse effects on neurological, physiological, behavioral, and reproduction success of many birds (Figure 12.4: Luke et al., 1989; Stronkhorst, 1992).

12.4 X-PRESS PEARL MARITIME DISASTER

While it was anchored about 10 nautical miles away from Colombo, the cargo vessel X-Press Pearl caught fire on May 20, 2021 and burned for almost two weeks (Figure 12.1), recording one of the worst environmental catastrophes in Sri Lankan history. At the time of burning, X-Press Pearl had 1486 shipping containers on board, with 81 of them categorized as dangerous cargo, containing hazardous chemicals such as 1040 tons of caustic soda (sodium hydroxide), 25 tons of nitric acid, 210 tons of methanol, several metals, including copper, 187 tons of lead, and aluminum, as well as lithium batteries, and an estimated 78 tons of and low-density polyethylene (LDPE) pellets, or commonly known as plastic nurdles (Rubesinghe et al., 2022; James et al., 2022; Karthik et al., 2022; Sewwandi et al., 2022).

The adverse environmental effects in relation to maritime and coastal birds caused by this incident can be categorized into three major groups. The first wave of effects was the air pollution from the burning chemicals and releasing gases, and the heat, flames, and chemical spill around the wreck. The area of impact is long range as the Yaas cyclone (Sewwandi et al., 2022) was active at that time with a wind speed of 60–70 km/h blowing inland; hence the fumes, chemicals, and debris dispersed into air and water. The second wave of environmental effects included billions of LDPE plastic pellets, various hazardous chemicals, oils, and materials from the shipwreck that entered into the water and eventually hit the beaches alongside the western coast. After the event, hundreds of dead turtles, fish, and other marine fauna washed up on shore, and tons of trash filled the beaches alongside the impact on the nearby communities. The third wave of effects is the long-term impacts on the marine environment and nearby terrestrial environments such as beaches, coastal lowlands, and coastal wetlands.

Considering the air pollution due to the X-Press Pearl incident, the National Building Research Organization (NBRO) determined that a 120 km^2 area was particularly susceptible to exposure, with an estimated 8000–13,000 tons of air pollutants being released into the air (James et al., 2022; Karthik et al., 2022; Rubesinghe et al., 2022; Sewwandi et al., 2022b). These pollutants may have contained a mixture of toxic substances, nitrogen oxides, sulfur dioxide, carbon monoxide, a variety of hydrocarbons, dioxins, heavy metals, and furans, depending on the assortment of materials that were on board the ship. The cargo that spilled out of the containers and was discovered along the beaches is mainly responsible for coastal and marine pollution.

A tier II oil spill warning was issued on May 25 (Rubesinghe et al., 2022) due to the 348 tons of bunker fuel oil carried by the X-Press Pearl at that moment. The actual amount of oil released into the ocean is unknown; however, a good portion of it must have been burned during the intense burning of the wreck. The ship also contained lubricating oil and gear oil, both of which could leak even during a later salvaging effort.

12.5 THE AFFECTED AREA AND IMMEDIATE IMPACT ON BIRDS

The X-Press Pearl accident took place on the western continental shelf between the Dickowita and Negombo harbors at about 10 km offshore (Figure 12.1). The affected shoreline spans approximately 250 km from Chilaw to Mirissa alongside the western and southwestern coastal areas (Figure 12.5). The Chilaw Sand Spits, which is a prominent point jutting out of the general contour of the coast, is a popular sea-watching site for birders where numerous species of seabirds can be aggregated due to their geological position. Similarly, Mirrissa is a prominent point where the continental shelf comes closer to the shore; hence sea-watching and whale-watching activities are conducted. Both these sites are oceanographically important locations as the western currents (at Chilaw sand spits) and southwestern sea currents (at Mirrissa) come close to the shore and rapidly divert into the deeper ocean. The Negombo Lagoon, two large estuaries formed by two main rivers, sandy beaches including two of the main tourist destinations of Sri Lanka, coral reefs along the western and south-western coast, mangrove forests, and coastal shrublands are the types of habitats affected by the accident (Figure 12.5). The Southwestern monsoons and the Yaas tropical storm of 2021 intensified the coastal currents that are moving mostly toward the south during the months of May and June (Pattiaratchi et al., 2022). Toxic fumes were blown directly into the inland and fogged the western coast near the accident site. Taking these factors into account, the species of birds recorded from Chilaw (51 km north of the accident site along the coastline) to Mirissa (170 km south of the accident site along the coastline) were used in the analysis, totaling 2431 km^2 from the 220 km transect consisting of 1 km inland and 10 km into the western horizon (Figure 12.5).

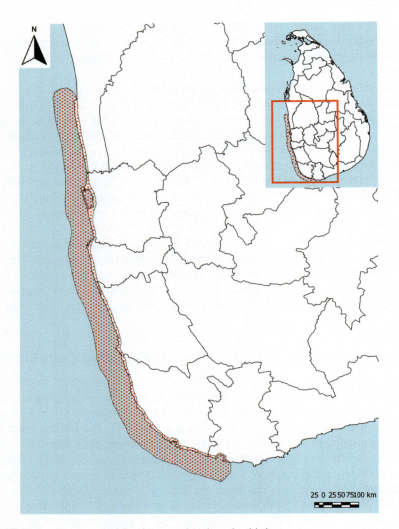

FIGURE 12.5 Area covered for the coastal and marine bird surveys.

12.5.1 Classification of Coastal and Marine Birds Based on Habitat Use

The birds that are associated with the coast can be classified into six categories:

Coastal grassland/shrubland birds: Species that are found in coastal scrublands, plantations, and grasslands. Most of them are medium to small insectivorous birds. They belong to the order Passeriformes (perching birds).
Coastal waterbirds: Species that are found in mangrove forests, intertidal mudflats, coastal canals, and such water-logged habitats. Storks, rails, and ducks are some examples.
Coastal raptors: Birds of prey found in the above two habitats are included here. Most of them depend on coastal fish, mammals, lizards, and amphibians as their prey.

Coastal shorebirds: Birds that are specially adapted to live on the shoreline. They forage and roost on the shore and therefore are particularly vulnerable to marine pollution. Some of these birds are migrants; hence they can carry some of these chemicals into inland habitats.

Near-shore seabirds: Seabirds that are foraging in the near shore are called near-shore seabirds. They feed fish near the shore or at fishing ports, and they roost onshore or on rocky outcrops near the shore and perform most of the activities except breeding in these affected waters.

Pelagic seabirds: Seabirds that are found in deep oceans but forage and migrate along the western and southern shoreline fall into this category. The number of pelagic birds especially goes up in the southwestern monsoon season as the strong southwest winds blow these birds closer to the western shore. During that time a large number of these birds come near-shore for shelter and foraging in the upwelling zone at the shelf-break.

12.6 PARAMETERS OF MARINE AND COASTAL BIRDS SUBJECTED TO INVESTIGATION

12.6.1 Foraging Ecology

Birds that feed in coastal and marine ecosystems can be exposed to chemical pollution through a variety of routes: direct consumption of toxic or other harmful chemicals, accident debris, and burned particles such as plastic pellets scattered in the environment, or ingesting indirectly through contaminated food or from the contaminated food web (Gallo et al., 2018; Huang et al., 2021; Islam & Tanaka, 2004; Pattiaratchi et al., 2022; Rochman, Manzano, et al., 2013; Sharma & Chatterjee, 2017).

Most seabirds and coastal raptors are at the top of the marine food web; therefore they are particularly vulnerable to bioaccumulation of chemical and biological pollutants. Shorebirds feed on marine invertebrates, which are filter feeders. The contamination in the lower levels of the food web, therefore, is directly affecting the food of shorebirds. Some coastal seabirds are scavengers; they feed on carcasses of large invertebrates and marine vertebrates. This wide spectrum of food types makes coastal and marine birds particularly vulnerable to marine pollution. Birds also drink water from coastal waterbodies and oceans. Therefore, water pollution can directly harm them as well.

12.6.2 Breeding Ecology

The majority of the coastal grassland birds, shorebirds, water birds, and seabirds breed on the ground, near water. Therefore, marine pollutants, contaminated water, washed-off debris, and contaminated food can affect nest building, the development of the egg, incubation, feeding young in a contaminated environment, contamination of the food web that could affect the developing chicks, and alteration of parental behavior due to certain contaminants. The depletion of food stocks (e.g. smaller fish) and the reduced productivity of the marine environment could be a consideration.

12.6.3 MIGRATION, ANNUAL MOLTING, AND METABOLISM

Most of the coastal shorebirds and seabirds in Sri Lanka are migrants. Some of them that are visiting the affected area are long-distance migrants (Panagoda et al. 2022a,b,c). To complete such an annual cycle, some of these birds had to fly over 20,000 km from Sri Lanka to the Arctic (Panagoda et al. 2022a). Therefore, it is an energetically expensive activity that requires feeding at critical periods. Migratory birds evolved to adjust their migratory routes over thousands of years. They may not be able to adjust in a year or two if the X-Press Pearl accident had caused drastic negative effects on their migratory routes, overwintering sites, or feeding areas.

Marine pollutants can alter molting cycles as well; they can damage young feathers and could drown molting birds (Figures 12.3 and 12.4; Camphuysen & Leopold 2004). Alteration of metabolism could lead to malnutrition and the depletion of fat reserves (Clark 1984), which in turn cause hypothermia while the bird is on the water.

12.7 DATA SOURCE AND SAMPLING METHOD

12.7.1 ON-FOOT BEACHED BIRD SURVEYS

During the month of May and June, a party of four experienced birders/biologists led by the authors conducted on-foot beach bird surveys along the entire length of the Dikkowita-Negombo beach up to the mouth of the Negombo Lagoon. The total length of the transect is 22 km.

Researchers walked along the beach looking for dead, injured, or distressed birds on the beach and near shore. Any unusual mortality of fish, marine invertebrates, marine mammals, and marine fauna was recorded as well.

Only two carcasses were recovered, including:

- Bridled tern *Onychoprion anaethetus* (1 specimen recovered from Dikkowita Beach)
- Sooty tern *Onychoprion fuscata* (1 specimen from Bopitiya Beach)

Specimens were collected for later laboratory analysis at the Laboratory for Molecular Ecology and Evolution, Department of Zoology and Environment Sciences, University of Colombo.

12.7.2 ONSHORE SEABIRD COUNTS

Onshore seabird counts were done from four locations to estimate the diversity and density of live pelagic seabirds in the area between 27 May and 15 June 2021 (Table 12.1). Specific raised platforms near carefully picked locations were used for this stationary survey for maximum visibility and effectiveness. Experienced seabird watchers conducted the survey using tripod-mounted Swarovski (80X20-60), Pentax (50X20-60), Nikon (80X20-40), and Vortex (80X20-60) spotting scopes.

TABLE 12.1
Diversity and Abundance of Seabirds from Negombo to Mt. Lavinia Coast Counted between 27 May and 15 June 2021

Seabird species	Estimated abundance
Sooty Tern	2100
Bridled Tern	14,400
Little Tern	900
Common Tern	600
Great Crested Tern	3000
Lesser Crested Tern	12,000
Whiskered Tern	840
Flesh-footed Shearwater	600
Wedge-tailed Shearwater	3300
Shearwater sp.	1800
Parasitic Jaeger	120
Yellow-billed Tropicbird	60
Total seabirds observed during the burning of the ship	**39,720**

12.8 BIRD DATA OBTAINED FROM A GLOBAL DATA REPOSITORY (PERIOD: JANUARY 2019 TO APRIL 2021)

Detailed bird observation data deposited as checklists by volunteer birders are stored in a global bird data repository called eBird (www.ebird.org). Data from January 1, 2019, to April 30, 2021, were used from eBird database to get the background bird numbers prior to the accident.

In total, 1,274,350 individual birds representing 159,155 observations were available for the transect. Of that, the bird observations from Chilaw Sand Spits to Mirissa, representing 1 km inland to 10 km offshore, were used, which accounted for 120,696 birds representing 15,087 observations, including 118 species of birds. The birds were categorized into coastal grassland, coastal wetland, coastal shorebirds, coastal raptors, coastal seabirds, and pelagic seabirds as described above (Figure 12.5).

12.9 BIRD DATA OBTAINED FROM A GLOBAL DATA REPOSITORY (PERIOD: MAY TO OCTOBER 2021)

To get the bird numbers that were directly relevant to the X-Press Pearl accident–related period, data from May 1 to October 31, 2021, were used from eBird database (www.ebird.org).

In total, 207,646 individual birds representing 21,891 observations were available for Puttalam, Gampaha, Colombo, Kalutara, and Galle Districts. Of that, the bird observations from Chilaw Sand Spits to Mirissa representing 1 km inland to 10 km offshore were used, which accounted for 191,608 birds representing 18,156 observations, including 129 species of birds (Tables 12.2 and 12.3, Figures 12.6, 12.7, 12.8, and 12.9).

TABLE 12.2
Birds Observed before (2019–2021) and during (in May to Oct 2021) X-Press Pearl Incident

District	Area	Before the event January 2019 to April 2021 Number of birds	Total Sp.	During the event May to Nov 2021 Number of birds	Total Sp.
Puttalam	Chilaw	2648	55	1690	20
Puttalam	Madampe	4992	59	2574	54
Puttalam	Wennappuwa	376	30		
Gampaha	Negombo	9456	81	89494	72
Gampaha	Ja-Ela	6224	48	2735	45
Gampaha	Wattala	33096	68	20309	68
Colombo	Colombo	4776	105	6044	58
Colombo	Dehiwala-Mt. Lavinia	3936	50	41803	65
Colombo	Moratuwa	0		405	25
Kalutara	Panadura	2480	61	594	24
Kalutara	Kalutara	5616	52		
Kalutara	Beruwala	7768	36		
Galle	Balapitiya	7176	89		
Galle	Ambalangoda	17176	45	4344	47
Galle	Hikkaduwa	1336	57	833	26
Galle	Galle	9960	61	20437	54
Galle	Habaraduwa	3680	48	346	38
Total		**120696**	**118**	**191608**	**129**

Source: www.ebird.org

TABLE 12.3
Different Categories of Birds Observed in the Affected Area during the Accident Period (May to November 2022)

Category	Total	Number of species
Coastal Grassland birds	70668	42
Coastal raptors	1488	10
Waterbirds	20382	35
Shorebirds	2295	18
Seabirds	95288	11
Pelagic seabirds	1487	13
TOTAL	191608	129

Impact on Coastal and Marine Birds 285

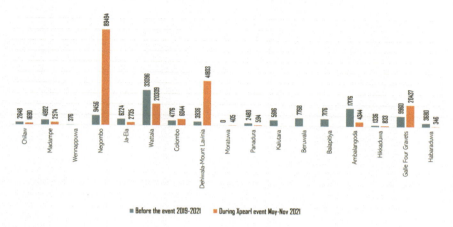

FIGURE 12.6 Bird numbers along the western and southern coast from Chilaw to Mirissa before (2019 January to 2021 April) and during the X-Press Pearl accident (May to November 2021).

Source: www.ebird.org

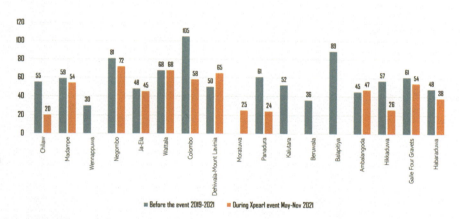

FIGURE 12.7 Bird species along the western and southern coast from Chilaw to Mirissa before (2019 January to 2021 April) and during the X-Press Pearl accident (May to November 2021).

Source: www.ebird.org

12.10 DEATH TOLL OBSERVED AFTER THE X-PRESS PEARL ACCIDENT

Only two carcasses of birds had been recovered from the immediate area of the accident site. The coastline of 220 km (Figure 12.5) did not yield dead or stressed seabirds. The recorded two specimens included a subadult bridled tern *O. anaethetus* (recovered from Dikkowita Beach) and a subadult sooty tern *O. fuscata* (from Bopitiya Beach). The specimens were collected for later laboratory analysis at the Laboratory for Molecular Ecology and Evolution, University of Colombo. The birds did not show any signs of injury, oil or plastic pellets in the gut. The fat levels (as measured from the subcutaneous

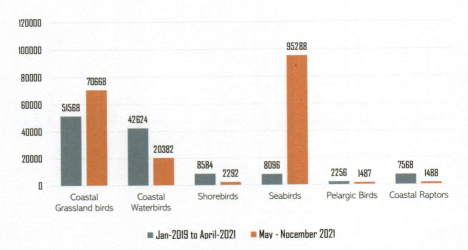

FIGURE 12.8 Birds from six categories distributed along the western and southern coast from Chilaw to Mirissa before (2019 January to 2021 April) and during the X-Press Pearl accident (May to November 2021).

Source: www.ebird.org

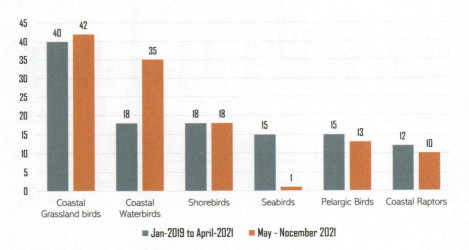

FIGURE 12.9 Bird species of different categories distributed along the western and southern coast from Chilaw to Mirissa before (2019 January to 2021 April) and during the X-Press Pearl accident (May to November 2021).

Source: www.ebird.org

fat deposits in the breast and belly; Kalnins et al. 2022) were very low in both specimens, suggesting that they were staved and probably stressed. The probable cause of death for those two birds could be the adverse weather caused by Yaas tropical storm.

Onshore sea watching at four locations in the morning hours recorded 39,720 seabirds, including sooty tern, bridled tern, little tern, common tern, great crested tern, lesser crested tern, whiskered tern, shearwater species, parasitic jaeger, and yellow-billed tropicbirds, during the burning of X-Press Pearl between 27 May and 15 June 2021 in the western sea border from Negombo to Mt. Lavinia (Table 12.1). Considering the diversity and density of seabirds in the area during the accident (Table 12.1), the lack of reported mortality (two birds) is unusual considering the previous maritime disasters outside the region.

12.11 A CRITICAL REVIEW OF THE LOW NUMBER OF BIRD DEATHS

Compared to the other marine accidents, the reasons behind the very low mortality (or no mortality) seen among birds during the X-Press Pearl incident are unclear. However, numerous species of fish, turtles, dolphins, and whales were found dead and beached carcasses were recorded around the island immediately after the accident.

The absence of a major oil spill might have prevented many Procellariiformes tubenose birds from immediate mortality. The rough sea due to the Yaas tropical storm and the southwestern monsoons could have prevented birds from landing on the water near the accident site for resting, which might have prevented them from direct contact with debris, oil, and other pollutants. Smoke, noise, and flames might have diverted birds from direct contact with the toxins in the vicinity. Again, the strong currents and wind blowing into the land might have kept the fumes and other pollutants directed toward the land and away from foraging marine birds.

In the other maritime disasters of similar magnitude, marine birds such as auks (family Alcidea), gulls (family Laridae), pelicans (family Pelecanidae), loons and grebes (family Podicipedidae), and sea ducks (family Anatidae) are the major victims. However, the members of these families were not present in southwestern coastal Sri Lanka, where the accident had taken place. The absence of such vulnerable families of birds in the region might have played a role in low mortality as well.

It has to be noted that only a very small percentage of carcasses can be recovered in a marine environment after a maritime accident. Therefore, the low mortality observed can be a complete oversight of the actual immediate impact of the accident on birds. Carcass floatation depends on the state of the body size of the bird (large birds are more likely to get detected), water temperature, and the direction and the force of the wind (Bibby 1981). The temperature of the water at sea can influence the fate of drifting avian carcasses, since carcasses in warmer waters may be scavenged or decompose quickly (Martin et al 2019). The overall recovery rate was 9.8% (59/600), with no significant variation between dropping sites in the Gulf of Mexico (Bibby 1981). In another study, only 11–16% of carcasses deployed in the nearshore were recovered. When deployed in the deep sea, both dummies and carcasses 100% were considered lost at sea (Wiese 2001, Martin et al. 2019). Bibby (1981) showed that bodies moved at about 4–12% of the wind speed.

Using a mark-recapture model to estimate the beaching probability of seabirds killed in nearshore waters, Himes Boor and Ford (2019) demonstrated that the carcasses in the Gulf of Mexico—or likely other relatively warm waters—begin to disappear immediately and continue to disappear with a roughly constant probability of 0.1468 per day. Thus, in warm waters, there would be only a probability of 0.269 that a carcass would remain floating for a week (i.e., surviving at sea for 7 days, $S^7 = 0.829^7$; Himes Boor & Ford 2019). Degradation or sinking can vary depending on the season, currents, and/or the distance a carcass would have to drift before reaching the shoreline (Van Pelt & Piatt, 1995; Varela & Zimmerman, 2019; Himes Boor & Ford, 2019).

12.12 IMPACTS DUE TO PLASTIC NURDLES AND OTHER CHEMICAL POLLUTANTS RELEASED FROM THE SHIPWRECK

Ingestion of plastic debris which is entangled among food sources (Awabdi et al., 2013a) such as grass buds and coastal seaweed where these birds forage could cause choking and malnutrition. Ingested plastic debris may effectively block the alimentary tract and/or reduce the feeding stimulus with lethal or sublethal effects (Caron et al., 2018a, 2018b). During the surveys, fish with nurdles in the gills and gut were reported. Birds too could have a similar fate.

Certain plastics and chemical residues of plastics can go through the food web (through coastal plants, invertebrates, insects, smaller vertebrates etc.) and biomagnify (Cecilia & Harry, 2003; Hernandez-Gonzalez et al., 2018a; Valderrama et al., 2018). Such magnifications can alter reproduction, molting, and stress physiology and ultimately reduce the fitness of the affected organism. The effects could be long term and yet to be seen.

Plastics laden with DDT (chlorinated hydrocarbon) and its metabolites cause reproductive failure in coastal birds (Jehl, 1973a). Lack of weight gain and chronic toxicity of PAHs cause mortality in larger predatory birds such as crows (Harris et al., 2000). PAHs can enter the circulation, leading to tissue and plasma contamination and causing long-term chronic effects on flight and molting (Troisi et al., 2006). Microplastics (high-density polyethylene [HDPE], polyethylene vinyl acetate [PEVA]) and additives have been identified in muscles in commercial fish in varying quantities (≤3–300 μm), suggesting that they can transfer into bird diet (Akhbarizadeh et al., 2018; Zitouni et al., 2020). Since most coastal birds forage on smaller fish, this accumulation of chemicals could contaminate them. Some of these chemicals can advance the degree of tissue changes (DTC) in liver (Karami et al., 2016; Rochman et al., 2013; Rochman et al., 2013). DEHP and DBP can alter the activity of enzymes involved in the synthesis of endogenous steroid hormones and their metabolism in fish, effects that can lead to the alteration of the balance between endogenous estrogens and androgens, and thus alterations in their reproductive physiology (Thibaut & Porte, 2004). The true effect of these contaminated fish on a bird's diet is yet to be established.

X-Press Pearl reportedly carried 1680 tons of plastic pellets (roughly 84 billion pellets). Based on previous plastic spills (Figure 12.2) and preliminary modeling, it is observed that they could affect coastlines in the regions extending from Indonesia and Malaysia to Somalia (James et al., 2022; Karthik et al., 2022; Rubesinghe et al., 2022; Sewwandi et al., 2022a).

12.13 MITIGATION AND LESSONS LEARNED

Oil spills have decreased due to a variety of preventive measures taken by the shipping industry, regulatory laws, and enforcement (Doerffer, 1992; EPA, 1999; Rubesinghe et al., 2022; Tran et al., 2014). However, there are more and more hazardous and noxious substances (HNS) being transported at sea, which might result in spills that are more difficult from both risk assessment and mitigation standpoints (Astiaso Garcia et al., 2013; Cho et al., 2013; Kim et al., 2019). The HNS definition applies to a number of the compounds transported on board X-Press Pearl. HNS spills are different from oil spills in that other chemicals can sink, producing hazardous, moving underwater plumes or float on the subsurface, whereas oil is assumed to float on the surface (Cho et al., 2013; Kim et al., 2019). These factors make the risk assessment more difficult, particularly in situations like X-Press Pearl, where it is unclear exactly what pollutants had escaped the disaster site.

By 25 May, 2021, according to the Marine Environment Protection Authority (MEPA) in Sri Lanka, cleaning operations had been carried out in around 250 locations from Mannar (Northwest) to Kirinda (South). Even though 899 tons of waste had been collected by August 4, 2021, the geographic spread of plastic pellets had been reported to be the largest on record and is reportedly increasing (James et al., 2022; Karthik et al., 2022; Rubesinghe et al., 2022; Sewwandi et al., 2022a). In some locations, the plastic pellets have reportedly accumulated to levels of 2 meters. Massive cleanup efforts won't be able to completely remove the pellets from the environment, and long-lasting effects are yet to be anticipated. The physical effects of the plastic pellets also include absorbed chemicals and metals, as well as additives for plastics.

A disaster of this scale and complexity is unavoidable with many unknowns since the cargo contained a complex mixture of pollutants. The risks would remain difficult to predict and uncertain throughout a few decades as we see from the previous such disasters that have happened to the marine and coastal ecosystems. This X-Press Pearl catastrophe along Sri Lanka's coast will undoubtedly continue for a very long time with mid- to long-term effects.

Many shipping regulations have been aimed at preventing and mitigating such disastrous oil spills, considering that today's container ships transport increasingly complex chemical and plastic mixtures, which can undoubtfully act as pollutants, should be regulated, and the laws and regulations should be more enforced compared to the old rules in order to avoid these ecological disasters. International agreements such as Safety of Life At Seas (SOLAS) Convention should be adequately updated to today's conditions along with setting up strategies for monitoring pollutants after plastic and chemical spills to provide early advice on appropriate restrictions and conservation measures.

12.14 RESTORATION

Beach cleaning to remove all possible plastic debris and other pollutants is essential as they tend to blow further inland and contaminate coastal ecosystems beyond the shoreline. Penetration of floating debris and contaminants via inland waterways such as canals should be prevented. And shores of such waterways up to several kilometers should be sampled and cleaned. Coastal waterways (examples such as Negombo Lagoon, Muthurajawela, Bolgoda Lagoon system, Maduganga Estuary, and Mahamodara Lagoon) should be monitored and shores of such waterways up to several kilometers inland should be sampled. Continued cleaning is essential as waves would carry material deposited from the sunken ship from the depths of the ocean in the next several monsoon seasons.

The containers with chemicals and other materials should be removed as soon as possible from the seabed to prevent further leakage of material into the ocean. Intense monitoring of sandy beaches, brackish and freshwater bodies for pollutants, and carcasses of birds should be carried out for chemical pollutants, ingested plastic nurdles, and other debris. The marine and wildlife protection agencies such as Marine Environment Protection Authority (MEPA) and Department of Wildlife Conservation (DWLC) together with Coast Conservation Department (CCA) should coordinate such action with experts in both the public and private sectors. Scientists in the university sector of Sri Lanka should be given funding and other logistics to carry out both short-term and long-term monitoring.

REFERENCES

Abelson, P. H. (1971). Marine pollution. *Science,* 171(3966), 21. https://doi.org/10.1126/science. 171.3966.21

Akhbarizadeh, R., Moore, F., & Keshavarzi, B. (2018). Investigating a probable relationship between microplastics and potentially toxic elements in fish muscles from northeast of Persian Gulf. *Environmental Pollution, 232,* 154–163. https://doi.org/10.1016/j.envpol. 2017.09.028

Albers, P. H. (1980). Transfer of crude oil from contaminated water to bird eggs. *Environmental Research, 22*(2), 307–314. https://doi.org/10.1016/0013-9351(80)90143-7

Astiaso Garcia, D., Cumo, F., Gugliermetti, F., & Rosa, F. (2013). Hazardous and Noxious Substances (HNS) Risk Assessment along the Italian Coastline. *Chemical Engineering Transactions, 32,* 115–120. https://doi.org/10.3303/CET1332020

Awabdi, D. R., Siciliano, S., & di Beneditto, A. P. M. (2013a). First record of killer whales (*Orcinus orca*) contents of juvenile green turtles, Chelonia mydas, in Rio de Janeiro, south-eastern Brazil. *Marine Biodiversity Records, 6,* e5. https://doi.org/10.1017/S1755267212001029

Bibby, C. J. (1981). An experiment on the recovery of dead birds from the North Sea. *Ornis Scandinavica,* 12, 261–265. www.jstor.org/stable/3676091

Bursian, S. J., Alexander, C. R., Cacela, D., Cunningham, F. L., Dean, K. M., Dorr, B. S., Ellis, C. K., Godard-Codding, C. A., Guglielmo, C. G., Hanson-Dorr, K. C., Harr, K. E., Healy, K. A., Hooper, M. J., Horak, K. E., Isanhart, J. P., Kennedy, L. v., Link, J. E., Maggini, I., Moye, J. K., ... Tuttle, P. L. (2017). Reprint of: Overview of avian toxicity studies for the Deepwater Horizon Natural Resource Damage Assessment. *Ecotoxicology and Environmental Safety, 146,* 4–10. https://doi.org/10.1016/j.ecoenv.2017.05.014

Camphuysen, C. J., & Heubeck, M. (2001). Marine oil pollution and beached bird surveys: The development of a sensitive monitoring instrument. *Environmental Pollution, 112*(3), 443–461. https://doi.org/10.1016/S0269-7491(00)00138-X

Camphuysen, C. J., & Leopold, M. F. (2004). The Tricolor oil spill: Characteristics of seabirds found oiled in The Netherlands. *Atlantic Seabirds*, *6*(3), 109–128.

Caron, A. G. M., Thomas, C. R., Berry, K. L. E., Motti, C. A., Ariel, E., & Brodie, J. E. (2018a). Ingestion of microplastic debris by green sea turtles (*Chelonia mydas*) in the Great Barrier Reef: Validation of a sequential extraction protocol. *Marine Pollution Bulletin*, *127*, 743–751. https://doi.org/10.1016/J.MARPOLBUL.2017.12.062

Caron, A. G. M., Thomas, C. R., Berry, K. L. E., Motti, C. A., Ariel, E., & Brodie, J. E. (2018b). Ingestion of microplastic debris by green sea turtles (Chelonia mydas) in the Great Barrier Reef: Validation of a sequential extraction protocol. *Marine Pollution Bulletin*, *127*, 743–751. https://doi.org/10.1016/J.MARPOLBUL.2017.12.062

Rubesinghe, C., Brosché, S., Withanage, H., Pathragoda, D., & Karlsson, T. (2022) X-Press Pearl, a 'new kind of oil spill' consisting of a toxic mix of plastics and invisible chemicals. International Pollutants Elimination Network (IPEN). https://ipen.org/sites/default/files/documents/ipen-sri-lanka-ship-fire-v1_2aw-en.pdf

Cecilia, E., & Harry, B. (2003). Origins and biological accumulation of small plastic particles in fur seals from Macquarie Island. *AMBIO: A Journal of the Human Environment*, *32*(6), 380–384.

Cho, S.-J., Kim, D.-J., & Choi, K.-S. (2013). Hazardous and noxious substances (HNS) risk assessment and accident prevention measures on domestic marine transportation. *Journal of the Korean Society of Marine Environment & Safety*, *19*(2), 145–154. https://doi.org/10.7837/KOSOMES.2013.19.2.145

Clark, R. B. (1984). Impact of oil pollution on seabirds. *Environmental Pollution. Series A, Ecological and Biological*, *33*(1), 1–22. https://doi.org/10.1016/0143-1471(84)90159-4

Croxall, J. P., Butchart, S. H. M., Lascelles, B., Stattersfield, A. J., Sullivan, B., Symes, A., & Taylor, P. (2012). Seabird conservation status, threats and priority actions: A global assessment. *Bird Conservation International*, *22*(1), 1–34. https://doi.org/10.1017/S0959270912000020

Day, R. H., Murphy, S. M., Wiens, J. A., Hayward, G. D., Harner, E. J., & Smith, L. N. (1997). Effects of the Exxon Valdez oil spill on habitat use by birds in Prince William Sound, Alaska. *Ecological Applications*, *7*(2), 593–613.

Dempsey, P. S. (1985). Compliance and enforcement in International Law – oil pollution of the marine environment by ocean vessels. *Northwestern Journal of International Law & Business*, *6*(2), 459–561.

Doerffer, J. W. (1992). Oil spill response in the marine environment. In *Oil Spill Response in the Marine Environment*. Elsevier. https://doi.org/10.1016/c2009-0-11195-9

EPA. (1999). *Understanding Oil Spills and Oil Spill Response – Ch 5: Wildlife And Oil Spills 5*. https://books.google.lk/books?hl=en&lr=&id=px5SAAAAMAAJ&oi=fnd&pg=PA1&dq=Understanding+Oil+Spills+And+Oil+Spill+Response.+Retrieved+from+&ots=cJ9SpPfmQq&sig=3CNA7z_GMZeOX_9tbgC0BBs2E18&redir_esc=y#v=onepage&q=Understanding Oil Spills And Oil Spill Response

Eppley, Z. A. (1992). Assessing indirect effects of oil in the presence of natural variation: the problem of reproductive failure in South Polar skuas during the Bahia Paraiso oil spill. *Marine Pollution Bulletin*, *25*(9–12), 307–312.

Eppley, Z., & Rubega, M. (1990). Indirect effects of an oil spill: reproductive failure in a population of South Polar skuas following the "Bahia Paraiso" oil spill in Antarctica. *Marine Ecology Progress Series*, *67*(1), 1–6. https://doi.org/10.3354/meps067001

Faksness, L.-G., Brandvik, P. J., Daling, P. S., Singsaas, I., & Sørstrøm, S. E. (2016a). The value of offshore field experiments in oil spill technology development for Norwegian waters. *Marine Pollution Bulletin*, *111*(1–2), 402–410.

Faksness, L. G., Brandvik, P. J., Daling, P. S., Singsaas, I., & Sørstrøm, S. E. (2016b). The value of offshore field experiments in oil spill technology development for Norwegian waters. *Marine Pollution Bulletin, 111*(1–2), 402–410. https://doi.org/10.1016/j.marpolbul.2016.07.035

Fallon, J. A., Smith, E. P., Schoch, N., Paruk, J. D., Adams, E. A., Evers, D. C., Jodice, P. G. R., Perkins, C., Schulte, S., & Hopkins, W. A. (2018). Hematological indices of injury to lightly oiled birds from the Deepwater Horizon oil spill. *Environmental Toxicology and Chemistry, 37*(2), 451–461. https://doi.org/10.1002/etc.3983

Focardi, S., Bargagli, R., & Corsolini, S. (1995). Isomer-specific analysis and toxic potential evaluation of polychlorinated biphenyls in Antarctic fish, seabirds and Weddell seals from Terra Nova Bay (Ross Sea). *Antarctic Science, 7*(1), 31–35. https://doi.org/10.1017/S095410209500006X

Gallo, F., Fossi, C., Weber, R., Santillo, D., Sousa, J., Ingram, I., Nadal, A., & Romano, D. (2018). Marine litter plastics and microplastics and their toxic chemicals components: the need for urgent preventive measures. In *Environmental Sciences Europe* (Vol. 30, Issue 1). Springer Verlag. https://doi.org/10.1186/s12302-018-0139-z

Goldsworthy, S. D., Gales, R. P., Giese, M., & Brothers, N. (2000). Effects of the Iron Baron oil spill on little penguins (*Eudyptula minor*). I. Estimates of mortality. *Wildlife Research, 27*(6), 559–571. https://doi.org/10.1071/WR99075

Gross, T. S., Arnold, B. S., Sepúlveda, M. S. & McDonald, K. (2003). Endocrine disrupting chemicals and endocrine active agents. In *Handbook of Ecotoxicology*, 2nd Edition (pp. 1033–1098). Lewis Publishers: Boca Raton, FL.

Haney, J. C., Geiger, H. J., & Short, J. W. (2014). Bird mortality from the Deepwater Horizon oil spill. I. Exposure probability in the offshore Gulf of Mexico. *Marine Ecology Progress Series, 513*, 225–237. https://doi.org/10.3354/meps10991

Harris, M. P., Wanless, S., & Webb, A. (2000). Changes in body mass of common guillemots *Uria aalge* in southeast Scotland throughout the year: Implications for the release of cleaned birds. *Ringing and Migration, 20*(2), 134–142. https://doi.org/10.1080/03078698.2000.9674235

Helm, R. C., Carter, H. R., Glenn Ford, R., Michael Fry, D., Moreno, R. L., Sanpera, C., & Tseng, F. S. (2015). Overview of Efforts to Document and Reduce Impacts of Oil Spills on Seabirds. In *Handbook of Oil Spill Science and Technology* (pp. 429–453). Wiley Online Library. https://doi.org/10.1002/9781118989982.ch17

Hernandez-Gonzalez, A., Saavedra, C., Gago, J., Covelo, P., Santos, M. B., & Pierce, G. J. (2018a). Microplastics in the stomach contents of common dolphin (*Delphinus delphis*) stranded on the Galician coasts (NW Spain, 2005–2010). *Marine Pollution Bulletin, 137*, 526–532. https://doi.org/10.1016/J.MARPOLBUL.2018.10.026
https://doi.org/10.1016/J.MARPOLBUL.2018.10.026

Himes Boor, G. K., & Ford, R. G. (2019). Using a mark-recapture model to estimate beaching probability of seabirds killed in nearshore waters during the Deepwater Horizon oil spill. *Environmental Monitoring and Assessment, 191*(4), 1–12. https://doi.org/10.1007/s10661-019-7919-9

Hoffman, D.J. (1990). Embryotoxicity and Teratogenicity of Environmental Contaminants to Bird Eggs, in G.W. Ware (ed.), *Reviews of Environmental Contamination and Toxicology: Continuation of Residue Reviews*. Springer: New York, NY, pp. 39–89.

Huang, W., Song, B., Liang, J., Niu, Q., Zeng, G., Shen, M., Deng, J., Luo, Y., Wen, X., & Zhang, Y. (2021). Microplastics and associated contaminants in the aquatic environment: A review on their ecotoxicological effects, trophic transfer, and potential impacts to human health. *Journal of Hazardous Materials, 405*. https://doi.org/10.1016/j.jhazmat.2020.124187

Irons, D. B., Kendall, S. J., Erickson, W. P., McDonald, L. L., & Lance, B. K. (2000). Nine years after the Exxon Valdez oil spill: Effects on marine bird populations in Prince William Sound, Alaska. *Condor*, *102*(4), 723–737. https://doi.org/10.2307/1370300

Islam, M. S., & Tanaka, M. (2004). Impacts of pollution on coastal and marine ecosystems including coastal and marine fisheries and approach for management: A review and synthesis. *Marine Pollution Bulletin*, 48, Issues 7–8, pp. 624–649. https://doi.org/10.1016/j.marpolbul.2003.12.004

James, B. D., de Vos, A., Aluwihare, L. I., Youngs, S., Ward, C. P., Nelson, R. K., Michel, A. P. M., Hahn, M. E., & Reddy, C. M. (2022). Divergent forms of pyroplastic: Lessons learned from the M/V X-Press Pearl ship fire. *ACS Environmental Au*, *2*(5), 467–479. https://doi.org/10.1021/acsenvironau.2c00020

Jehl, J. R. (1973a). The distribution of marine birds in Chilean Waters. *The Auk*, *90*(1), 114–135. www.jstor.org/stable/4084021

Joseph R. Jehl, (1973). Studies of a Declining Population of Brown Pelicans in Northwestern Baja California, *The Condor*, 75(1), 69–79. https://doi.org/10.2307/1366536

Kalnins, L., Krüger, O., & Krause, E. T. (2022). Plumage and fat condition scores as well-being assessment indicators in a small passerine bird, the zebra finch (*Taeniopygia guttata*). *Frontiers in Veterinary Science*, *9*, 791412.

Karami, A., Romano, N., Galloway, T., & Hamzah, H. (2016). Virgin microplastics cause toxicity and modulate the impacts of phenanthrene on biomarker responses in African catfish (*Clarias gariepinus*). *Environmental Research*, *151*, 58–70. https://doi.org/10.1016/j.envres.2016.07.024

Karthik, R., Robin, R. S., Purvaja, R., Karthikeyan, V., Subbareddy, B., Balachandar, K., Hariharan, G., Ganguly, D., Samuel, V. D., Jinoj, T. P. S., & Ramesh, R. (2022). Microplastic pollution in fragile coastal ecosystems with special reference to the X-Press Pearl maritime disaster, southeast coast of India. *Environmental Pollution*, *305*, 119297. https://doi.org/10.1016/j.envpol.2022.119297

Katsanevakis, S. (2008). Marine Debris, a Growing Problem: Sources. *Distribution, Composition, and Impacts in: Hofer, Tobias N.(Hg.), Marine Pollution New Research, New York*, 53–100.

Kim, Y. R., Lee, M., Jung, J. Y., Kim, T. W., & Kim, D. (2019). Initial environmental risk assessment of hazardous and noxious substances (HNS) spill accidents to mitigate its damages. *Marine Pollution Bulletin*, *139*, 205–213. https://doi.org/10.1016/J.MARPOLBUL.2018.12.044

Luke, B. G., Johnstone, G. W., & Woehler, E. J. (1989). Organochlorine pesticides, PCBs and mercury in Antarctic and subantarctic seabirds. *Chemosphere*, *19*(12), 2007–2021. https://doi.org/10.1016/0045-6535(89)90024-6

Martin, N., Varela, V. W., Dwyer, F. J., Tuttle, P., Ford, R. G., & Casey, J. (2019). Evaluation of the fate of carcasses and dummies deployed in the nearshore and offshore waters of the northern Gulf of Mexico. *Environmental Monitoring and Assessment*, *191*(4), 1–13. https://doi.org/10.1007/s10661-019-7923-0

Murphy, S. M., Day, R. H., Wiens, J. A., & Parker, K. R. (1997). Effects of the Exxon Valdez oil spill on birds: Comparisons of pre-and post-spill surveys in Prince William Sound, Alaska. *The Condor*, *99*(2), 299–313.

Panagoda, P. A. B. G., Meng, F., Zhang, B., Mundkur, T., Balachandran, S., Kotagama, S. W., Lei, C., & Seneviratne, S. S. (2022a). Tracking the lesser-known migratory routes of Heuglin's Gulls (Larus heuglini), from Sri Lanka to the Arctic. 78th Annual Scientific Sessions of the Sri Lanka Association for the Advancement of Science, Sri Lanka.

Panagoda, P. A. B. G., Meng, F., Zhang, B., Mundkur, T., Balachandran, S., Kotagama, S. W., Lei, C., & Seneviratne, S. S. (2022b). Mannar island; a centre for the movement of waterbirds within the Palk Bay and Gulf of Mannar. Proceedings of the Postgraduate Institute of Science Research Congress, Sri Lanka: 28th–29th October 2022, Sri Lanka.

Panagoda, P. A. B. G., Meng, F., Zhang, B., Mundkur, T., Balachandran, S., Kotagama, S. W., Lei, C., & Seneviratne, S. S. (2022c). Overflying the Himalayas; the Northward Migration of Sri Lankan-wintering Brown-headed Gulls. Proceedings of the International Research Conference of the Sir John Kothalawala Defense University, Sri Lanka: September 2022, Sri Lanka.

Pattiaratchi, C., van der Mheen, M., Schlundt, C., Narayanaswamy, B. E., Sura, A., Hajbane, S., White, R., Kumar, N., Fernandes, M., & Wijeratne, S. (2022). Plastics in the Indian Ocean-sources, transport, distribution, and impacts. In *Ocean Science* (Vol. 18, Issue 1, pp. 1–28). Copernicus GmbH. https://doi.org/10.5194/os-18-1-2022

Rubesinghe, C., Brosche, S., Withanage, H., Pathragoda, D., & Karlsson, T. (2022). X-Press Pearl: a "New Kind of Oil Spill." *International Pollutants Elimination Network (IPEN), February.* https://ipen.org/sites/default/files/documents/ipen-sri-lanka-ship-fire-v1_2aw-en.pdf

Piatt, J. F., Lensink, C. J., Butler, W., & Nysewander, D. R. (1990). Immediate Impact of the "Exxon Valdez" Oil Spill on Marine Birds. *The Auk, 107*(2), 387–397. https://doi.org/10.2307/4087623

Prince, R. C., & Bragg, J. R. (1997). Shoreline bioremediation following the Exxon Valdez oil spill in Alaska. *Bioremediation Journal, 1*(2), 97–104. https://doi.org/10.1080/10889869709351324

Roberts, J. (2005). Protecting sensitive marine environments: The role and application of ships' routeing measures. *International Journal of Marine and Coastal Law, 20*(1), 135–159. https://doi.org/10.1163/157180805774851599

Rochman, C. M., Hoh, E., Kurobe, T., & Teh, S. J. (2013). Ingested plastic transfers hazardous chemicals to fish and induces hepatic stress. *Scientific Reports, 3*(1), 1–7. https://doi.org/10.1038/srep03263

Rochman, C. M., Manzano, C., Hentschel, B. T., Simonich, S. L. M., & Hoh, E. (2013). Polystyrene plastic: A source and sink for polycyclic aromatic hydrocarbons in the marine environment. *Environmental Science and Technology, 47*(24), 13976–13984. https://doi.org/10.1021/es403605f

Sewwandi, M., Amarathunga, A. A. D., Wijesekara, H., Mahatantila, K., & Vithanage, M. (2022a). Contamination and distribution of buried microplastics in Sarakkuwa beach ensuing the MV X-Press Pearl maritime disaster in Sri Lankan sea. *Marine Pollution Bulletin, 184*, 114074. https://doi.org/10.1016/j.marpolbul.2022.114074

Sewwandi, M., Amarathunga, A. A. D., Wijesekara, H., Mahatantila, K., & Vithanage, M. (2022b). Contamination and distribution of buried microplastics in Sarakkuwa beach ensuing the MV X-Press Pearl maritime disaster in Sri Lankan sea. *Marine Pollution Bulletin, 184*. https://doi.org/10.1016/j.marpolbul.2022.114074

Sharma, S., & Chatterjee, S. (2017). Microplastic pollution, a threat to marine ecosystem and human health: A short review. *Environmental Science and Pollution Research, 24*(27), 21530–21547. https://doi.org/10.1007/s11356-017-9910-8

Smith, R., Baker, K., Fraser, W., Hofmann, E., Karl, D., Klink, J., Quentin, L., Prezelin, B., Ross, R., Trivelpiece, W., & Vernet, M. (1995). The Palmer LTER: A long-term ecological research program at Palmer Station, Antarctica. *Oceanography, 8*(3), 77–86. https://doi.org/10.5670/oceanog.1995.01

Stronkhorst, J. (1992). *Thesis for MSc. Ecotoxicology of Natural Populations, University of Reading, Reading, United Kingdom.*

Thibaut, R., & Porte, C. (2004). Effects of endocrine disrupters on sex steroid synthesis and metabolism pathways in fish. *Journal of Steroid Biochemistry and Molecular Biology, 92*(5), 485–494. https://doi.org/10.1016/j.jsbmb.2004.10.008

Tran, T., Yazdanparast, A., & Suess, E. A. (2014). Effect of oil spill on birds: A graphical assay of the Deepwater Horizon oil spill's impact on birds. *Computational Statistics, 29*(1–2), 133–140. https://doi.org/10.1007/s00180-013-0472-z

Troisi, G. M., Bexton, S., & Robinson, I. (2006). Polyaromatic hydrocarbon and PAH metabolite burdens in oiled common guillemots (*Uria aalge*) stranded on the East Coast of England (2001–2002). *Environmental Science and Technology, 40*(24), 7938–7943. https://doi.org/10.1021/es0601787

Tunnell, J. W., Dunning, K. H., Scheef, L. P., & Swanson, K. M. (2020). Measuring plastic pellet (nurdle) abundance on shorelines throughout the Gulf of Mexico using citizen scientists: Establishing a platform for policy-relevant research. *Marine Pollution Bulletin, 151*, 110794. https://doi.org/10.1016/j.marpolbul.2019.110794

Valderrama, S. P., Avila, A. H., Méndez, J. G., Martínez, O. M., Rojas, D. C., Azcona, H. F., Hernández, E. M., Aragón, H. C., Alcolado, P. M., Pina-Amargós, F., González, Z. H., Pantoja, L. E., & Farrat, L. F. R. (2018). Marine protected areas in Cuba. *Bulletin of Marine Science, 94*(2), 423–442. https://doi.org/10.5343/bms.2016.1129

Van Pelt, T. I., & Piatt, J. F. (1995). Deposition and persistence of beach cast seabird carcasses. *Marine Pollution Bulletin, 30*(12), 794–802.

Varela, V. W., & Zimmerman, G. S. (2019). Persistence of avian carcasses on sandy beaches and marsh edges in the northern Gulf of Mexico. *Environmental Monitoring and Assessment, 191*(4), 1–18.

Vidal, M., Domínguez, J., & Luís, A. (2011). Spatial and temporal patterns of polycyclic aromatic hydrocarbons (PAHs) in eggs of a coastal bird from northwestern Iberia after a major oil spill. *Science of the Total Environment, 409*(13), 2668–2673. https://doi.org/10.1016/j.scitotenv.2011.03.025

Vikas, M., & Dwarakish, G. S. (2015). Coastal pollution: A review. *Aquatic Procedia, 4*, 381–388.

Walter, S. T., Carloss, M. R., Hess, T. J., & Leberg, P. L. (2014). Demographic trends of brown pelicans in Louisiana before and after the Deepwater Horizon oil spill. *Journal of Field Ornithology, 85*(4), 421–429. https://doi.org/10.1111/jofo.12081

Weis, J. S. (ed.) (2014). Physiological, developmental and behavioral effects of marine pollution. In *Physiological, Developmental and Behavioral Effects of Marine Pollution.* Springer. https://doi.org/10.1007/978-94-007-6949-6

Whitton, B. A., Phillips, D. J. H., & Rainbow, P. S. (1994). Biomonitoring of trace aquatic contaminants. *The Journal of Applied Ecology, 31*(3), 595. https://doi.org/10.2307/2404456

Wiens, J. A., Crist, T. O., Day, R. H., Murphy, S. M., & Hayward, G. D. (1996). Effects of the Exxon Valdez oil spill on marine bird communities in Prince William Sound, Alaska. *Ecological Applications, 6*(3), 828–841.

Wiese, F. K., Montevecchi, W. A., Davoren, G. K., Huettmann, F., Diamond, A. W., & Linke, J. (2001). Seabirds at risk around offshore oil platforms in the North-west Atlantic. *Marine Pollution Bulletin, 42*(12), 1285–1290. https://doi.org/10.1016/S0025-326X(01)00096-0

Wright, D. A. (2004). The Empire of the St. Lawrence: A study in commerce and politics (review). *The Canadian Historical Review, 85*(3), 555–558. https://doi.org/10.1353/can.2004.0142

Wright, S. K., Allan, S., Wilkin, S. M., & Ziccardi, M. (2022). Oil spills in the Arctic. In Tryland, M. (Ed.), *Arctic One Health: Challenges for Northern Animals and People* (pp. 159–192). Cham: Springer International Publishing.

Zitouni, N., Bousserrhine, N., Belbekhouche, S., Missawi, O., Alphonse, V., Boughatass, I., & Banni, M. (2020). First report on the presence of small microplastics (≤ 3 μm) in tissue of the commercial fish *Serranus scriba* (Linnaeus. 1758) from Tunisian coasts and associated cellular alterations. *Environmental Pollution, 263*, 114576. https://doi.org/10.1016/j.envpol.2020.114576

13 Beach and Marine Microplastics
Physiochemical Removal Techniques Targeting Marine Disasters

Kalani Imalka Perera, Madushika Sewwandi,
Indika Hema Kumara Wijerathna,
Sujitra Vassanadumrongdee, and
Meththika Vithanage

13.1 OCCURRENCE AND DISTRIBUTION OF MARINE MICROPLASTICS

13.1.1 Sources and Pathways

The presence of microplastics in the marine environment has become a massive problem to date due to the widespread consumption and extensive and improper management of plastic waste. Marine plastic waste can be categorized as anthropogenic and natural flotsam. Since all the water systems end up in the sea, the anthropogenic load of plastic in marine environments is increasing day by day (Anderson et al., 2015). Thus plastic waste brought in by effluent from stormwater runoff, wastewater treatment plants, and drainage systems pollutes the marine ecosystem. Plastic waste and microplastics on land can also enter the marine system through storms and flash floods (Zhang et al., 2022). Plastic waste discharged by harbors and marine industries also causes the ubiquitous presence of plastic waste in the marine ecosystem. Plastic waste is abundant and widespread in coastal areas due to the disposal of plastic litter and washed-off plastic waste from the sea (Paul et al., 2022). The adjoining oceanic water with the waste disposal of surrounding industrialized regions also causes the common presence of microplastics on beaches. Atmospheric deposition of suspended microplastics also contributes the microplastic pollution in coastal zones (Liu et al., 2019; Szewc et al., 2021). Metrological factors such as air humidity, precipitation height, air mass routes, and wind speed have been recognized as major influencers for marine microplastic contamination (Szewc et al., 2021). The deposition of microplastics in the ocean during nighttime is higher than daytime deposition (Liu et al., 2019). The long-term existence of plastic waste on beaches or in the sea leads to the formation of microplastics undergoing biological, chemical, and physical weathering (Corcoran, 2020).

DOI: 10.1201/9781003314301-13

Plastic litter on beaches undergoes rapid solar UV weathering since coastal areas are rich in atmospheric oxygen and UV irradiation (Corcoran, 2022). Weathering causes the discoloration of the plastic debris, making it weaker and brittle during advanced stages of degradation.

Embrittled and highly weathered plastic debris is easily fragmented with any mechanical force such as animal bite, wind, wave, and human activities and ultimately causes the quick formation of microplastics in the marine environment (Bajt, 2021). The light-induced degradation of secondary plastics is enhanced by the UV and heat stabilizers and antioxidants used as additives. Nevertheless, depending on the factors, polymer type, temperature, mechanical force, salinity, biofouling, hydrostatic pressure, and the presence of oil in seawater, the degradation rate of the marine plastics and thereby microplastic formation is varied.

Other than the derivation of microplastics from plastic waste in the marine environment, microplastics can be added to the sea due to maritime accidents of ships (Annette and Gaëlle, 2020; Schumann et al., 2019). Since the plastic industry has become popular for decades, most countries largely produce plastic items for various purposes. The worldwide distribution of primary plastic beads, nurdles, fiber, or pellets mostly occurs through marine transport. Different types of issues, collisions, storms, fires, technical failures, and damages in engines cause maritime accidents (Annette and Gaëlle, 2020; Häkkien et al., 2013). The accidental spillages of microplastics cause the distribution of them on the beach area after they wash upon the beach with continuous sea waves (Figure 13.1). Massive maritime accidents have been reported worldwide and the microplastic pollution caused by them in marine environments in the North Sea, South Africa, New Zealand, Hong Kong, and Durban has been largely discussed in the 20th century (Annette and Gaëlle, 2020). The most recent incredible incident has been reported in Sri Lanka after a fire caught in the X-Press Pearl cargo vessel while transporting 31.9 tons of polystyrene pellets, 81.3 tons of plastic pellets, 19.3 tons of polymeric beads, and 9700 tons of potentially toxic epoxy resins (Hassan et al., 2021). Other than the spillage of primary microplastic nurdles from the fired ship vessel, partially burned large plastic debris was also added to the sea during the shipwreck. That large plastic debris was fragmented with the waves forming of partially burnt microplastics (Sewwandi et al., 2022b).

13.1.2 FATE AND ENVIRONMENTAL IMPACTS FROM ACCIDENTAL SPILLAGE OF MICROPLASTICS

13.1.2.1 Distribution

The massive and complex ocean environment greatly influences the fate of microplastics, changing their destination in the ocean. Spilled microplastics into the sea are not restricted to a specific sea region as they start to scatter everywhere since ocean currents, breezes, and stream discharges have a substantial impact on their transport (Gündoğdu et al., 2022). Vertically shearing oceanic currents and wind-occurring sea environments cause the transport of nurdles into, out of, and within a particular sea region (Kanhai et al., 2022; Lentz and Fewings, 2011). Since microplastics are lightweight and float in seawater, their distribution is very fast

Beach and Marine Microplastics 299

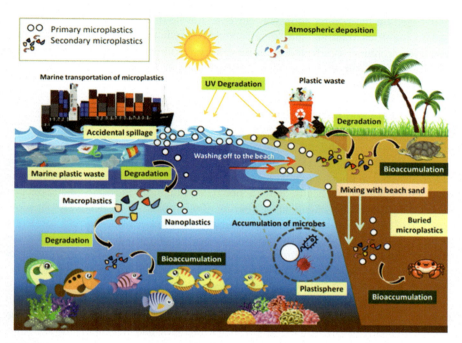

FIGURE 13.1 The sources, possible pathways, and fate of microplastics in marine environments.

with wave action. Therefore, wave and wind-induced microplastic transport greatly causes the higher abundances of microplastics on beaches (Gündoğdu et al., 2022). Accordingly, microplastics are accumulated in the nearby coastal areas after a few days from the accidental spillage. Several examples of the scattering of spilled microplastics in marine environments can be picked up in the literature.

A shipping accident that happened in the Friesland region in 2019 caused the spill of a large number of plastic pellets, and after a few months, nurdle pollution has been reported in approximately 700 locations along the Norwegian coast and a few locations in Denmark and Sweden (Annette and Gaëlle, 2020). Further, the microplastic nurdles spilled from the X-Press Pearl vessel to the Sri Lankan Sea were distributed to many coastal areas around the island and deposited over the beaches to two to three days after the disaster (Sewwandi et al., 2022b). Since the incident occurred during the season of high waves and strong winds, the highest nurdle accumulation was reported on beaches Pamunugama and Sarakkuwa which are directly onshore of the shipwreck (Hassan et al., 2021). The first appearance of nurdles was reported along the southwest coast (approximately 300 km stretch) of Sri Lanka and then to the east coast due to the prevailing southwest monsoon during those days. According to the predicted pathways of nurdles through a nurdle distribution modeling based on monsoon wind and currents, the nurdles have a higher chance of transporting toward the east impacting Indonesia (Pattiaratchi and Wijeratne, 2021).

Thus based on the reverse monsoon currents, nurdle distribution to the beaches in western countries Maldives and Somalia could be predicted. Consequently, the entire marine environment is affected by those disasters without limiting it to a particular area. Eventually, the wide spread of microplastics and their existence disturb the marine environment.

13.1.2.2 Fate and Environmental Impact

The preference of microbes in water and soil to live on microplastics has been confirmed with the soundings on the plastisphere through several studies (Yang et al., 2020; Zhang et al., 2021). Therefore, floating microplastics in the seawater would be an excellent home for pathogenic marine microbes, causing bad consequences for the environment and animal health. Wave action, UV irradiation, and atmospheric oxygen trigger the physical and chemical weathering of microplastics, respectively (Bandow et al., 2017). The long-term presence of microplastics in seawater would accelerate their degradation, causing the origination of nanoplastics which can severely impact environmental pollution and animal health than microplastics since the environmental complexity of nanoplastics is distinct from microplastics (ter Halle and Ghiglione, 2021). Apart from that, microplastics washed up on the beach tend to be buried in beach sand, causing beach microplastic pollution. Microplastics can easily be ground down by the different sizes of grains in beach sand. The depositional behavior of the beach with the dynamics of the beach, seasonal changes, and anthropogenic activities on the beach cause the rapid burial of microplastics, increasing the incorporation of highly dense microplastics in the sand. After the X-Press Pearl maritime disaster, the plastic nurdles and partially burned microplastics deposited on Sarakkuwa Beach were buried in the beach over 1 m depth (Sewwandi et al., 2022a).

The potentially harmful effects of microplastics on a diverse range of aquatic fauna and humans have raised concerns around the world. Microplastics pose a pervasive and preventable threat to the health of marine ecosystems because they are persistent and common pollutants in the water. There are many different types, sizes, and shapes of microplastics, all of which have different physical and chemical characteristics as well as toxicological effects. Marine organisms are likely to ingest microplastics through water, food, or inhalation. (Arias et al., 2019; Cox et al., 2019). The tiny pieces that float in the seawater would mislead small fishes with their food. Microplastics are trophic transferred through the food chain from primary producers to upper predators and bioaccumulated in them. Therefore, the presence of microplastics in marine environments directly affects fish health, lowering feeding intensity, reducing reproducibility, and inducing gill dysfunction and immunosuppression (Mallik et al., 2021). Trophic transfer of microplastics can enhance the accumulation of toxic plastic-derived chemicals such as bisphenols and phthalates and in fish (Capó et al., 2022). Marine organisms at higher trophic levels are more exposed to microplastics and related chemicals than organisms with low trophic levels (Hasegawa et al., 2021). The surface and buried microplastics also would badly impact beach organisms (de Souza Machado et al., 2019; Fernandino et al., 2015). The long-term presence of microplastics on the beach also would harm the coastal vegetation. The burial of microplastics in turtle-nesting beach environments would endanger the life of turtles, influencing their health (Duncan et al., 2018). Ultimately, humans would be exposed to marine microplastics either directly or indirectly via marine environments or seafood, respectively (Vital et al., 2021).

13.2 MICROPLASTIC REMOVAL TECHNOLOGIES

Microplastic removal techniques are categorized by type and location of the treatment. They can be either in situ (removal of microplastics in the natural environment) or ex situ (treated in an artificial setting, laboratories, and water treatment plants). Based on the mode of treatment, both in situ and ex situ methods can further be classified into physical, chemical, and biological methods. Most of the literature available is based on ex situ methods that are used in wastewater treatment plants and laboratories to reduce the contamination of microplastics in water bodies. A limited amount of literature related to microplastic removal in marine water and sediments was found. The majority of them used techniques such as filtration which are physical methods (Badola et al., 2021). Since each method consists of both advantages and disadvantages, the focus should be on the better technique (Table 13.1).

13.2.1 Physical Removal

One of the biggest challenges with microplastics is capturing them from the environment to recycle or incinerate. Microplastics in beach environments have been largely removed through physical removal methods since they are extremely practical.

13.2.1.1 Removal of Microplastics in Seawater

Filtration and sedimentation are the most popular physical removal methods for microplastics in seawater. For the laboratory-scale experiments, plankton or neuston nets are used to collect the particles floating on the surface water and water column (Marialilokyee; Setälä et al., 2016). Accordingly, to remove the microplastics in marine surface water, wet sieving filtration can be used. Filtration methods can be introduced as the most suitable for the removal of microplastics in marine water since microplastics can be collected on the sieve, passing unwanted liquid directly through the filters. The pore size of the filter should be designed according to the sizes of the microplastics mixed with seawater. Though in-field net filtration seems an easy method to remove microplastics in marine water, it might be problematic when the contaminated area and the volume of marine water to be treated are large. In those cases, bulk filtration can be applied under two types. Microplastics in marine water can be removed by passing pumped seawater through a filter set up in a ship or boat (Stuart, 2021). Through a separate outlet, the filtered water can be removed. The net sieving might harm marine life, namely, sea turtles, pelagic fishes, and marine mammals unintentionally (Stuart, 2021). Hence, filtration is not a sustainable solution for the removal of marine microplastics in terms of the well-being of marine organisms.

When the in-field bulk filtration is inapplicable, the seawater can be carried through large pumps from the sea to the treated plant and conducted the filtration. In line with the grab sampling method for microplastic extraction available in the literature for laboratory scale, filtering the pumped polluted surface seawater can be performed

TABLE 13.1
Comparison of the Removal Methods that Were Utilized in the Removal of Marine Microplastic from Sediments and Waters

Technique	Treated Source	Advantages	Disadvantages	Reference
Physical – plankton and neuston net	Water	• In-field filtration system • Easy to remove microplastics on site • Simple equipment and low cost for it	• Hard to use on large scale • Labor intensive • Operation is costly • Regular cleaning or changing the net is required	Setälä et al. (2016)
Physical and chemical – filters installed in scrubbers of large ships	Water	• After the installation, the filtration process happens automatically • Large quantities of microplastics can be removed • Not labor intensive • All possible ship routes can be cleared	• The high initial cost of filter and installation • Technical advancements needed • Regular cleaning required	Magloff (2022)
Bulk filtration	Water	• Can be used in larger volumes • Harm to marine life can be minimized • In-situ removal can be done	• High energy and labor-intensive • Costly • Requires regular cleanups	Stuart (2021)
Density separation	Beach sediment	• 'Quicksand' is a simple mix of water and sand. • Microplastics can be separated by the density separation method • Low cost • Can be applied to PVC removal as well	• Hard to handle large quantities • Labor intensive • Time-consuming process	Bayley (2019b)
Electrical hand-sieving tool	Beach sediment	• A simple design of a sieve mounted between poles • A mobile device can be carried easily • One or two persons can handle the sieve for the operation • Low cost	• Time-consuming • Hard to handle larger quantities • Regular cleanups are needed for the sieve	Ward (2015)

TABLE 13.1 (Continued)
Comparison of the Removal Methods that Were Utilized in the Removal of Marine Microplastic from Sediments and Waters

Technique	Treated Source	Advantages	Disadvantages	Reference
Beach cleaning robot	Beach sediment	• Used to filter the top layer on sand beaches • Can be adjusted for larger particle sizes as well • Not labor intensive	• The only top layer of sand is filtered • In-depth filtration is hard • Depends on the sand conditions	Wells (2021)

(Barrows et al., 2017). Nevertheless, no evidence exists in the literature for the bulk removal of marine microplastics via the grab sampling technique. Continuous filtration can cause an effective removal of microplastics. However, the water level of the sea should be balanced without disturbing the life of marine organisms. As well, this bulk filtration is also not applicable to the deep-sea regions where marine animals are living.

With the use of oil and magnet, a viable method for the removal of microplastics in ocean water has been developed by a teenager in Ireland (Jacobo, 2019). Once the microplastic particles migrated and agglomerated in the oil phase, a strong magnet was used to remove the fluid layer. Though this method has been experienced in the laboratory, it is essential to scale this up to an industrial level. Sedimentation of microplastics with different coagulants has also been experienced preliminary in small scales. An elegant solution was demonstrated by Liu and coworkers to remove microplastics in wastewater with the use of the bacteria *Pseudomonas aeruginosa* (Quaglia, 2021). A biofilm was designed to incorporate the floating microplastics in the water due to the sticky property of bacteria. After trapping the microplastics, the microbe nets sunk to the bottom of the water, and thereafter the biofilm dispersal genes were used to unlatch the microplastics. Then, recycling can be carried out for the separated microplastic blob.

13.2.1.2 Removal of Microplastics in Beach Sand

Dry sieving, density separation, and other new methods were used for the removal of microplastics in beach environments (Table 13.1). The microplastics mixed with beach sand and sediment can also be removed through filtration after they are mixed with water or high-dense solution than plastics. However, microplastics like polyvinyl chloride which are denser than water do not float in the water. In that case, filtration is not successful since they remain in the sand, whereas other light-dense microplastics float. Accordingly, there should be an alternative solution to overcome these limitations. Dr. Richard Coultan invented a highly dense solution by mixing sand and water to float highly dense microplastic particles and thereafter filter them through conventional filtration methods (Bayley, 2019b). The resultant liquid was 'quicksand' and it was optimized in a sand skimmer consisting of a separator and tray to remove and collect floating microplastics. A few novel sieving techniques were also designed to remove the microplastics from the beaches.

An electrical hand-sieving tool was designed with a mesh screen to remove the marine microplastics. To enhance the sieving process through electrostatic interactions, beside the mesh screen two opposing poles were placed (Ward, 2015). A beach cleaning robot, BeBot, was designed to remove the hidden plastic trash in the top layer of beach sand through the cocontribution of two companies, Poralu Marine and 4ocean (Wells, 2021). The BeBot was a green cleaner and lesser landscape disruptor, unlike other beach cleaning machines. Since BeBot was made for large plastic debris, a little change to the sieve size is required for the removal of microplastics. Another machine has been invented and designed by the collaboration of the Hawaii Wildlife Fund, the University of Hawaii at Hilo, and the University of de Sherbrook in Québec to clean the beaches in Hawaii for the removal of beach microplastics (Sherwood, 2019). Nevertheless, the way of physical removal of microplastics should be matched with the level of pollution and beach properties. Since the absence of a large-scale standard method to remove marine microplastics physically, future studies are encouraged to design or propose economically and ecologically friendly removal methods.

13.2.1.3 Case Study: Nurdles and Plastic Debris Removal from Sarakkuwa Beach, Sri Lanka

After the recent maritime disaster happened in Sri Lanka, the deposited microplastics, nurdles, and other partially burned plastic fragments on the beach were removed using a simply designed machine and manual sieving (Hassan et al., 2021). The initial cleanup in Sarakkuwa Beach was done with the contribution of volunteer communities. The removal process for the deposited nurdles and partially burned plastic fragments on the beach after the ship disaster was not easy due to the continuous receival of nurdles with the waves. Consequently, the requirement for nonstop cleaning was raised. The beach cleaning process was based on physical removal techniques. An automated machine and manual sieving tools were parallel used for the removal of microplastics. A Sri Lankan innovator, Mr. Chinthaka Waragoda, designed a machinery plant, a blue treatment facility, to mitigate the devastation that happened (Sewwandi et al., 2022b). The removal of plastic fragments and nurdles through the blue treatment facility was involved in the bulk removal, whereas the manual sieving was limited to an area per day.

Manual dry sieving was performed with extensive manpower and manually operated machinery. As illustrated in Figures 13.2a and b, beach sand was handsieved by women who live in the Sarakkuwa village. Some foreign and local volunteer communities and Sri Lankan Army forces supported the beach cleaning process by offering new manual machines (Figures 13.2c and d). The dry sieving was limited for the surface beach sand or up to approximately 1 m depth. Continuous manpower was required for input and sieving of sand to the hand sieve and the rolling sieve. The cost behind the utilization of manpower was high, whereas the efficiency per day of the cleaning process was less than that of the blue treatment facility. Since the required manpower for each step of the cleaning process, digging the beach, carrying and putting sand to the sieves, and sieving sand was high enough to cover a large

Beach and Marine Microplastics

FIGURE 13.3 Physical removal method that was followed by the blue treatment facility to remove the nurdles mixed with Sarakkuwa Beach sand after the X-Press Pearl ship disaster in Sri Lanka.

area of the beach per day, the cost for their payments was also high. This was the biggest disadvantage of this manual dry-sieving technique. Besides, there was no systematic digging or refilling of beach sand. This would disturb the beach profile, missing its natural beauty. Utilization of hand rolling sieve system was comparatively more effective than the hand dry sieving, but the limitations were the same as the hand sieving (Figures 13.2c and d). Accordingly, the carried out manual sieving process for the microplastic removal at Sarakkuwa Beach consumed more time and was expensive; therefore those techniques were not much successful. Since both types of dry sieving techniques were limited to surface sand treatment, the removal of buried nurdles and partially burned plastic debris from the beach was also unsuccessful.

The technique used in the blue treatment facility seems to be the enlarged version of the microplastic extraction procedure through density separation that applies to environmental solid samples (Figure 13.3). The technique of the blue treatment process was based on wet sieving. As the first step of the blue treatment facility, microplastics-mixed sand was excavated from the beach and carried in bulk to the treatment plant. The excavation was carried out up to 2 m depth in the beach. An excavator bucket of sand was added to well water containing a big vessel and thoroughly mixed under an automated mixer. Since the plastics have a lesser density to the water, plastics materials come to the surface of the water during the mixing process, whereas sand goes and settles at the bottom of the vessel. Water was passed through a set of sieves to separate various plastic wastes of different sizes. The mechanically

FIGURE 13.2 Dry sieving (a and b: Hand sieving; c and d: machine sieving) of microplastics and other plastic fragments at Sarakkuwa Beach by manual and mechanical techniques after the X-Press Pearl maritime disaster in Sri Lanka. (Photo courtesy: Marine Environment Protection Authority (MEPA), Sri Lanka.)

separated nurdles and other plastic debris were removed through several outlets based on their particle size. The nurdles in the water were manually removed by using nets. The small seashells removed from the outlets were collected separately.

Cleaned sand was mixed with those separated seashells and put onto the beach again, letting them undergo the natural settling. Approximately 80 excavator buckets were treated with 80000 L of well water per day, removing 420 kg of nurdles, 60 kg of burnt fragments, and 2000 L of waste sludge. The water was reused for the removal process after filtering them. Therefore, the wastage of water can be controlled to some extent. Throughout a systematic sand excavation and refilling in the beach, the removal of microplastics in bulk can be managed successfully, providing an economically friendly and time-saving treatment facility.

13.2.2 Chemical Removal Methods

13.2.2.1 Coagulation and Agglomeration

Wastewater treatment plants use coagulation and agglomeration processes to form larger contaminant particles, making them easier to separate (Hu et al., 2012; Lee et al., 2012; Shirasaki et al., 2016). Chemical processes might be employed as the first stage of a more intricate filtration system to get rid of microplastics (Abuwatfa et al., 2021). Past studies reveal that chemical methods were used most of the time to change the properties of microplastics present in the water body such as making them float, or as an adhesive, etc. Various chemical coagulates are used in water treatment for the removal of microplastics, such as different iron salts ($Fe_2(SO_4)_3 \cdot 9H_2O$, $FeCl_3 \cdot 6H_2O$) and aluminum salts ($KAl(SO_4)_2 \cdot 12H_2O$, $AlCl_3 \cdot 6H_2O$, $Al_2(SO_4)_3 \cdot 18H_2O$) to capture dissolved solids in wastewater by forming flocculants and to make them settle in the bottoms of coagulation tanks (Dey et al., 2021).

Furthermore, parameters such as the concentrations of pollutants, surface charge, and the pH of the wastewater are found to be associated with the process (Jamal et al., 2019). Also in experiments, the performance of aluminum-based coagulants with polyacrylamide incorporation was evaluated to be better than the performance of iron-based coagulants (Ma et al., 2019; Triebskorn et al., 2019). Coagulation and agglomeration have been researched to eliminate microplastics in laboratories. The type of coagulant, pH, the chemical composition of the media, and the concentrations found as factors affecting the efficiency. Although It has been noted that the salinity of the media and the presence of a light source can change the rate at which microplastic constructions degrade in the aquatic environment, degradation processes haven't been extensively researched (Padervand et al., 2020). So it was observed that chemical methods of microplastic removals are tested in wastewater treatment plants and laboratories rather than in the real environment. The reason might be that even though these methods are found to be effective in the removal of small microplastics in controllable operational conditions using simple mechanical devices, In the real marine environment utilization is not practical as the addition of chemicals to the media resulting in contaminations and creating toxicities to the ecosystem (Herbort et al., 2018; Ma et al., 2019).

13.2.2.2 Adhesion

Adhesion is a physicochemical process that occurs due to the van der Waals force or ion exchange that causes adsorbate to bind to the surface of adsorbents (Mrvčić et al., 2012). Recent studies have found that the marine algae *Fucus vesiculosus* (brown algae) might adsorb microplastics due to the presence of alginic acids in its cell wall (Sundbæk et al., 2018). In these brown algae, on alginate polymers, the availability of the carboxylic functional group is said to be responsible for the plastic binding capacity of the adsorbents. Other than that Red Sea giant clam (*Tridacna Maxima*) (Arossa et al., 2019), Antarctic krill (*Euphausia Superba*) (Dawson et al., 2018), and some corals (Corona et al., 2020; Martin et al., 2019) have been studied for their capability of microplastic adsorption even though the reported efficiency was very low. As discussed before while microalgae are identified as a biological tool for removing fine microplastics (Cunha et al., 2020), corals also could be identified as natural sinks for microplastics in the ocean as tested in the laboratory (Martin et al., 2019). According to research by Corona et al., in a laboratory context (Corona et al., 2020), the microplastics removal effectiveness of mushroom coral taken from the reef of the island of Magoodhoo, Faafu Atoll, Republic of Maldives, exhibited an efficiency of 97% for virgin + biofouled plastic of size 200–1000 μm. According to this experiment, corals are a significant microplastic sink in the seas. Retention, adhesion, and ingestion are identified to be the primary causes of microplastic sinking in these species (Arossa et al., 2019). Since the removal efficiency of those methods is comparably low compared to the physical removal methods, large-scale utilization is not applicable. Nevertheless, the long-term mitigation of microplastic pollution in marine environments can be acceptable through the sinking process of corals.

13.3 COMPARISON OF MICROPLASTIC REMOVAL TECHNIQUES IN MARINE ENVIRONMENTS

13.3.1 Selecting a Better Removal Technique : Factors to Consider

There are currently no widely established standards for the removal process since the field of microplastic large-scale removal is still in its infancy (Miller et al., 2020). Overall, the properties of the microplastic occurrence in a particular place, such as type, size, shape, quantity, accessibility, etc., determine the technique to be followed in the removal of microplastics. Accurate measurements of microplastic occurrence in the environment and identification of plausible sources are necessary to comprehend the scope of the problem and establish the highest mitigation priorities (Hidalgo-Ruz et al., 2012).

Because microplastic contamination of beach sand occurs through rivers, sea currents, anthropogenic activities along the coast, and transfer by water movement or by air emissions, the choice of removal method affects the amount and types of polymers observed. Depending on the thickness of the beach sand layer going to be treated, the beach excavating method should be decided. The selection should be less harmful to the coastal vegetation and animals in beach environments. However, the lack of studies on developing methodologies for microplastic removal remains uncertain and researchers tend to follow the most suitable methods as per the site conditions. Since the existing beach cleaning procedures were established to collect trash on the beach, such as cigarette ends, pieces of plastic, and seaweed, to improve the appearance of the beaches, the sifting equipment of them is general. They take the top layer of sand and filter it through 10 mm holes (Bayley, 2019a). However, because the microplastics are less than 5 mm in size, they fall through the holes and remain on the beach. Therefore, it is better to adjust the mesh screen of that equipment, enabling the removal of microplastics.

The weather condition of the particular area should be considered before the implementation of the microplastic removal processes. The techniques should be designed to match the wind and current conditions. Since the complexity of the ocean, the removal of microplastics in marine water needs to be handled very carefully. Buoyancy is a crucial factor in determining the fate of microplastics in marine water when contrary to beaches (Bråte et al., 2014). Thus, filter feeders, zooplanktons, and planktivores can easily access buoyant plastics with densities below those of water. Depending on the material density, microplastics can be buoyant in seawater in either a positive or negative manner (Li et al., 2018). Examples of positive buoyant polymers are polypropylene, polyethylene, and polystyrene and negative buoyant plastics are polystyrenes, polyvinyl chloride, and polyethylene (Khatmullina and Chubarenko, 2019). Therefore, the removal method should be selected depending on the polymer types of microplastics that caused the pollution. Removal of microplastics accumulated in the seabed is a challenging task as the range of organisms that would be affected is large. Disturbing marine organisms and destroying their nests would create a greater issue than the long-term existence of microplastics. Consequently, the percentage risk behind the existence of microplastics in the marine environment should be estimated and compared with the risk on the marine organisms during the removal process, before entering the removal stage. However, similarly to the beach sediments, not much information or past studies are available on the specific factors to be considered in selecting a microplastic removal technique in marine waters.

13.4 CHALLENGES AND FUTURE DIRECTIONS IN MICROPLASTIC REMOVAL IN MARINE ENVIRONMENTS

Since the removal of microplastics has not been discussed largely, there are many issues faced during this research and drawbacks in techniques. As the name implies, dealing with microlevel particles spread across a wide area at different states and conditions is not an easy task. The establishment of a standard method for the removal of microplastics remains a challenge due to variations in their characteristics and uncertainty in complete eradication from the environment. Therefore, the main challenges in removal could be briefed as the identification, collection, separation, and disposal of microplastics, minimizing the damage to the environment (Heloisa and Amira, 2017). As discussed, the majority of the removal techniques were studied in wastewater treatment plants, laboratories, and controlled environments; the effectiveness and efficiencies of their practical applications are questionable. Even comparison among the technologies is difficult as the studies focus on different types of microplastics and different marine environments considering much variability. Besides, most of the time those studies are case specific. Microplastics can be successfully removed from wastewater using existing technologies, but they can be expensive, challenging to install in existing facilities, and only employed when there is a necessity. High energy requirements in some technologies, which result in greater operating costs, is another disadvantage (Richards et al., 2021). Inventing a method to remove marine microplastics with zero harm to the natural environment is challenging. Since the focus always should be on the minimum harmfulness, the cost, time, and techniques should be arranged accordingly. Tides and wave currents would disturb the removal of microplastics from marine water. Since beaches undergo habitual natural mixing the vertical distribution of microplastics can be increasingly challenging the removal. Besides, the low energy waves cause to make more mixed beaches. Accordingly, the receival of low-energy tides can also be a big problem for the removal of microplastics as they enhance the vertical distribution of beach microplastics.

Even if an excellent removal technology exists, the absence of policies, laws, and standardization in removal and handling of microplastic is a major problem in microplastic removal processes. Specifically, when considering the X-Press Pearl disaster in Sri Lanka, the absence of standard policies and regulations created a difficult situation for the government and the related responsible institutes to manage the crisis, estimate the level of damage, and take necessary actions (de Vos et al., 2022).

In conclusion, there is a requirement for an efficient method/protocol or system for the treatment of microplastics and a policy implication throughout the world to control microplastic contamination in the environment including the marine environment. Robust policies and regulatory approaches are required from governments, industries, and International Maritime Organization (IMO) to prevent microplastic pollution at all stages of the supply chain. To prevent and manage disasters such as X-Press Pearl, there is a timely need for a strict protocol for nurdle shipment and handling, along with the guidelines for managing accidental spillages, to improve the disaster responses in the events of major spillages. The upcoming United Nations Global Plastic Treaty would be an ideal platform to pave the pathway towards establishing policies, regulations, protocols, and standards to prevent and manage microplastic pollution on a national, regional, and international basis.

REFERENCES

Abuwatfa, W.H., Al-Muqbel, D., Al-Othman, A., Halalsheh, N. and Tawalbeh, M. 2021. Insights into the removal of microplastics from water using biochar in the era of COVID-19: A mini review. Case Studies in Chemical and Environmental Engineering 4, 100151. https://doi.org/10.1016/j.cscee.2021.100151

Anderson, A., Andrady, A., Arthur, C., Baker, J., Bouwman, H., Gall, S., Hildalgo-Ruz, V., Köhler, A., Lavender Law, K., Leslie, H.A. and Kershaw, P., 2015. Sources, fate and effects of microplastics in the environment: a global assessment. *GESAMP Reports & Studies Series* (90).

Annette, G. and Gaëlle, H. 2020. Plastic giants polluting through the backdoor: The case for a regulatory supply-chain approach to pellet pollution. Surfrider Foundation Europe. https://surfrider.eu/wp-content/uploads/2020/11/report-pellet-pollution-2020.pdf

Arias, A.H., Ronda, A.C., Oliva, A.L. and Marcovecchio, J.E. 2019. Evidence of microplastic ingestion by fish from the Bahía Blanca Estuary in Argentina, South America. Bulletin of Environmental Contamination and Toxicology 102(6), 750–756. https://doi.org/10.1007/s00128-019-02604-2

Arossa, S., Martin, C., Rossbach, S. and Duarte, C.M. 2019. Microplastic removal by Red Sea giant clam (Tridacna maxima). Environmental Pollution 252, 1257–1266. https://doi.org/10.1016/j.envpol.2019.05.149

Badola, N., Bahuguna, A., Sasson, Y. and Chauhan, J.S. 2021. Microplastics removal strategies: A step toward finding the solution. Frontiers of Environmental Science & Engineering 16(1), 7. https://doi.org/10.1007/s11783-021-1441-3

Bajt, O. 2021. From plastics to microplastics and organisms. FEBS Open Bio 11(4), 954–966. https://doi.org/10.1002/2211-5463.13120

Bandow, N., Will, V., Wachtendorf, V. and Simon, F.-G. 2017. Contaminant release from aged microplastic. Environmental Chemistry 14(6), 394–405. https://doi.org/10.1071/EN17064

Barrows, A.P.W., Neumann, C.A., Berger, M.L. and Shaw, S.D. 2017. Grab vs. neuston tow net: A microplastic sampling performance comparison and possible advances in the field. Analytical Methods 9(9), 1446–1453. https://doi.org/10.1039/C6AY02387H

Bayley, S. 2019a. How we can rid our beaches of microplastics, the inventor coming up with a solution, *Evening Standard*.

Bayley, S. 2019b. How we can rid our beaches of microplastics, the inventor coming up with a solution, Evening Standard, 04/11/2022. https://www.standard.co.uk/futurelondon/theplasticfreeproject/microplastics-beaches-inventor-a4244051.html

Bråte, I.L., Halsband, C., Allan, I. and Thomas, K. 2014. Report made for the Norwegian Environment Agency: Microplastics in marine environments; Occurrence, distribution and effects.

Capó, X., Alomar, C., Compa, M., Sole, M., Sanahuja, I., Soliz Rojas, D.L., González, G.P., Garcinuño Martínez, R.M. and Deudero, S. 2022. Quantification of differential tissue biomarker responses to microplastic ingestion and plasticizer bioaccumulation in aquaculture reared sea bream *Sparus aurata*. *Environmental Research* 211, 113063. https://doi.org/10.1016/j.envres.2022.113063

Corcoran, P.L. 2022. Degradation of Microplastics in the Environment, in *Handbook of Microplastics in the Environment*, T. Rocha-Santos, M.F. Costa, and C. Mouneyrac, Editors. 2022, Springer International Publishing: Cham, pp. 531–542

Corona, E., Martin, C., Marasco, R. and Duarte, C. 2020. Passive and active removal of marine microplastics by a mushroom coral (Danafungia scruposa). *Frontiers in Marine Science* 7. https://doi.org/10.3389/fmars.2020.00128

Cox, K.D., Covernton, G.A., Davies, H.L., Dower, J.F., Juanes, F. and Dudas, S.E. 2019. Human consumption of microplastics. *Environmental Science & Technology* 53(12), 7068–7074. https://doi.org/10.1021/acs.est.9b01517

Cunha, C., Silva, L., Paulo, J., Faria, M., Nogueira, N. and Cordeiro, N. 2020. Microalgal-based biopolymer for nano- and microplastic removal: a possible biosolution for wastewater treatment. *Environment Pollution* 263(Pt B), 114385. https://doi.org/10.1016/j.envpol.2020.114385

Dawson, A.L., Kawaguchi, S., King, C.K., Townsend, K.A., King, R., Huston, W.M. and Bengtson Nash, S.M. 2018. Turning microplastics into nanoplastics through digestive fragmentation by Antarctic krill. *Nature Communications* 9(1), 1001. https://doi.org/10.1038/s41467-018-03465-9

de Souza Machado, A.A., Lau, C.W., Kloas, W., Bergmann, J., Bachelier, J.B., Faltin, E., Becker, R., Görlich, A.S. and Rillig, M.C. 2019. Microplastics can change soil properties and affect plant performance. *Environmental Science & Technology* 53(10), 6044–6052. https://doi.org/10.1021/acs.est.9b01339

de Vos, A., Aluwihare, L., Youngs, S., DiBenedetto, M.H., Ward, C.P., Michel, A.P.M., Colson, B.C., Mazzotta, M.G., Walsh, A.N., Nelson, R.K., Reddy, C.M. and James, B.D. 2022. The M/V X-Press Pearl nurdle spill: Contamination of burnt plastic and unburnt nurdles along Sri Lanka's beaches. *ACS Environmental Au* 2(2), 128–135. https://doi.org/10.1021/acsenvironau.1c00031

Dey, T.K., Uddin, M.E. and Jamal, M. 2021. Detection and removal of microplastics in wastewater: evolution and impact. *Environmental Science and Pollution Research* 28(14), 16925–16947. https://doi.org/10.1007/s11356-021-12943-5

Duncan, E.M., Arrowsmith, J., Bain, C., Broderick, A.C., Lee, J., Metcalfe, K., Pikesley, S.K., Snape, R.T.E., van Sebille, E. and Godley, B.J. 2018. The true depth of the Mediterranean plastic problem: Extreme microplastic pollution on marine turtle nesting beaches in Cyprus. *Marine Pollution Bulletin* 136, 334–340. https://doi.org/10.1016/j.marpolbul.2018.09.019

Fernandino, G., Elliff, C.I., Silva, I.R. and Bittencourt, A.C. 2015. How many pellets are too many? The pellet pollution index as a tool to assess beach pollution by plastic resin pellets in Salvador, Bahia, Brazil. *Revista de Gestão Costeira Integrada-Journal of Integrated Coastal Zone Management* 15(3), 325–332.

Gündoğdu, S., Ayat, B., Aydoğan, B., Çevik, C. and Karaca, S. 2022. Hydrometeorological assessments of the transport of microplastic pellets in the Eastern Mediterranean. *Science of The Total Environment* 823, 153676. https://doi.org/10.1016/j.scitotenv.2022.153676

Häkkien, J., Posti, A., Weintrit, A. and Neumann, T. 2013. Overview of Maritime Accidents Involving Chemicals Worldwide in the Baltic Sea. CRC Press.

Hasegawa, T., Mizukawa, K., Yeo, B.G., Sekioka, T., Takada, H. and Nakaoka, M. 2021. Trophic Transfer of Microplastics Enhances Plastic Additive Accumulation in Fish. bioRxiv.

Hassan, P., Camille, L., Stephane, L.F. and Luigi, A. 2021. X-Press Pearl maritime disaster Sri Lanka-report of the UN environmental advisory mission JULY 2021, UN environmental advisory mission, UN. https://wedocs.unep.org/20.500.11822/36608

Heloisa, W. and Amira, A. 2017. Water Challenges of an Urbanizing World. Matjaž, G. (ed), p. Ch. 5, IntechOpen, Rijeka. https://doi.org/10.5772/intechopen.71494

Herbort, A.F., Sturm, M.T., Fiedler, S., Abkai, G. and Schuhen, K. 2018. Alkoxy-silyl induced agglomeration: A new approach for the sustainable removal of microplastic from aquatic systems. *Journal of Polymers and the Environment* 26(11), 4258–4270. https://doi.org/10.1007/s10924-018-1287-3

Hidalgo-Ruz, V., Gutow, L., Thompson, R.C. and Thiel, M. 2012. Microplastics in the marine environment: a review of the methods used for identification and quantification. *Environmental Science & Technology* 46(6), 3060–3075.

Hu, C., Liu, H., Chen, G. and Qu, J. 2012. Effect of aluminum speciation on arsenic removal during coagulation process. *Separation and Purification Technology* 86, 35–40. https://doi.org/10.1016/j.seppur.2011.10.017

Jacobo, J. 2019. Irish teen invents method to remove microplastics from ocean, wins $50K Google Science Fair prize. ABC News. https://abcnews.go.com/Technology/irish-teen-invents-method-remove-microplastics-ocean-wins/story?id=64731771

Jamal, M., Razeeb, K.M., Shao, H., Islam, J., Akhter, I., Furukawa, H. and Khosla, A. 2019. Development of tungsten oxide nanoparticle modified carbon fibre cloth as flexible pH sensor. *Scientific Reports* 9(1), 4659. https://doi.org/10.1038/s41598-019-41331-w

Kanhai, L.D.K., Asmath, H. and Gobin, J.F. 2022. The status of marine debris/litter and plastic pollution in the Caribbean Large Marine Ecosystem (CLME): 1980–2020. *Environmental Pollution* 300, 118919. https://doi.org/10.1016/j.envpol.2022.118919

Khatmullina, L. and Chubarenko, I. 2019. Transport of marine microplastic particles: why is it so difficult to predict? 1. Anthropocene Coasts 2, 293–305. https://doi.org/10.1139/anc-2018-0024

Lee, K.E., Morad, N., Teng, T.T. and Poh, B.T. 2012. Development, characterization and the application of hybrid materials in coagulation/flocculation of wastewater: A review. *Chemical Engineering Journal* 203, 370–386. https://doi.org/10.1016/j.cej.2012.06.109

Lentz, S.J. and Fewings, M.R. 2011. The wind- and wave-driven inner-shelf circulation. *Annual Review of Marine Science* 4(1), 317–343. https://doi.org/10.1146/annurev-marine-120709-142745

Li, J., Zhang, K. and Zhang, H. 2018. Adsorption of antibiotics on microplastics. *Environmental Pollution* 237, 460–467. https://doi.org/10.1016/j.envpol.2018.02.050

Liu, K., Wu, T., Wang, X., Song, Z., Zong, C., Wei, N. and Li, D. 2019. Consistent transport of terrestrial microplastics to the ocean through atmosphere. *Environmental Science & Technology* 53(18), 10612–10619. https://doi.org/10.1021/acs.est.9b03427

Ma, B., Xue, W., Hu, C., Liu, H., Qu, J. and Li, L. 2019. Characteristics of microplastic removal via coagulation and ultrafiltration during drinking water treatment. *Chemical Engineering Journal* 359, 159–167. https://doi.org/10.1016/j.cej.2018.11.155

Magloff, L. 2022. Ships' scrubbers used to remove ocean microplastics, Spring Wise, Italy, 11/11. www.springwise.com/innovation/sustainability/a-new-filter-system-for-ocean-microplastics/

Mallik, A., Xavier, K.A.M., Naidu, B.C. and Nayak, B.B. 2021. Ecotoxicological and physiological risks of microplastics on fish and their possible mitigation measures. *Science of the Total Environment* 779, 146433. https://doi.org/10.1016/j.scitotenv.2021.146433

Martin, C., Corona, E., Mahadik, G.A. and Duarte, C.M. 2019. Adhesion to coral surface as a potential sink for marine microplastics. *Environmental Pollution* 255, 113281. https://doi.org/10.1016/j.envpol.2019.113281

Miller, M.E., Hamann, M. and Kroon, F.J. 2020. Bioaccumulation and biomagnification of microplastics in marine organisms: A review and meta-analysis of current data. *PLoS One* 15(10), e0240792. https://doi.org/10.1371/journal.pone.0240792

Mrvčić, J., Stanzer, D., Solić, E. and Stehlik-Tomas, V. 2012. Interaction of lactic acid bacteria with metal ions: Opportunities for improving food safety and quality. *World Journal of Microbiology and Biotechnology* 28(9), 2771–2782. https://doi.org/10.1007/s11274-012-1094-2

Padervand, M., Lichtfouse, E., Robert, D. and Wang, C. 2020. Removal of microplastics from the environment. A review. *Environmental Chemistry Letters* 18(3), 807–828. https://doi.org/10.1007/s10311-020-00983-1

Pattiaratchi, C. and Wijeratne, S. 2021. X-Press Pearl disaster: An oceanographic perspective. Groundviews. 14/11/2021.

Paul, P., Mohan, A.K., Dev, G. and PG, S.K. 2022. Microplastics as contaminants in marine environment. *Sustainability, Agri, Food and Environmental Research* 10(1). (ISSN: 0719-3726)

Quaglia, S. 2021. Scientists find way to remove polluting microplastics with bacteria. *The Guardian*. www.theguardian.com/science/2021/apr/28/scientists-find-way-to-remove-polluting-microplastics-with-bacteria

Richards, S., Rao, L., Connelly, S., Raj, A., Raveendran, L., Shirin, S., Jamwal, P. and Helliwell, R. 2021. Sustainable water resources through harvesting rainwater and the effectiveness of a low-cost water treatment. *Journal of Environmental Management* 286, 112223. https://doi.org/10.1016/j.jenvman.2021.112223

Schumann, E.H., MacKay, C.F. and Strydom, N.A. 2019. Nurdle drifters around South Africa as indicators of ocean structures and dispersion South African *Journal of Science* 115, 1–9. http://dx.doi.org/10.17159/sajs.2019/5372

Setälä, O., Magnusson, K., Lehtiniemi, M. and Norén, F. 2016. Distribution and abundance of surface water microlitter in the Baltic Sea: A comparison of two sampling methods. *Marine Pollution Bulletin* 110(1), 177–183. https://doi.org/10.1016/j.marpolbul.2016.06.065

Sewwandi, M., Amarathunga, A.A.D., Wijesekara, H., Mahatantila, K. and Vithanage, M. 2022a. Contamination and distribution of buried microplastics in Sarakkuwa beach ensuing the MV X-Press Pearl maritime disaster in Sri Lankan sea. *Marine Pollution Bulletin* 184, 114074. https://doi.org/10.1016/j.marpolbul.2022.114074

Sewwandi, M., Hettithanthri, O., Egodage, S.M., Amarathunga, A.A.D. and Vithanage, M. 2022b. Unprecedented marine microplastic contamination from the X-Press Pearl container vessel disaster. *Science of the Total Environment* 828, 154374. https://doi.org/10.1016/j.scitotenv.2022.154374

Sherwood, L. (2019) UH Hilo analyzes data from machine designed to remove beach microplastics. University of Hawai'i News. www.hawaii.edu/news/2019/05/07/uh-hilo-microplastic/

Shirasaki, N., Matsushita, T., Matsui, Y. and Marubayashi, T. 2016. Effect of aluminum hydrolyte species on human enterovirus removal from water during the coagulation process. *Chemical Engineering Journal* 284, 786–793. https://doi.org/10.1016/j.cej.2015.09.045

Stuart, R. 2021. Scooping plastic out of the ocean is a losing game. Hakai Magazine. https://hakaimagazine.com/features/scooping-plastic-out-of-the-ocean-is-a-losing-game/

Sundbæk, K.B., Koch, I.D.W., Villaro, C.G., Rasmussen, N.S., Holdt, S.L. and Hartmann, N.B. 2018. Sorption of fluorescent polystyrene microplastic particles to edible seaweed *Fucus vesiculosus*. *Journal of Applied Phycology* 30(5), 2923–2927. https://doi.org/10.1007/s10811-018-1472-8

Szewc, K., Graca, B. and Dołęga, A. 2021. Atmospheric deposition of microplastics in the coastal zone: Characteristics and relationship with meteorological factors. *Science of the Total Environment* 761, 143272. https://doi.org/10.1016/j.scitotenv.2020.143272

ter Halle, A. and Ghiglione, J.F. 2021. Nanoplastics: A complex, polluting terra incognita. *Environmental Science & Technology* 55(21), 14466–14469. https://doi.org/10.1021/acs.est.1c04142

Triebskorn, R., Braunbeck, T., Grummt, T., Hanslik, L., Huppertsberg, S., Jekel, M., Knepper, T.P., Krais, S., Müller, Y.K., Pittroff, M., Ruhl, A.S., Schmieg, H., Schür, C., Strobel, C., Wagner, M., Zumbülte, N. and Köhler, H.-R. 2019. Relevance of nano- and microplastics for freshwater ecosystems: A critical review. *TrAC Trends in Analytical Chemistry* 110, 375–392. https://doi.org/10.1016/j.trac.2018.11.023

Vital, S.A., Cardoso, C., Avio, C., Pittura, L., Regoli, F. and Bebianno, M.J. 2021. Do microplastic contaminated seafood consumption pose a potential risk to human health? *Marine Pollution Bulletin* 171, 112769. https://doi.org/10.1016/j.marpolbul.2021.112769

Ward, M. 2015. Marine microplastic removal tool, Patent, U.S., US 8,944.253 B2. https://patentimages.storage.googleapis.com/b1/5f/0c/11c3c799777321/US8944253.pdf

Wells, A. 2021. New Beach-cleaning Robot Removes Microplastics Hidden under Sand. Thomas. www.thomasnet.com/insights/new-beach-cleaning-robot-removes-microplastics-hidden-under-sand/

Yang, Y., Liu, W., Zhang, Z., Grossart, H.-P. and Gadd, G.M. 2020. Microplastics provide new microbial niches in aquatic environments. *Applied Microbiology and Biotechnology* 104(15), 6501–6511. https://doi.org/10.1007/s00253-020-10704-x

Zhang, K., Xu, S., Zhang, Y., Lo, Y., Liu, M., Ma, Y., Chau, H.S., Cao, Y., Xu, X., Wu, R., Lin, H., Lao, J., Tao, D., Lau, F.T.K., Chiu, S.-c., Wong, G.T.N., Lee, K., Ng, D.C.M., Cheung, S.-G., Leung, K.M.Y. and Lam, P.K.S. 2022. A systematic study of microplastic occurrence in urban water networks of a metropolis. *Water Research* 223, 118992. https://doi.org/10.1016/j.watres.2022.118992

Zhang, X., Li, Y., Ouyang, D., Lei, J., Tan, Q., Xie, L., Li, Z., Liu, T., Xiao, Y., Farooq, T.H., Wu, X., Chen, L. and Yan, W. 2021. Systematical review of interactions between microplastics and microorganisms in the soil environment. *Journal of Hazardous Materials* 418, 126288. https://doi.org/10.1016/j.jhazmat.2021.126288

14 Navigating Container Ship Impacts

Xiaokai Zhang and Mona Wells

14.1 THE ENVIRONMENTAL HAZARDS OF THE X-PRESS PEARL

The one-year anniversary of the sinking of the X-Press Pearl has passed, seemingly with little global fanfare considering the seriousness of the disaster. Attendant to the usual complex anatomy of a disaster's occurrence (Pallardy 2022; Swuste et al. 2020; Willmsen and Mapes 2017; Sewwandi et al., 2022a), for the X-Press Pearl debacle there is an even more complex postdisaster environmental effects envelope. Many aspects of the disaster contribute to complexity. The disaster was not, strictly speaking, localized. Between the first report of fire aboard ship on May 20, 2021, within the jurisdictional limits of Sri Lanka's Colombo Port, to the eventual sinking of the ship in 21 m of water on June 2, 2021, northwest of the port and approximately 4.5 nautical miles off the coast, roughly abeam Negombo Lagoon, the ship was actively shedding containers into the sea (Pattiaratchi and Wijeratne 2021). At the end of April 2022, *Container News* reports that 300 items of debris were cleared from a region outside a 1 km radius of the wreck and deeper than 10 m of water (Li 2022; Thassim 2022). Clearly many containers were washed into shallower coastal waters inasmuch as eight damaged containers were recovered ashore in the immediate aftermath of the beginning of the ship's disintegration, and efforts to map the affected area and recover debris from shallow waters continue (Li, 2022; Partow 2021; Thassim 2022). The ship's manifest identified more than 80 containers with known and diverse hazardous cargo (Wamsley 2021); however, one expert involved in advisement on the disaster estimated the number of containers whose contents might reasonably be expected to cause environmental impacts to be 862 (Wells 2021). It is not clear that all cargo was accurately declared or described (see, for example Donahue 2022), and this list omits cargo that, subsequent to the disaster, had a clear short-term hazardous effect with long-term potential, a prime example of which is plastic nurdles (Karthik et al. 2022; Sewwandi et al., 2022a, b; Reuters 2022; Rubesinghe et al. 2021). Added to the complexity of the spatial distribution of impact and the diversity of hazardous substances among the ship's cargo is the significantly complicating feature of transformation of toxic or even benign substances into toxic or even more toxic substances under the influence of the shipboard fire and explosions. As if all of this is not sufficient, the environment in which this disaster occurred has particular ecological diversity values (CBD 2022; Gunatilleke et al. 2008), as well as, e.g. economic value to the people of

Sri Lanka who make their livelihoods off the sea (Rodrigo 2021) and tourism value via the natural beauty of the environs (Ranasinghe and Sugandhika 2018).

In contemplating analysis of the X-Press Pearl disaster, its aftermath, and what lessons might be learned, one is reminded of the old adage concerning how to eat an elephant—the task is massive and may only be accomplished one bite at a time. Accordingly, this chapter considers the X-Press Pearl disaster and others like it that will inevitably occur with greater frequency in future, from the context of impact assessment, specifically, the need to link impacts to antecedent hazards. Within this focus, however, it is also useful to consider some background considerations within the general approach to impact assessment that need to be in place in order for effective effects assessment to be supported. A detailed analysis would be far beyond the scope of a book chapter; hence the information herein is orientational. This seems appropriate given the circumstances that, as the disaster unfolded, significant issues arose in terms of coordinating an Environmental Impact Assessment (EIA) response (Wells 2021), and, if nothing else, the event offers much scope for reflection and preparation for dealing with similar events in the future. We start with background commentary on EIA and decision support, segue to monitoring planning and commentary on the needs for integrated monitoring, and close with a discussion of approaches to link measurement of chemical stressors with effects, an exercise required in order to assess impacts.

14.2 BACKGROUND—ENVIRONMENTAL IMPACT ASSESSMENT AND DECISION SUPPORT

The ultimate goal of EIA is to assess impacts caused by environmental changes or events (planned or actual) for the purposes of Environmental Decision Support in developing effective management and/or recovery interventions (e.g. Leknes 2001). While the actual practice of EIA is distinct and different from that of Environmental Risk Assessment (ERA), starting early in the development of environmental practice similarities in the needs of the two have been recognized (Carpenter 1995; Go 1987; Suter et al. 1987; Verhoeven et al. 2012). This includes an increasing emphasis on a weight-of-evidence (WoE) approach (Suter et al. 2017; US EPA 2016) involving the integration and evaluation of multiple and heterogeneous pieces of evidence (e.g. causes of biological impairment based on evidence from laboratory analysis, biomarkers, biological surveys, model projections, etc.). Figure 14.1 shows a schematic representation of the evidentiary needs of EIA cast in terms of ERA, which begins with hazard identification. In the case of the X-Press Pearl, hazard identification is clear—its genesis is the disaster itself. It may not be, however, possible to identify all stressors to the environment resulting from the hazard event, e.g. because accuracy of the container manifests is at issue, multiple stressor effects may be in play, or short-term transformation of hazards is in play due to the fire and explosions on the ship. It is still necessary to scope issues that might result in impacts. Ideally, identification of stressors, through monitoring, enables exposure assessment—if analysis reveals a persistence of a contaminant in the environment, monitoring can confirm exposure. Concomitant to that, toxicological information may be obtained from prior literature (e.g. Genus Mean Acute or Chronic Values for any given hazard)

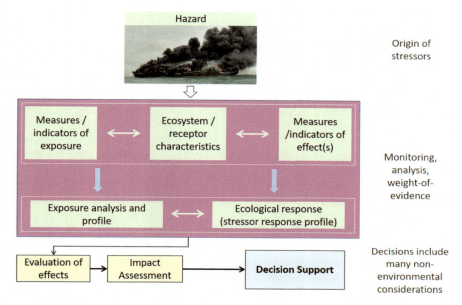

FIGURE 14.1 Schematic representation of the process behind Environmental Impact Assessment with emphasis on the underpinnings of weight-of-evidence.

Source: Picture of X-Press Pearl from Zhang et al. (2023), CC BY 4.0.

or via measurement, in order to understand the consequences of exposure (ecological response/stressor-response behavior). These two, exposure and response to exposure, enable quantitative assessment of effects, leading to impact assessment. In the absence of information concerning specific stressors, it is nonetheless possible to measure adverse effects directly, for instance, via presence and absence of effects in comparison to a control site or area (vide infra). In any case, however, there must be a clear association between a stressor or stress event and adverse effects to fulfill requirements of EIA.

Figure 14.1 also provides an indication of how the EIA result in turn informs decision making. Impact characterization via EIA is an ineluctable prerequisite to cogent and evidentiary environmental decision support for environmental management decisions and is specifically aimed at the aspects of decision-making that are to be informed by biophysical evidence, i.e. natural science (DiMento and Ingram 2005, Janssen et al. 2005, Leknes 2001). The WoE approach entails that linkages from impetus (hazard) to outcome (impacts) are quantitative and thus associated with probabilities/confidence. Often, standards of evidence in science are aligned with that of legal jurisprudence, such that demonstration of impact is required in order to receive relief for a claim of damage (Yang 2019). Legal considerations are balanced by technicofeasibility of scenarios for environmental damage recovery, which in turn also require consideration of social and economic cost-benefit (Dietz 2003, van der Sluijs et al. 2008). An alternate way to view the process in Figure 14.1 is that the entire and substantial weight of decision making rests on the head of a pin, i.e.

small failures in the process that lead to EIA may result in unintended and adverse decisions, propagating rather than ameliorating environmental damage.

A case in point involves the monitoring program itself. As the complex instrumentation needed to conduct sophisticated environmental monitoring, and the skills needed to support same, become increasingly available, it may be argued that the mania for monitoring is more intense than ever. The defensibility of monitoring data, however, is often lacking (Batley 1999). While a large body of guidance documentation on best practice in monitoring plans has been available for quite some time (e.g. Quevauviller 1995), it is not clear to what extent these guidance documents have been effective (Batley 1999; Chunlong et al. 2014). A monitoring program should be designed with Data Quality Objectives (DQOs) in mind to identify what decision support is needed, specifying data quality requirements (i.e. identifying and navigating uncertainties) and developing a defensible sampling and analysis plan for monitoring (US EPA 2006). Prior to further consideration of basic tenets of environmental best practice, it is worthwhile to ensure that consistent terminology is in use to avoid confusion. In one document sourced during the writing of this chapter, hazard was conflated with risk and monitoring was conflated with impact assessment (Partow et al. 2021). It seems reasonable to follow the lead of prominent agencies (Hannah et al. 2020) and begin this discussion with a list of terminology (Table 14.1). While some of these terms are not within the primary focus of this chapter, the list itself serves to provide an idea of the need to communicate concepts explicitly.

14.3 WHY MONITORING PLANNING AND INTEGRATED MONITORING IS CRITICAL

Integrated monitoring is much the goal as a method to support assessment/EIA (Davies and Vethaak 2012). Increasingly, it is not enough to monitor and demonstrate the presence of stressors; biological effects that can be demonstrated to be associated with or caused by stressors are needed (e.g. WoE, Suter et al. 2017). One advantage of an integrated approach that has special relevance to the X-Press Pearl disaster is that this approach produces explicit data to address situations wherein contaminants that might be present, but have not been detected, or even assessed as specific stressors due to the inability to identify them structurally (e.g. from the shipboard fire and explosions), are nonetheless identifiable as causing impacts via observation of adverse effects (Connon et al. 2012). As such, integrated monitoring at its heart is consistent with the Pathways of Effects (PoE) conceptual approach taken by Fisheries and Oceans Canada (Hannah et al. 2020). While such conceptual frameworks are ever-maturing, the general concepts have been around for some time in various contexts (e.g. Teeguarden et al. 2016; Vaughan 1964). Figure 14.2 is a representative illustration of what a partial/abbreviated conceptual PoE model looks like for the X-Press Pearl disaster. Ten stressors are specifically shown in Figure 14.2, each of which is associated with known impacts described in detail by Hannah et al. (2020); however, many more are applicable under "chemicals in cargo". Each particular stressor may be associated with various forms of effects, classed in this diagram in the three highly aggregated and interrelated categories of change in habitat, change in fitness, and change in mortality, which effects in turn are evidenced via

TABLE 14.1
List of Terminology Related to Concepts Addressed in this Chapter

Term	Usage	Reference
Acceptance/ exceedance/ assessment criteria	A set of terminologies related to environmental quality standards and generally used to indicate quantitative criteria for different endpoint evaluations as being acceptable or unacceptable. For disaster events, there may be a comparison of two criteria such as "good" versus "impacted".	Hauge et al. (2011); OSPAR (2014)
Effects	The result or outcome of the stimulus of one or more stressors, i.e. always the result of a cause. Effects may be positive or negative and are not necessarily measurable, however, must produce measurable changes in order to be assessed with confidence.	Boehlert and Gill (2010)
Endpoint	Explicit expressions of the actual environmental value that is to be protected, operationally defined by an ecological entity (e.g. fish, bivalve), and/or its measurable attributes. Endpoints are chosen to support environmental decision making and management based on ecological relevance, susceptibility, and relevance to management goals.	US EPA (1998)
Environmental hazard	Environmental hazards are defined as constituting threats to people and the things that people value and are normally thought of as occurring as a consequence of the interaction between natural systems and humans. As a result, hazards are often associated with causal agents and may include disasters that cause pollution and environmental degradation.	Cutter (2005)
Environmental risk	Variously defined by many organizations, usually characterized as a quantity based on the bivariate of probability of an adverse effect occurring and the degree or seriousness of consequence.	Galatchi (2006)
Impacts	Unlike effect, this term does not indicate simply a consequence or result. Impacts are effects that, with some level of confidence, rise to the level of deleterious ecological significance.	Boehlert and Gill (2010)
Integrated monitoring	Monitoring that anticipates and plans for assessment of individual determinand contaminants at different sites across matrices. Matrices may include, e.g. sediment, fish, and shellfish and other biota, and assessment is against specified environmental criteria associated with resultant biological effects.	Davies and Vethaak (2012)
Stressor	A physical, chemical, or biological stimulus that, at some given level of intensity, has the potential to exert effects on an ecosystem or one or more of its components	O et al. (2015)

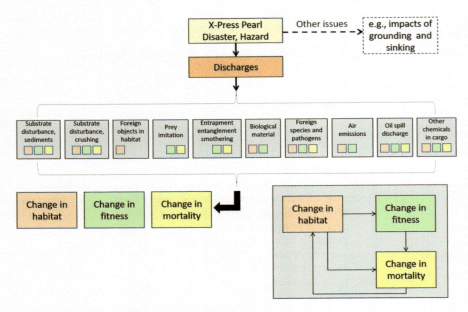

FIGURE 14.2 Diagram representing linkages in the Pathways of Effects model for one aspect of the X-Press Pearl disaster that is associated with impacts (discharge, see Hannah et al., 2020 for more details). Other aspects of the disaster are not shown but are also potentially applicable. The inset in the lower right shows interrelationships between the three highly aggregated categories of effect (change in habitat, change in fitness, change in mortality).

one to many specific or generic pieces of measurable evidence. Evidence endpoints enable assessment of impact.

The inset to Figure 14.2 (lower right) shows how even in this highly aggregated representation of effects, the interrelationships of effects ensure that few will be highly separable in terms of apportioning results of the action of one stressor from another with reference to overall outcome. Organisms are dependent upon their habitat (nature and quality of substrate, benthos, water column, etc.), changes which affect fitness. Changes in fitness may be complex and difficult to measure, spanning all aspects of the physiological condition of an organism and its ability to grow, survive, reproduce, and rear offspring, and include diverse aspects such as communication, predation avoidance and predation, susceptibility to stress, disease, state of immunosuppression, changes to nervous system, changes to endocrine system, mutagenic effects, and many other nonlethal effects (Hannah et al. 2020; Heugens et al., 2001). The indirect outcome of these nonlethal effects can nonetheless result in mortality. For instance, plastics, ranging from microplastics to nurdles, are a form of debris associated with various stressor effects (e.g. foreign object change in habitat, smothering, prey imitation) that may be reflected in measurable endpoints associated with, e.g. marine microorganisms, marine reptiles, marine fishes, marine mammals, marine birds, and others (Hannah et al. 2020; Heugens et al. 2001). As Figure 14.2 is associated with

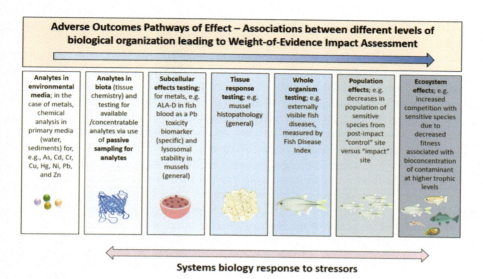

FIGURE 14.3 Schematic showing how multiple measurements, from chemical analysis and across different levels of biological organization, contribute to a weight-of-evidence approach to measurement of effects that are needed to assess environmental impact.

a continuum of potential assessment endpoints and criteria, it is necessary to incorporate into a monitoring plan a range of suitable assessment measurements (Connon et al. 2012; Davies and Vethaak 2012; Suter et al. 2017; US EPA 2014).

Figure 14.3 shows details regarding how a suite of measurements are used to measure effects and forge links between stressors and biological effect at different levels of biological organization ranging from subcellular to organism and ecosystem levels; thus, for instance, toxic metals might be analyzed chemically in exposure media, while at the same time being measured in tissues of organisms. Associated investigations might include a biomarker (subcellular response), analysis of a tissue-level response such as macroscopic liver neoplasms in fish, and one or more whole organism responses, and, if severe enough, effects in these categories would be measurable in measurements at the population level (e.g. changes in population growth rate) or ecosystem level (in the case of metals for instance, via bioaccumulation in the food chain and associated effects across trophic levels). The shift of emphasis from risk and impact assessment oriented to individual substances or classes of substances to multiple substance exposure, evidenced by multiple integrated levels of biological effects, has prominently been driven by the need for cost-effective monitoring programs in the face of increasingly prevalent occurrence of complex exposure scenarios (Connon et al. 2012; Davies and Vethaak 2012; Suter et al. 2017).

The foregoing comments serve to illustrate the complexity, ergo difficulty, of developing a fit-for-purpose monitoring program for the X-Press Pearl disaster and the importance of an integrated approach for monitoring in such cases. As early as the 1990s, the US EPA called for an integrated approach, noting that problem formulation (understanding the dimensions that require action, developing an assessment plan) is, or should be, an iterative process, taking into account available information on stressors, likely exposure routes, ecosystem's characteristics, integrated ecological effects, data quality needed or obtainable, and feasibility (US EPA 1998). For cases wherein data are unavailable, unobtainable, or infeasible to obtain, such a process-based approach serves to delineate what information can be obtained and should be collected. This iterative process is a form of project scoping whereby, as the scoping proceeds, information quality, applicability, and integrability coalesce into a cogent monitoring sampling and analysis plan (SAP), as shown in Figure 14.4.

A number of leading environmental entities have issued guidance documents and/or templates in order to assist with project planning and capacity building. A notable case with respect to the development of SAPs for monitoring is the SAP guidance and template of the US EPA (2014). While this template was not designed with projects of the scope of the X-Press Pearl in mind, it nonetheless has much to offer. Early in the document, it calls for the organizational framework around the SAP (e.g. needed to support EIA) to be explicitly specified, including the agency with overarching responsibility, other agencies involved, a clearly stated delineation for roles and responsibility of each (which will lead in turn to responsibility for project deliverables), also calling for a project organizational chart with lines of communication within the project shown. The document calls for specification of contractors, their duties, responsibilities, and their lines of communication as well. In the case of the X-Press Pearl disaster, for instance, in a document outlining advice on EIA for the disaster (Partow et al. 2021), 40 stakeholders were listed as being part of consultation with one team of international experts, and

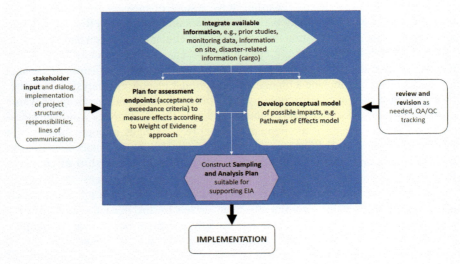

FIGURE 14.4 Schematic illustrating a process for developing an organized Sampling and Analysis Plan that foresees the needs of EIA. The process inherently incorporates a system of problem formulation and is consistent with the tenets of best practice in project management.

these stakeholders were not listed by role, but rather by agency. Other international experts who provided input and/or advice, as well as other Sri Lankan government agencies and other entities who were actively involved, were not listed (see Wells 2021 and commentary therein), i.e. there was a systemic level of coordination lacking. It is notable that the team of experts whose report is readily available to the public (Partow et al. 2021) listed their team members according to organization and role within the team, a sort of tacit recognition of the importance of establishing an organizational framework early in a project. Since the time of the expert reporting in the immediate aftermath of the X-Press Pearl disaster, transparency has not emerged. Aside from the document cited, no SAP, EIA scope, or other such detailed information, i.e. official centralized government reporting on the program of work conducted to date, was found during the writing of this chapter.

In addition to organizational framework, the US EPA SAP guidance and template also establishes that an early consideration in planning is to understand what environmental data is already available and to engage in SAP planning with DQOs in view (US EPA 2006), with a suitable quality assurance and quality control (QA/QC) plan in place to ensure that data collected are suitable to support evidentiary needs, for instance, for EIA. The guidance document suggests that there should not only be a QA/QC plan, also providing guidance on how that might look, but that an independent project team member should be assigned to review in a context of continuous improvement and ensure that the plan is appropriately followed throughout a project lifetime. The guidance document points out that such planning applies to both laboratory analytical and field operations. This approach of development of DQOs within a framework of appropriate QA/QC helps to ensure that uncertainties in data to be collected are anticipated and managed to best effect. This includes review of the measurements to be performed, whether these are likely to meet needs, and indeed, whether integration of lines of investigation is possible (per Figure 14.3). The importance of establishing an organizational framework and following the best tenets of project management from the planning stage cannot be overemphasized.

While there is both guidance and templates on development of SAPs, the issue of linking physical, chemical, and biological stressors to effects and outcomes is difficult. Guidance exists on this as well; in particular, the International Council for the Exploration of the Sea (ICES) guidance documentation is the result of decades of work and coordination and is aimed at marine applications; therefore it is ideal for consideration vis-à-vis the X-Press Pearl disaster. Casting this sort of guidance into a template framework is more difficult for a number of reasons. The next section will offer some commentary concerning approaches to stressor-effects linkages. Physical stressors, and to a lesser extent, biological stressors, are well described elsewhere (e.g. Hannah et al. 2020), and the commentary below will focus on chemical stressors, which undoubtedly constitute a large part of the damage envelope associated with the X-Press Pearl disaster. The content should be regarded as commentary only, as a full discussion of such has been the topic of more than one book. The purpose of the exercise, therefore, is not to be complete, but to consider what types of available approaches are logistically feasible for a disaster such as the X-Press Pearl.

14.4 LINKING CHEMICAL STRESSORS TO BIOLOGICAL EFFECTS

14.4.1 INTEGRATED SAMPLING AND TISSUE CHEMISTRY

Integrated sampling is, fortunately, reasonably straightforward, with a primary consideration being to sample water, sediments, and biota at the same time and at the same stations for analytes of interest. Various guidance documents (Al-Alam et al. 2020; Davies and Vethaak 2012; Hannah et al. 2020; Menchaca et al. 2014) suggest a standard suite of analytes for toxic metals, polycyclic aromatic hydrocarbons (PAHs), and polychlorinated biphenyls (PCBs). Other classes of contaminants associated with the X-Press Pearl have been identified and are discussed in other chapters herein. A fundamental problem of analysis in the primary medium of water for marine chemical contamination such as that from the X-Press Pearl is that contaminant concentration in the water column may be too low for practical measurement and/or associated with high uncertainties in analytical method due to low levels (Lepom et al. 2009). A problem with sediment analysis is that in this complex medium there may be a great difference between available contaminant and total contaminant (Ehlers and Luthy, 2003; Zhang et al. 2022). Available contaminants may be measured as bioavailability (for discussion here, availability to biota, i.e. that which is not bioavailable, is not toxic; see Wells 2012; Zhang et al. 2017). One well-known method for measuring bioavailability is the use of caged mussels (ASTM 2022, 1987; Kiddon et al. 2020). These organisms are particularly useful for water column measurements as the organisms bioconcentrate many contaminants, thereby offering one way to solve the issue of low water column contaminant concentrations. The advisory document cited above concerning the X-Press Pearl disaster that was released shortly after the event recommended caged mussel biomonitoring as "a relatively simple and cost-effective option" (Partow et al. 2021). This description of the technique may be accurate for environmental practitioners already familiar with the approach. For example, practitioners who already have (a) identified local sources of the best species for the region, (b) infrastructure for the depuration and biological assessment of organism condition (size/growth status, spawning state, etc.) in place, (c) the necessary QA/QC procedures in place, etc. Certainly, one advantage of the approach is that standards of practice exist (e.g. ASTM 2022). The approach is arguably not ideal, however, for events such as the X-Press Pearl, where the environmental practitioners responsible for implementing a SAP are not already engaged in this type of biomonitoring and are therefore not fully familiar with the considerable requirements and preparation needed to make effective use of the technique.

It is perhaps with the considerations of these issues around biomonitoring in mind that the ICES (2012) instead emphasizes passive sampling via an approach that has been referred to as chemoavailability, i.e. availability of contaminants to chemical proxies (Li et al. 2022; Wells 2012). In some cases, such approaches may correlate well with bioavailability; however, there is general agreement that such approaches measure relatively labile contaminants (Akkanen et al., 2007) and are therefore useful in risk assessment. The primary generic limitation of such an approach is simultaneously its advantage: in being inherently abiotic, it cannot accurately be said to reflect "availability to biota". This also entails, however, that in using such approaches, practitioners are freed from the vagaries of organisms' physical state that need to be accounted for and controlled in biomonitoring, issues that may also produce results that do not meet QA/QC standards and are thus inadequate to support DQOs. While eliminating many variables inherent and difficult to control in biomonitoring, as for the technique of using caged mussels, appropriate guidance documentation on the use of passive samplers in monitoring for both organic and inorganic contaminants is readily available (ISO 2011; Smedes and Booij 2012). As part of integrated monitoring, it is expected that analysis of contaminants in tissues of biota is measured in tandem with water and sediment media to which these biota are exposed.

14.4.2 BIOLOGICAL EFFECTS

Examples of ways to measure effects are listed below according to different levels of biological organization. It is imperative to protect life at the biological levels of population and ecosystem. Because chemical contaminants first begin to exert their effects at the molecular level, however, it is important to design SAPs for gathering data that will bridge the gap between the relatively short-term (acute) effects that can realistically be quantified in laboratory or field experiments and longer-term (chronic) ecological effects (Hansel et al. 2015; Mueller et al. 2020). The use of biomarkers in ecotoxicology, combined with chemical analysis, is becoming prevalent for the detection of sublethal effects that are linkable to population/ecosystem effects (Connon et al. 2012; Dagnino et al. 2007). Ultimately, the most defensible EIA is based on measurement of effects that, according to adverse outcome pathways (AOP, Figure 14.3), link mechanistic responses on the cellular level with whole organism, population, and ecosystems' communities, covering a broad range of effects in a WoE approach. It should be noted, however, that at the present state of the art, this goal is typically ambitious.

14.4.3 EXAMPLES OF SUBCELLULAR EFFECTS MEASUREMENTS

A few representative commonly used subcellular effects measurements, including biomarkers, are as follows:

- Cytochrome P450 1A (CYP1A or "EROD") activity is a biomarker of cytosolic aryl hydrocarbon (Ah) receptor/ARNT complex activation/inactivation (ISO 2007; Stagg et al. 2000; Stegeman et al. 1992) strongly associated with biotransformation of planar compounds such as PAHs, PCBs, and dioxins, typically measured by enzyme-linked immunoassay (ELISA) or mRNA abundance.
- Metallothionein in fish and mussels is typically associated with metal sequestration and/or regulation of copper and zinc, though also affected by oxidative stress processes; a variety of analytical methods exist (Davies and Vethaak 2012), which may be interpretively difficult.
- Vitellogenin protein is measured in the blood plasma of male fish and is considered a biomarker of exposure to xenoestrogens; measured by ELISA or mRNA abundance (García-Reyero et al. 2004; ISO 2013; Roesijadi and Robinson 1994).

- Delta-aminolevulinic acid dehydratase activity (ALA-D) is a biomarker for monitoring lead toxicity (Ciğerci et al. 2008; Hylland 2004) that is in increasing use due to being a good indicator of effects and being measured colorimetrically.
- The Micronucleus Assay (Davies and Vethaak 2012; ISO 2006) is a visual microscopic assay of genotoxicity/accumulated genetic damage using hemolymph and gill cells of mollusks and peripheral blood cells of fish.
- DNA adducts or covalent modification of DNA as part of carcinogenesis is a generic indicator of genotoxicity; however, it is often associated with PAH contamination, typically measured in benthic-feeding fish. The most common technique for measurement involves a P^{32}-labeling assay (Gupta et al. 1982; Pfau 1997).
- DNA strand breaks (Comet Assay) are a measure of genotoxicity (Singh et al. 1988) applicable to mussels and fish, measured using single-cell gel electrophoresis. Significant confounding issues exist (Davies and Vethaak 2012).
- Inhibition of acetylcholinesterase (AChE) activity is used as a measure of neurotoxicity and applies to a wide spectrum of organic and inorganic contaminants in fish and mussels, measured using a colorimetric microplate method (Bocquené and Galgani 1998; Wheelock et al. 2005).
- Lysosomal stability in mussels and fish is a measure of cellular integrity (Allen and Moore 2004; Borenfreund and Puerner 1985; Moore et al. 2006; Mosmann 1983; Ringwood et al. 2004), and represents a generic damage indicator associated with pollutant impacts, commonly determined using neutral red retention assay in a microplate format.
- PAH metabolites in fish bile are a measurement of PAH exposure and may be determined via fluorescence spectroscopy (screening) or alternately high-performance liquid chromatography (HPLC) and gas chromatography–mass spectrometry (GCMS) (quantitation) (Aas et al. 2000; Ariese et al. 2005).
- "Omics" techniques, including genomics, proteomics, and metabolomics, are techniques with tremendous potential to link the molecular/biochemical state of organisms to contaminant exposure (Connon et al. 2012 and references therein) and, however, require highly specialized analytical capacity (analysts, instruments, databases).

Throughout the list above, there is an interplay between quality of evidence, complexity of analysis, and feasibility of implementation in terms of the capacity needed to perform analysis and/or the ready availability of commercial labs to perform analysis. Genomic approaches, or "omics", for example, have proliferated in recent times as a result of advances in relevant areas of biology, coupled with increasingly advanced instrumentation available. In addition to being potentially very powerful, these omics approaches are further attractive in being implemented in or amenable to high-throughput analytical techniques. These approaches could not reasonably be said, however, to be practical for implementation in response to the needs of impact assessment for a disaster such as the X-Press Pearl due to their complexity and lack of standard methods and guidance for such an application. This said, already the measurement of some well-known biomarkers such as CYP1A and vitellogenin may be conducted at the protein level using ELISA or using mRNA abundance, the former of which is comparatively simple and the latter of which represents an omics approach. Of the list above, commonly used methods that are reasonably feasible for rapid implementation and or support via commercial laboratories include the Micronucleus Assay, the Comet Assay, lysosomal stability, AChE activity, and to a lesser extent, CYP1A, vitellogenin, and ALA-D.

14.4.4 Examples of Tissue-Response Effects Measurements

Some representative tissue-response effects measurements are as follows:

- The ICES has developed five categories by which to characterize <u>histopathological liver lesions</u> for use in general and PAH-specific effects monitoring (Feist et al. 2004; OSPAR 2008).
- Guidance documentation regarding practice also exists for the determination of <u>macroscopic liver neoplasms</u>/nodules (Bucke et al. 1996; Davies and Vethaak 2012; OSPAR 2008) in fish for use in general and PAH-specific effects monitoring.
- <u>Mussel histopathology</u> has been associated with contaminants including PCBs, dioxins, PAHs, and heavy metals since the early days of the Mussel Watch program (Kim et al. 2006). While the histopathological methods used are not complicated per se, experience is needed to differentiate contaminant- from pathogen-related histological changes.
- The incidence of <u>imposex</u> in gastropods (OSPAR 2008) has been used to demonstrate endocrine disrupting effects, which may be associated with a large array of contaminants and contaminant mixtures; this approach may be viewed and/or implemented at various levels of biological organization, including population effects, as below.

At the tissue-response level, tests for biological effects may be conducted in a variety of schemes, including use of standardized conditions in laboratories, to testing tissue-level, organism-level, etc. tests in actual field conditions (control versus impact sites, i.e. per the scheme of using caged mussels in a field setting). Mesocosm testing is becoming more prominent and enables some level of control that is afforded in laboratory tests while also introducing a level of reality associated with field testing (see discussion in Zhang et al. 2020). In any case, the focus of biological testing must be on ecological relevance in the context of the hazard and associated stressors. One issue with the ICES methods for tissue response (e.g. histopathological liver lesions) and some of the whole organism effects below is that this guidance has been developed and largely tested in the Atlantic, specifically a limited region of the North Atlantic. This said, the species studied have analogues in many regions, and thus the guidance may nonetheless prove useful/applicable. In contrast, the US National Oceanic and Atmospheric Administration's (Kim et al. 2006) guidance on mussel histopathology has seen wide use across a large range of regional environments.

Increasingly, *in vitro* tests can be effective because of demonstrated links between different *in vitro* tests and one or more modes of biological action, e.g. endocrine disruption. A drawback with these tests at present is the lack of demonstrated connectivity to effects at population and community levels of biological organization. *In vitro* tests in increasing use include the Chemical Activated Luciferase Gene Expression (CALUX, Leusch et al. 2010; van der Linden et al. 2008) and Yeast Estrogen Screen (YES, Routledge and Sumpter 1996), both of which are based on a reporter gene approach and respond to substances that bind specific receptors.

14.4.5 Examples of Whole-Organism Effects Measurements

Some representative whole-organism effects measurements are as follows:

- A fish disease index (FDI) has been developed to quantify externally visible fish diseases (ICES 2012, 2018; Lang and Wosniok 2008; Lang et al. 2017).
- Guidance is available for testing chronic toxicity of sediments to mussel and sea urchin (echinoid) embryos (ASTM 2021a, b).
- Mussel scope for growth testing has been in common use for decades and has been shown to be mechanistically linked to protein turnover, lysosomal stability, and larval viability while also being a correlate of contaminant exposure, i.e. it is a test with high ecological relevance (Widdows and Staff 2006).
- The incidence of imposex in gastropods has been used to demonstrate endocrine disrupting effects; this approach may be viewed and/or implemented at various levels of biological organization, including population effects, as below OSPAR (2008).

At the whole-organism level, the approach often entails using various well-known acute (short-term response) or chronic (longer-term response) toxicity tests, many of which are well known and standardized. As with *in vitro* tests, such *in vivo* tests may be conducted using sample materials in a laboratory framework or in mesocosm and field settings. Bioassay battery testing (Di Paolo et al., 2016) is used when feasible, based on the concept of a simple food chain using at least three species from different trophic levels (usually a primary producer, a detritivore or filter feeder, and a consumer such as a larval fish). This sort of testing is considerably more demanding in terms of expertise and infrastructure and hence is impractical in many settings. Toxicity identification evaluation (TIE) protocols (Wheelock et al. 2008) should also be mentioned, which involve methods to detect toxicity in whole organisms under realistic exposure conditions. This approach may also suffer from feasibility issues in the case of the X-Press Pearl disaster and additionally may be insensitive with regard to some types of effects. For an excellent discussion of *in vivo* and *in vitro* assays relevant to the marine environment, as well as some discussion of acceptance/exceedance criteria, see Davies and Vethaak (2012).

14.4.6 Population and Community/Ecosystem Effects

With regard to EIA, it has been stated, "*The weighing of heterogeneous evidence such as conventional laboratory toxicity tests, field tests, biomarkers, and community surveys is essential to environmental assessments*" (Suter et al. 2017). On first examination, this statement seems at once rational and intuitive, yet there is a gap between such assertions, increasingly common in EIA, and the body of guidance documentation by which to effect such a WoE approach that incorporates population and community effects, particularly with respect to the marine domain. Measurements for population effects would include, for example, population growth rate(s) (Chabanne et al. 2017; Dolbeth 2021). Stress on stress in mussels is another population-applicable test that measures the time to mortality subsequent to applying a stress, with mortality having been demonstrated to be shorter in impacted populations; appropriate guidance documentation exists for this test (Hellou and Law 2003; Thain et al. 2019; Viarengo et al. 1995). Other tests include impact-associated changes in population sex ratios (ISO 2015) and population changes in viability (e.g. for egg-laying species, egg viability, and emergence success, Morris et al. 1999). With respect to community and ecosystem stress, a common focus is on intra- and interspecific interactions (e.g. competition, predation). The challenges in assessing effects at higher system levels of biological organization have been outlined in a constructive way by Rouphael et al. (2011), who discuss two frameworks and approaches commonly used to monitor biodiversity in marine environments. Of these, one involves use of "true controls", which would not have been possible for the X-Press Pearl disaster. This leaves a framework whereby determination of population- and ecosystem-level changes is based on how one or more response variables differ from impact locations in relation to nearby/comparable control locations. The cause of impact at such distal levels can be inferred using multiple lines of evidence, i.e. which could include connectivity to effects demonstrated at lower levels of biological organization (Beyers 1998; Fabricius and De'ath 2004). Modeling is also recognized as having much value at an ecosystem level; however, it requires significant specialist expertise (Suter 2006).

One serious conundrum of population and community effects measurement is that effects may be, on the one hand, subtle, certainly sublethal, and at the same time compromising ecosystems over long periods, therefore being difficult to detect and not recognized for a long time. This is at the heart of the concern with the X-Press Pearl and future X-Press Pearl disasters. By the time the true effects are known, it may be impossible to ascribe them to the disaster, relegating them to a general decline in environmental quality, or even "change happens naturally".

14.5 THE X-PRESS PEARLS OF THE FUTURE

By any standard, the disaster of the X-Press Pearl is a tragedy, one component in the ongoing cycle of human destruction of the environment that is now being referred to as the sixth great extinction (Shivanna 2020). To a great extent, such events will increasingly be a certainty. Even if greater care could be taken, there is no way to avoid human error. The increasing demand for goods shipped all over the world (Lun et al. 2010), the cost-efficiency of marine shipping (Ameln et al. 2020), and the resultant ever-increasing size and number of container ships (Allianz Global 2022; Jungen et al. 2021) ensures not one but many future X-Press Pearls, and in locations that are in delicate ecosystems, and/or far from the most advanced response capability that the OECD has to offer. The website *g-Captain*, for instance, lists the top ten worst containership disasters in recent history (up to 2018, Schuler 2018), of which only one of the ten occurred in proximity to the full force of trained response capability and capacity (United Kingdom).

In the aftermath of the X-Press Pearl disaster, various organizations offered assistance and/or provided one or more experts to provide technical support (Wells 2021). Certain countries, including Sri Lanka, are eligible for participation in technical cooperation projects, some of which involve capacity building to improve the ability to cope with an event such as the X-Press Pearl (Wells 2021; UNECE 2022). The concept is laudable. Rather than give a person a fish that they might eat that day, one teaches a person to fish that they might eat always. Let us consider the matter from a different perspective. Sri Lanka is not in the OECD and, therefore, is not a member of the organization whose member countries, according to a UNICEF report (UNICEF 2022), are contributing disproportionately to the destruction of children's environments in other parts of the world; pollution is specifically a major contributor to said destruction. In addition to not being one of the richest countries globally, Sri Lanka has less than 0.3% of the total global population. Is it reasonable that Sri Lanka and countries like it should need to mount their own response to disasters such as that of the X-Press Pearl so that persons in the countries of the world who can readily afford perfumes, carpets, cars and car parts, and plastic toys should be able to enjoy vastly cheaper ones? Those same countries whose citizens are more likely to have the disposable income enabling a visit to Sri Lanka to enjoy its stunning beauty and ecosystem diversity?

Accepting that it is not reasonable for small non-OECD countries to deal with such events such as the X-Press Pearl and that EIA rationally requires an explicit organizational framework around it (vida supra and Figure 14.4) is to accept that more needs to be done at a global level, in a centralized and coordinated manner. If we as a society, and this includes all of us as individuals, really wish to do more to reduce the number of future events of this magnitude and the cumulative impact on world ecosystems, it would arguably help if goods shipped were priced at the real value of ecosystem's damage incurred. It would also help if countries such as Sri Lanka would have access to palpable and ongoing assistance for response, paid for (in advance) by goods shippers (i.e. ultimately consumers). While this chapter has provided orientational context for and content on EIA from the standpoint of practice, and with a focus on feasibility for non-OECD countries, it is neither possible nor practical to upskill a workforce into the complexity of what is required in short order. While the best manner in which to deal with such weighty issues as the next X-Press Pearl disasters of the future is not to be found in these pages, by any stretch of the imagination, the idea that small countries such as Sri Lanka should be expected to have or develop capacity equal to the level of dealing with the X-Press Pearl disaster is munificently unreasonable. Equally unreasonable is that the value of cheap(er) perfumes, carpets, cars and car parts, and plastic toys, i.e. items in the X-Press Pearl manifest, should be valued above any country's natural heritage. It is to be hoped that the disaster, as reported in this book, will form the basis of a call to action.

REFERENCES

Aas E, Beyer J, Goksøyr A 2000. Fixed wavelength fluorescence (FF) of bile as a monitoring tool for polyaromatic hydrocarbon exposure in fish: An evaluation of compound specificity, inner filter effect and signal interpretation. *Biomarkers*, 5: 9–23.

Akkanen J, Sormunen A, Lyytikäinen M, Leppänen M, Pehkonen S, Kukkonen JVK 2007. Does Tenax® Extraction-based Desorption Measure (Bio)Availability of Sediment-associated Contaminants? ICES CM 2007/J:14. *ICES (International Council for the Exploration of the Sea)*: Copenhagen.

Al-Alam J, Baroudi F, Chbani A, Fajloun Z, Millet M 2020. A multiresidue method for the analysis of pesticides, polycyclic aromatic hydrocarbons, and polychlorinated biphenyls in snails used as environmental biomonitors. *J. Chromatogr. A*, 1621: 461006.

Allen JI, Moore MN 2004. Environmental prognostics: Is the current use of biomarkers appropriate for environmental risk evaluation. *Mar. Environ. Res.*, 58: 227–232.

Allianz Global 2022. Loss Drivers in the Shipping Industry: Larger Vessels. Allianz Global News, www.agcs.allianz.com/news-and-insights/expert-risk-articles/shipping-safety-22-losses.html.

Ameln M, Fuglum JS, Thun K, Andersson H, Stalhane M 2020. A new formulation for the liner shipping network design problem. *Int. T. Oper. Res.*, 28: 638–659.

Ariese F, Beyer J, Jonsson G, Visa CP, Krahn MM 2005. Review of Analytical Methods for Determining Metabolites of Polycyclic Aromatic Compounds (PACs) in Fish Bile. *ICES Techniques in Marine Environmental Sciences*, 39. ICES: Copenhagen.

ASTM (American Society for Testing and Materials) 2022. Standard Guide for Conducting In-situ Field Bioassays with Caged Bivalves, Standard E2122-22. ASTM: West Conshohocken, Pennsylvania.

ASTM 2021a. Standard Guide for Conducting Short-Term Chronic Toxicity Tests with Echinoid Embryos, Standard E1563-21a. ASTM: West Conshohocken, Pennsylvania.

ASTM 2021b. Standard Guide for Conducting Static Short-Term Chronic Toxicity Tests Starting with Embryos of Four Species of Saltwater Bivalve Molluscs, Standard E724-21. ASTM: West Conshohocken, Pennsylvania.

ASTM 1987. "Mussel Watch"—Measurements of Chemical Pollutants in Bivalves as One Indicator of Coastal Environmental Quality, Symposia Paper STP28585S. ASTM: West Conshohocken, Pennsylvania.

Batley GE 1999. Quality assurance in environmental monitoring. *Mar. Pollut. Bull.*, 39: 23–31.

Beyers DW 1998. Causal inference in environmental impact studies. *J. North Am. Benthol. Soc.*, 17: 367–373.

Bocquené G, Galgani F 1998. *Biological Effects of Contaminants: Cholinesterase Inhibition by Organophosphate and Carbamate Compounds*. ICES Techniques in Marine Environmental Sciences, 22. ICES: Copenhagen.

Boehlert GW, Gill AB 2010. Environmental and ecological effects of ocean renewable energy development: A current synthesis. *Oceanography*, 23: 68–81.

Borenfreund E, Puerner JA 1985. Toxicity determined *in vitro* by morphological alterations and neutral red absorption. *Toxicol. Lett.* 24: 119–124.

Bucke D, Vethaak D, Lang T, Mellergaard S 1996. Common Diseases and Parasites of Fish in the North Atlantic: Training Guide for Identification. ICES No. 19. ICES: Copenhagen.

Carpenter RA 1995. Risk assessment. *Impact Assess.*, 13: 153–187.

CBD 2022. Convention on Biodiversity, Country Profile for Sri Lanka. www.cbd.int/countries/profile/?country=lk#:~:text=However%2C%20at%20present%2C%20Sri%20Lanka's,species%20and%2059%25%20of%20reptiles. Accessed July, 2022.

Chabanne DBH, Finn H, Bejder L 2017. Identifying the relevant local population for environmental impact assessments of mobile marine fauna. *Front. Mar. Sci.*, 4: 00148.

Chunlong Z, Mortimer MR, Mueller JF 2014. Quality Assurance & Quality Control of Environmental Field Sampling, pp. 8–24. Future Medicine: London.

Ciğerci İ. H. Korcan S. E. Konuk M. Öztürk S. 2008. Comparison of ALA-D activities of *Citrobacter* and *Pseudomonas* strains and their usage as biomarker for Pb contamination. *Environ. Monitor. Assess.*, 139: 41–48.

Connon RE, Geist J, Werner I 2012. Effect-based tools for monitoring and predicting the ecotoxicological effects of chemicals in the aquatic environment. *Sensors*, 12: 12741–12771.

Cutter SL 2005. The Origin of Environmental Hazards. In: Encyclopedia of Social Measurement. Elsevier: Amsterdam.

Dagnino A, Allen JI, Moore MN, Broeg K, Canesi L, Viarengo A 2007. Development of an expert system for the integration of biomarker responses in mussels into an animal health index. *Biomarkers*, 12: 155–172.

Davies IM, Vethaak AD 2012. Integrated Marine Environmental Monitoring of Chemicals and Their Effects. ICES Cooperative Research Report No. 315, Anderson ED (series ed). ICES: Copenhagen.

Dietz T 2003. What is a good decision? Criteria for environmental decision making. *Hum. Ecol. Rev.* 10: 33–39.

DiMento JFC, Ingram H 2005. Science and environmental decision making: The potential role of environmental impact assessment in the pursuit of appropriate information. *Nat. Resour. J.*, 45: 283–309.

Di Paolo C, Ottermanns R, Keiter S, Ait-Aissa S, Bluhm K, Brack W, Breitholtz M, Buchinger S, Carere M, Chalon C 2016. Bioassay battery interlaboratory investigation of emerging contaminants in spiked water extracts–towards the implementation of bioanalytical monitoring tools in water quality assessment and monitoring. *Water Res.*, 104: 473–484.

Dolbeth M 2021. Defining and Measuring a Marine Species Population or Stock. In: Life Below Water. Encyclopedia of the UN Sustainable Development Goals. Leal Filho W, Azul AM, Brandli L, Lange Salvia A, Wall T (eds). Springer: Cham.

Donahue B 2022. Unicorns and Explosives: A Burning Ship off Victoria's Coast Hints at the Dangerous Secrets of Cargo Carriers. The Narwhal, https://thenarwhal.ca/bc-zim-kingston-dangerous-goods/.

Ehlers LJ, Luthy RG 2003. Peer reviewed: Contaminant bioavailability in soil and sediment. *Environ. Sci. Technol.*, 37: 295A–302A.

Fabricius KE, De'ath G 2004. Identifying ecological change and its causes: A case study on coral reefs. *Ecol. Appl.*, 14: 1448–1465.

Feist SW, Lang T, Stentiford GD, Köhler A 2004. Biological Effects of Contaminants: Use of Liver Pathology of the European Flatfish Fab (*Limanda limanda* L.) and Flounder (*Platichthys flesus* L.) for Monitoring. ICES Techniques in Marine Environmental Sciences Report No. 38. ICES: Copenhagen.

Galatchi LD 2006. Environmental Risk Assessment. In: Chemicals as Intentional and Accidental Global Environmental Threats, Simeonov L, Chirila E (eds). NATO Security through Science Series. Springer: Dordrecht.

García-Reyero N, Raldúa D, Quirós L, Llaveria G, Cerdà J, Barceló D, Grimalt JO, Piña B 2004. Use of vitellogenin mRNA as a biomarker for endocrine disruption in feral and cultured fish. *Anal Bioanal. Chem.*, 378: 670–675.

Go F 1987. Environmental Impact Assessment: An EIA Guidance Document. Report No. 41, MARC/WHO (Monitoring and Assessment Research Centre King's College London/World Health Organization): London.

Gunatilleke N, Pethiyagoda R, Gunatilleke S 2008. Biodiversity of Sri Lanka. *J. Natn. Sci. Foundation Sri Lanka,* 36: 25–62.
Gupta RC, Reddy MC, Randerath K 1982. P-postlabelling analysis of nonradioactive aromatic carcinogen DNA adducts. *Carcinogenesis,* 3: 1081–1092.
Hannah L, Thornborough K, Murray CC, Nelson J, Locke A, Mortimor J, Lawson J 2020. Pathways of Effects Conceptual Models for Marine Commercial Shipping in Canada: Biological and Ecological Effects. *Can. Sci. Advis. Sec. Res.* Doc. 2020/077. Fisheries and Oceans Canada: Ottawa.
Hansel TC, Osofsky HJ, Osofsky JD, Speier A 2015. Longer-term mental and behavioral health effects of the Deepwater Horizon Gulf oil spill. *J. Mar. Sci. Eng.,* 3: 1260–1271.
Hauge S, Kråkenes T, Håbrekke G, Johansen G, Merz M, Onshus M 2011.Barriers to prevent and limit acute releases to sea – Environmental risk acceptance criteria and requirements to safety systems, SINTEF Report No. A20727, 2011.
Hellou J, Law RJ 2003. Stress on stress response of wild mussels, *Mytilus edulis* and *Mytilus trossulus*, as an indicator of ecosystem health. *Environ Pollut.,* 126: 407–416.
Heugens EHW, Hendriks AJ, Dekker T, van Straalen NM, Admiraal W 2001. A review of the effects of multiple stressors on aquatic organisms and analysis of uncertainty factors for use in risk assessment. *Crit. Rev. Toxicol.,* 31: 247–84.
Hylland K 2004. Biological Effects of Contaminants: Quantification of δ-aminolevulinic Acid Dehydratase (ALA-D) Activity in Fish Blood. ICES Techniques in Marine Environmental Sciences, No. 34. ICES: Copenhagen.
ICES 2018. Report of the Working Group on Pathology and Diseases of Marine Organisms CM 2018/ASG:01. ICES: Copenhagen.
ICES 2012. Report of the Working Group on Pathology and Diseases of Marine Organisms CM 2012/SSGHIE:03. ICES: Copenhagen.
ISO (International Organization for Standardization) 2015. Water Quality—Determination of Acute Lethal Toxicity to Marine Copepods (*Copepoda, Crustacea*). Draft International Standard ISO/DIS 14669, revised from 1999 standard. ISO: Geneva.
ISO 2013. Quality—Biochemical and Physiological Measurements on Fish— Part 3: Determination of Vitellogenin. ISO/TS 23893-3:2013(en). ISO: Geneva.
ISO 2011. Water quality—Sampling—Part 23: Guidance on Passive Sampling in Surface Waters. ISO 5667-23:2011(en). ISO: Geneva.
ISO 2007. Water Quality—Biochemical and Physiological Measurements on Fish— Part 2: Determination of Ethoxyresorufin-O-deethylase (EROD). ISO/TS 23893-2:2007(en). ISO: Geneva.
ISO 2006. Water Quality—Evaluation of Genotoxicity by Measurement of Induction of Micronuclei — Part 2: Mixed Population Method using the Cell Line V79. ISO 214273-2:2006(en). ISO: Geneva.
Janssen PHM, Petersen AC, van der Sluijs JP, Risbey JS, Ravetz JR 2005. A guidance for assessing and communicating uncertainties. *Water Sci. Technol.,* 52: 125–131.
Jungen H, Specht P, Ovens J, Lemper B 2021. The Rise of Ultra Large Container Vessels: Implications for Seaport Systems and Environmental Considerations. In: Dynamics in Logistics. Freitag M, Kotzab H, Megow N (eds). Springer: Cham.
Karthik R, Robin R, Purvaja R, Karthikeyan V, Subbareddy B, Balachandar K, Hariharan G, Ganguly D, Samuel V, Jinoj T 2022. Microplastic pollution in fragile coastal ecosystems with special reference to the X-Press Pearl maritime disaster, southeast coast of India. *Environ. Pollut.,* 305: 119297.
Kiddon JA, Sullivan H, Nelson WG, Pelletier MC, Harwell L, Nord M, Paulsen S 2020, Lessons Learned from 30 Years of Assessing U.S. Coastal Water. In: Water Quality— Science, Assessments and Policy, Summers K (ed). IntechOpen: London.

Kim Y, Ashton-Alcox A, Powell EN 2006. Histological techniques for marine bivalve molluscs: Update NOAA Technical Memorandum NOS NCCOS 27, Status and Trends "Mussel Watch" Program. *Mar. Environ. Res.*, 65: 101–127.

Lang T, Wosniok W 2008. The Fish Disease Index: A Method to Assess Wild Fish Disease Data in the Context of Marine Environmental Monitoring. ICES CM 2008/D:01. ICES: Copenhagen.

Lang T, Feist SW, Stentiford GD, Bignell JP, Vethaak AD, Wosniok W, 2017. Diseases of dab (*Limanda limanda*): Analysis and assessment of data on externally visible diseases, macroscopic liver neoplasms and liver histopathology in the North Sea, Baltic Sea and off Iceland. *Mar. Environ. Res.*, 124: 61–69.

Leknes E 2001.The roles of EIA in the decision-making process. *Environ. Impact Assess. Rev.*, 21: 309–334.

Lepom P, Brown B, Hanke G, Loos R, Quevauviller P, Wollgast J 2009. Needs for reliable analytical methods for monitoring chemical pollutants in surface water under the European Water Framework Directive. *J. Chromatogr. A*, 1216: 302–315.

Leusch FDL, de Jager C, Levi Y, Lim R, Puijker L, Sacher F, Tremblay L.A, Wilson VS, Chapman HF 2010. Comparison of five in vitro bioassays to measure estrogenic activity in environmental waters. *Environ. Sci. Technol.*, 44: 3853–3860.

Li B, Zhang X, Tefsen B, Wells M 2022. From speciation to toxicity: Using a "Two-in-One" whole-cell bioreporter approach to assess harmful effects of Cd and Pb. *Water Res.* 217: 118384.

Li M 2022. Cargo Debris Outside X-Press Pearl Wreckage Safety Zone Fully Recovered. Container News, https://container-news.com/cargo-debris-outside-x-press-pearl-wreckage-safety-zone-fully-recovered/.

Lun YHV, Lai KH, Cheng TCE. 2010. Shipping and Logistics Management. Springer: London.

Menchaca I, Rodríguez JG, Borja Á, Belzunce-Segarra MJ, Franco J, Garmendia JM, Larreta J 2014. Determination of polychlorinated biphenyl and polycyclic aromatic hydrocarbon marine regional Sediment Quality Guidelines within the European Water Framework Directive. *Chem. Ecol.*, 30: 693–700.

Moore MN, Allen JI, McVeigh A 2006. Environmental prognostics: An integrated model supporting lysosomal stress responses as predictive biomarkers of animal health status. *Mar. Environ. Res.*, 61: 278–304.

Morris W, Doak D, Groom M, Kareiva P, Fieberg J, Gerber L, Murphy P, Thomson D 1999. A Practical Handbook for Population Viability Analysis. The Nature Conservancy, Washington, D.C., USA.

Mosmann T 1983. Rapid colorimetric assay for cellular growth and survival: application to proliferation and cytotoxicity assays. *J. Immunol. Meth.*, 65: 55–63.

Mueller MT, Fueser H, Höss S, Traunspurger W 2020. Species-specific effects of long-term microplastic exposure on the population growth of nematodes, with a focus on microplastic ingestion. *Ecol. Indic.*, 118: 106698.

O M, Martone R, Hannah L, Greig L, Boutillier J, Patton S 2015. An Ecological Risk Assessment Framework (ERAF) for Ecosystem-based Oceans Management in the Pacific Region. *DFO Can. Sci. Advis. Sec. Res.* Doc. 2014/072. Fisheries and Oceans Canada: Ottawa.

OSPAR (OSPAR Commission) 2014. Levels and Trends in Marine Contaminants and Their Biological Effects – CEMP Assessment Report 2013. OSPAR Commission: London.

OSPAR 2008. JAMP Guidelines for Contaminant-Specific Biological Effects. Technical Annex 3: TBT-specific biological effects monitoring. OSPAR Agreement 2008-09. OSPAR Commission: London.

Pallardy R 2022. Deepwater Horizon Oil Spill. Encyclopedia Britannica, www.britannica.com/event/Deepwater-Horizon-oil-spill.

Partow H, Lacroix C, Le Floch S, Alcaro L 2021. X-Press Pearl Maritime Disaster, Sri Lanka, UNEP Report, UNEP (United Nations Environment Programme): Geneva.

Pattiaratchi C, Wijeratne S 2021. X-Press Pearl Disaster: An Oceanographic Perspective. Groundviews, https://groundviews.org/2021/06/08/x-press-pearl-disaster-an-oceanographic-perspective/.

Pfau W 1997. DNA adducts in marine and freshwater fish as biomarkers of environmental contamination. *Biomarkers*, 2: 145–151.

Quevauviller P (ed) 1995. Quality Assurance in Environmental Monitoring. VCH: Weinheim.

Ranasinghe R, Sugandhika MGP 2018. The contribution of tourism income for the economic growth of Sri Lanka. *J. Manag. Tour. Res.*, 1: 67–84.

Reuters 2022 Shipping Firm CMA CGM to Stop Transporting Plastic Waste. Reuters News, www.reuters.com/business/sustainable-business/shipping-firm-cma-cgm-stop-transporting-plastic-waste-2022-02-11/.

Ringwood AH, Hoguet J, Keppler C, Gielazyn M 2004. Linkages between cellular biomarker responses and reproductive success in oysters, *Crassostrea virginica*. *Mar. Environ. Res.*, 58: 151–155.

Rodrigo M 2021. X-Press Pearl Sinking Puts a Lens on Seafood Safety in Sri Lanka. EcoBusiness, www.eco-business.com/news/x-press-pearl-sinking-puts-a-lens-on-seafood-safety-in-sri-lanka/.

Roesijadi G, Robinson WE 1994. Metal Regulation in Aquatic Animals: Mechanisms of Uptake, Accumulation and Release. In: Aquatic Toxicology, Molecular, Biochemical and Cellular Perspectives, Malins DC, Ostrander GK (eds). Lewis Publishers: Chelsea, Michigan.

Rouphael AB, Abdulla A, Said Y 2011. A framework for practical and rigorous monitoring by field managers of marine protected areas. *Environ. Monit. Assess.*, 180: 557–72.

Routledge EJ, Sumpter JP 1996. Estrogenic activity of surfactants and some of their degradation products assessed using a recombinant yeast screen. *Environ. Sci. Technol.*, 15: 241–248.

Rubesinghe C., Brosché S., Withanage H., Pathragoda D., Karlsson T., 2021. X-Press Pearl, A "New Kind of Oil Spill" Consisting of a Toxic Mix of Plastics and Invisible Chemicals. International Pollutants Elimination Network (IPEN): Global.

Schuler M., 2018. The Worst Container Ship Disasters in Recent History. *gCaptain*, https://gcaptain.com/the-worst-containership-disasters-in-recent-history-in-photos/.

Sewwandi, M., Amarathunga, A.A.D., Wijesekara, H., Mahatantila, K., Vithanage, M., 2022a. Contamination and distribution of buried microplastics in Sarakkuwa beach ensuing the MV X-Press Pearl maritime disaster in Sri Lankan sea. *Mar. Pollut. Bull.* 184, 114074.

Sewwandi, M., Hettithanthri, O., Egodage, S.M., Amarathunga, A.A.D., Vithanage, M., 2022b. Unprecedented marine microplastic contamination from the Xpress pearl container vessel disaster. *Sci. Total Environ.* 828, 154374.

Shivanna K.R., 2020. The sixth mass extinction crisis and its impact on biodiversity and human welfare. *Resonance,* 25: 93–109.

Singh NP, McCoy MT, Tice RR, Schneider EL 1988. A simple technique for quantitation of low levels of DNA damage in individual cells. *Exp. Cell Res.* 175: 184–191.

Smedes F, Booij K 2012. Guidelines for Passive Sampling of Hydrophobic Contaminants in Water using Silicone Rubber Samplers. ICES Techniques in Marine Environmental Sciences No. 52. ICES: Copenhagen.

Stagg RM, Rusin J, McPhail ME, McIntosh AD, Moffat CF, Craft JA 2000. Effects of polycyclic aromatic hydrocarbons on expression of CYP1A in salmon (*Salmo salar*) following experimental exposure and after the Braer oil spill. *Environ. Toxicol. Chem.* 19: 2797–2805.

Stegeman JJ, Brouwer M, Richard TDG, Foerlin L, Fowler BA, Sanders BM, van Veld PA 1992. Molecular responses to environmental contamination: enzyme and protein systems as indicators of chemical exposure and effect. In: Biomarkers: Biochemical, Physiological and Histological Markers of Anthropogenic Stress. Huggett RJ, Kimerly RA, Mehrle PM Jr, Bergman HL (eds). Lewis Publishers: Chelsea, MI.

Suter GW 2006. Ecological Risk Assessment (2nd ed.). CRC Press: Boca Raton.

Suter GW, Barnhouse LW, O'Neill RV 1987. Treatment of risk in environmental impact assessment. *Environ. Manag.* 11: 295–303.

Suter G, Cormier S, Barron M 2017. A weight of evidence framework for environmental assessments: Inferring qualities. *Integr. Environ. Assess. Manag.* 13: 1038–1044.

Swuste P, van Gulijk C, Groeneweg J, Zwaard W, Lemkowitz S, Guldenmund F 2020. From Clapham Junction to Macondo, Deepwater Horizon: Risk and safety management in high-tech-high-hazard sectors: A review of English and Dutch literature: 1988–2010. *Saf. Sci.,* 121: 249–282.

Teeguarden JG, Tan YM, Edwards SW, Leonard JA, Anderson KA, Corley RA, Kile ML, Simonich SM, Stone D, Tanguay RL, Waters KM, Harper SL, Williams DE 2016. Completing the link between exposure science and toxicology for improved environmental health decision making: The aggregate exposure pathway framework. *Environ. Sci. Technol.,* 50: 4579–4586.

Thain J, Fernández B, Martínez-Gómez C 2019. Biological Effects of Contaminants: Stress on Stress (SoS) Response in Mussels. ICES Techniques in Marine Environmental Science Report No. 59. ICES: Copenhagen.

Thassim F 2022. Salvaging the X-Press Pearl: Progress so Far. Ceylon Today, https://ceylontoday.lk/2022/06/18/salvaging-the-x-press-pearl-progress-so-far/.

UNECE (United Nations Economic Commission for Europe) 2022. What is Technical Cooperation? https://unece.org/what-technical-cooperation. Accessed July, 2022.

UNICEF (United Nations Children's Fund) 2022. Places and Spaces: Environments and Children's Well-being. Innocenti Report Card 17. UNICEF Office of Research: Innocenti, Florence.

US EPA (United States Environmental Protection Agency) 2016. Weight of Evidence in Ecological Assessment. Report no. EPA/100/R-16/001. US EPA: Washington, DC.

US EPA 2014. Sampling and Analysis Plan Guidance and Template, Version 4. Document R9QA/009.1. US EPA: Washington, DC.

US EPA 2006. Guidance on Systematic Planning Using the Data Quality Objectives Process, QA/G-4. Document EPA/240/B-06/001. US EPA: Washington, DC.

US EPA 1998. Guidelines for Ecological Risk Assessment. Report no. EPA/630/R-95/002F. US EPA: Washington, DC.

van der Linden SC, Heringa MB, Man HY, Sonneveld E, Puijker LM, Brouwer A, van der Burg B 2008. Detection of multiple hormonal activities in wastewater effluents and surface water, using a panel of steroid receptor CALUX bioassays. *Environ. Sci. Technol.,* 42: 5814–5820.

van der Sluijs JP, Petersen AC, Janssen PHM, Risbey JS, Ravetz JR 2008. Exploring the quality of evidence for complex and contested policy decisions. *Environ. Res. Lett.,* 3: 024008.

Vaughan RD 1964. Use of the critical-path method in a pollution control project. *J. Am. Water Works Assoc.,* 56: 1092–1096.

Verhoeven JK, Bakker J, Bruinen de Bruin Y, Hogendoorn EA, de Knecht JA, Peijnenburg WJGM, Posthuma L, Struijs J, Vermeire TG, van Wijnen HJ, de Zwart D 2012. From Risk Assessment to Environmental Impact Assessment of Chemical Substances. Report 601353002, RIVM (Dutch National Institute for Public Health and the Environment): Bilthoven, Holland.

Viarengo A, Canesi L, Pertica M, Mancinelli G 1995. Stress on stress response: A simple monitoring tool in the assessment of a general stress syndrome in mussels. *Mar. Environ. Res.*, 39: 245–248.

Wamsley L 2021. Sri Lanka Faces An Environmental Disaster As A Ship Full Of Chemicals Starts Sinking. NPR (National Public Radio) USA, www.npr.org/2021/06/02/1002484499/sri-lanka-faces-environmental-disaster-as-ship-full-of-chemicals-starts-sinking.

Wells M 2021. Status Report Describing the Short-term Provision of Technical Assistance to Sri Lanka for the Mitigation of the X-Press Pearl Marine Ecological Disaster. Report RAS0081. IAEA (International Atomic Energy Agency): Vienna.

Wells, M., 2012. Polycyclic aromatic hydrocarbon (PAH) sensitive bacterial biosensors in environmental health. In: Biosensors and Environmental Health, Hunter RJ, Preedy VR (Eds). CRC Press: Boca Raton.

Wheelock CE, Eder KJ, Werner I, Huang H, Jones PD, Brammell BF, Elskus AA, Hammock BD 2005. Individual variability in esterase activity and CYP1A levels in Chinook salmon (*Oncorhynchus tshawytscha*) exposed to esfenvalerate and chlorpyrifos. *Aquat. Toxicol.*, 74: 172–92.

Wheelock CE, Phillips BM, Anderson BS, Miller JL, Miller MJ, Hammock BD 2008. Applications of carboxylesterase activity in environmental monitoring and toxicity identification evaluations (TIEs). *Rev. Environ. Contam. T.*, 117–178.

Widdows J, Staff F 2006. Biological Effects of Contaminants: Measurement of Scope for Growth in Mussels. ICES Techniques in Marine Environmental Science Report No. 40. ICES: Copenhagen.

Willmsen C, Mapes LV 2017. A Disaster Years in the Making. Seattle Times, https://projects.seattletimes.com/2017/west-point/.

Yang T 2019. The emergence of the Environmental Impact Assessment duty as a global legal norm and general principle of law. *Hastings Law J.*, 70, 525.

Zhang X, Jiang M, Zhu Y, Li B, Wells M 2023. The X-Press Pearl disaster underscores gross neglect in the environmental management of shipping: Review of future data needs. *Mar. Pollut. Bull.*, 189, 114728.

Zhang X, Li B, Schillereff DN, Chiverrell RC, Tefsen B, Wells M 2022. Whole-cell biosensors for determination of bioavailable pollutants in soils and sediments: Theory and practice. *Sci. Tot. Environ.*, 811: 152178.

Zhang X, Li B, Deng J, Qin B, Wells M, Tefsen B 2020. Regional-scale investigation of dissolved organic matter and lead binding in a large impacted lake with a focus on environmental risk assessment. *Water Res.*, 172: 115478.

Zhang X, Qin B, Deng J, Wells M 2017. Whole-cell bioreporters and risk assessment of environmental pollution: A proof-of-concept study using lead. *Environ. Pollut.*, 229: 902–910.

15 Management and Policy Recommendations
MV X-Press Pearl Ship Accident

Shakeela S. Bandara

15.1 OVERVIEW

15.1.1 BACKGROUND

Onboard fire occurred on May 20, 2021 at MV X-Press Pearl ship, which then continued for more than 5 days, and onboard explosion occurred on May 25, 2021, resulting in massive destruction of ship by releasing large amounts of containers into the sea. That fire continued for more than 7 days, and ship was sunk 9.5 nautical miles away from Port of Colombo. Due to the onboard explosion, cargo containers were damaged and dropped down to the sea during the fire. Both liquid and solid cargo entities were dispersed across the sea. Water-soluble materials were dissolved and were unable to be identified visually. Due to the fire incident, waste items were being washed up to several locations of the coastal line of the country, causing huge environmental and health threats.

Purview of this chapter is the management of waste including the shipwreck and analyzing the strengths and weaknesses of the existing state policies of Sri Lanka. Also this chapter tries to look back at the incident in a scientific manner after 1 year by paying more attention to the missed opportunities that could have been handled in a more environmentally favorable way.

After the onboard explosion, cargo containers started to drop down to the sea and released different types of cargo including liquids and solids. However, liquid cargo especially water soluble liquids and solid chemicals dissolved in seawater and it was not visible. That became an unseen tragedy to the sea animals. Hazardous and noxious substances (HNS) are defined as any substance other than oil, which if introduced into the marine environment is likely to create hazards to human health, to harm living resources and other marine life, to damage amenities, and/or to interfere with other legitimate uses of the sea (IMO, 2000). A huge amount of plastic pellets carrying containers were released to the sea and it was released from containers reaching the coastal areas of Sri Lanka, polluting more than 758 km from Kirinda to Mannar. Plastic pellets contaminated with various chemicals, partially burned plastic fragments, and aggregated plastic debris due to heat exposure polluted Sri Lanka's coastal areas, causing significant environmental harm, disrupting socio-economic activities, and posing health risks to humans (Figure 15.3). Figure 15.7 shows the fatal damage caused by the ingestion of plastic pellets by sea fish and Figures 15.8 and 15.9 show the harmfulness of chemical exposure to sea turtles in the sea around Sri Lanka. The full extent of the environmental damage remains largely unknown, still hidden underneath the sea, and most of the scattered plastic nurdles within the groynes will never be collected due to the hardship of collection. Figure 15.6 shows plastic pellets dispersal throughout the shore protective boulder layer at Bambalapitiya Beach area. All these wastes at the beach side of different areas of the county should be properly collected and transported to suitable intermediate storage facility which has adequate space to store until the final disposal in an environmentally safe manner.

Moreover, there were 1486 containers in the ship when this incident occurred. As per the cargo manifestos some nondangerous, general cargo items such as foods, industrial raw materials, tires, and furniture were contained in this vessel. At present, totally or partially burned waste washed up to the coast and the others were in the sea or in deck vessel. On the other hand, all the waste generated from the burned ship is considered hazardous waste due to the possibility of contamination with the chemicals that were in the same ship during the fire. At the same time, there is a necessity to clearly identify and categorize the items which were hazardous or nonhazardous. Therefore, a series of analytical tests of samples are needed to be performed to confirm their harmfulness before making the final decision on the method of disposal.

However, if any waste generated in the sea areas which belongs to Sri Lanka is to be disposed of in terrestrial areas, that should be regulated by National Waste Management regulations which are published by Central Environmental Authority, Sri Lanka, under National Environmental Act of 1980 and its amendments. If any waste is not clearly identified or categorized as hazardous waste but is mixed with identified hazardous wastes, that should be disposed of via the hazardous waste disposal criteria and it must be done in accordance with the Hazardous Waste Management Regulation No 1534/18 dated February 01, 2008 under the 1980 National Environmental Act no 47 and its further amendments.

15.2 STATE POLICIES AND REGULATIONS ON CHEMICAL AND WASTE MANAGEMENT IN SRI LANKA

National Environmental Act No. 47 of 1980 and its further amendments function as an umbrella-type cover-up on terrestrial environment of the country, and many other legal frameworks are adapted to the coastal, exclusive economic zone of the sea and its extensions. Coast Conservation Act, No. 57 of 1981 and its further amendments for defined coastal areas and Marine Pollution Prevention Act, No. 35 of 2008 for territorial waters of Sri Lanka or any other maritime zones function as basic legal frames. Accordingly, all these laws had to be implemented together with the mutual understanding and need to take collective responsibility at the state level. This incident is recorded as the worst marine disaster in recent history of the country and it was extremely difficult to practice those lawsuits which were limited to inscribed documents. It is even more difficult to implement those laws and policies in an integrated manner, and using such laws to obtain reasonable compensation for this type of catastrophe is more difficult for a developing country like Sri Lanka. Political interventions, bribes or underhand deals, and bad precedents of past drove this incident to an unfortunate end.

15.2.1 MARINE POLLUTION PREVENTION ACT, NO. 35 OF 2008

An act to provide for the prevention, control, and reduction of pollution in the territorial waters of Sri Lanka or any other maritime zone as well as the offshore and coastal zones of Sri Lanka and for matters connected therewith or incidental thereto (Long title of Act). Marine Environment Protection Authority was established by the act (Section 2) and it defines the duties and powers of the Authority as to: effectively safeguard and preserve the territorial waters of Sri Lanka or any other maritime zone, its foreshore and the coastal zone of Sri Lanka from pollution, conduct investigations and inquiries, and to institute legal actions and oversee all sea transport of oil and bunkering operations (Section 7). Sections 07 and 11 order the owner of any ship to take such steps as may be prescribed to prevent, mitigate, control, and clean up any pollution and detain any ship, if there is reasonable cause to believe that any oil, harmful substance, or other pollutants has been discharged.

The Authority shall provide reception facilities within or outside any port to discharge or deposit any residue of oil or other pollutants and direct the person in charge of all ports to provide adequate reception facilities. According to Section 21 of Marine Pollution Prevention Act, it should direct the person in charge of all ports to prepare a waste management plan and to carry out at prescribed intervals an EIA by a Classification Society approved by the MEPA. Releasing or permitting the discharge of oil, hazardous substances, or other pollutants into Sri Lanka's territorial waters, maritime zones, foreshores, or coasts constitutes a criminal offense. The owner of the vessel, offshore installation, or pipeline responsible for the discharge will be subject to a fine (4 million to 15 million LKR) (S.26). Moreover, if any person dumps any oil, harmful substances, or any other pollutant, they shall be guilty of an offense and shall be liable to a fine (4 million to 15 million LKR) (S.27).

The Act states in part 01 of sentence 34 on the liabilities of the pollution of territorial waters the following:

> Where any act referred to in Section 24 or Section 26, results in the pollution of the territorial waters of Sri Lanka or any other maritime zone, its fore-shore and the coastal zone of Sri Lanka, the owner or the operator of the ship or the owner or the person in charge of the apparatus or the owner or the occupier of the off-shore installation or the owner or occupier of the pipe line or the owner or the occupier of the place on land for the time being, as the case may be or the person carrying on the operation of exploration of natural resources including petroleum or the person in charge of such operation shall be liable for
>
> (a) any damage caused by the discharge, escape or dumping of any oil, harmful substances or other pollutant in to the territorial waters of Sri Lanka or any

other maritime zone, its fore-shore and the coastal zone of Sri Lanka under such Law or to the fore-shore or any interests related thereto;
(b) the costs of any measures taken for the purposes of preventing, reducing or removing any damage caused by the discharge, escape or dumping of any oil, harmful substance or pollutant into the territorial waters of Sri Lanka or any other maritime zone, its fore-shore and the coastal zone or any interests related thereto.

Furthermore, part 02 of the same section highlights interests pertaining to Sri Lanka's territorial waters or any other maritime zone, including its fore-shore:

(a) marine, coastal, port or estuarine activities including fisheries activities;
(b) the promotion of tourism and the preservation and development of tourist attractions in the territorial waters of Sri Lanka or any other maritime zone or on the fore-shore including beaches and coral reefs;
(c) the health of the coastal population and their wellbeing;
(d) the protection and conservation of living marine resources and wildlife.

Section 35 is based on the civil liability of the pollution and that stated as the liability in respect of any one incident under Section 34 shall be limited in accordance with the provisions of the International Convention on the Civil Liability for Pollution Damage, 1992 as may be incorporated into regulations made under this act.

Accordingly, Marine Environmental Prevention Authority, Sri Lanka, is the government Authority most responsible for mitigating, controlling, and cleaning up the pollutants discharged by any ship into the territorial sea of Sri Lanka. Moreover, MEPA has legal power to intervene in the MV X-Press Pearl ship accident as a regulatory and legislative arm of the State for damage recovery and get reasonable compensation. Eight months prior to this incident, on around September 3, 2020 MT New Diamond, a Greece-owned large crude oil tanker with 2 million barrels of oil (Bloomberg.com, 2020) caught fire off the western coast of the country. Compensation and damage estimation was carried out by MEPA expert committee and submitted to the ITOF. Though it was revealed on April 13, 2022 at the Committee on Public Enterprises (COPE) investigation that the estimated value under Section 34 of the Act was Rs. 3480 million, only Rs. 51.3 million had been received in response precisely for the ship's fire and oil spill (Committee on Public Enterprises (COPE), Parliament of Sri Lanka 2022).

These bad experiences have to be used to fix future marine disasters; nevertheless, another incident happened again after several months. Although such incidents occur in practice, the Act has empowered the legal framework with strong sentences in this regard. It is clearly stated in Section 38 of the Act that if any oil or other pollutant is discharged, escapes or is dumped for any reason, then the owner shall forthwith report all details to the Authority. If any person fails to do so, he shall be guilty of an offence and a fine (1 million–5 million) (Section 38). Also, where any act referred to in Section 24 or 26 results in pollution, the owner shall be liable for any damage caused or any interests related thereto (Section 34).

15.2.2 Coast Conservation Act, No. 57 of 1981 and Its further amendments and Coast Conservation and Coastal Resource Management Act

The long title of this act is, "The administration, control, custody and management of the Coastal Zone are hereby vested in the Republic" (S.2). "Coastal Zone" means that area lying within a limit of three hundred meters landwards of the Mean High water line and a limit of two kilometres seawards of the Mean Low Water line and in the case of rivers, streams, lagoons, or any other body of water connected to the sea either permanently or periodically, the landward boundary shall extend to a limit of two kilometres measured perpendicular to the straight baseline drawn between the natural entrance points thereof and shall include the waters of such rivers, streams and lagoons or any other body of water so connected to the sea. (S.42)

Moreover, this Act vests the administration, control, custody, and management of the Coastal Zone in the Republic (S.2), and upon receipt of an application for a permit, the Director may require the applicant to furnish an IEER or EIAR or both such reports. [S.16(1)]. If IEER is sufficient, copy to the Council [S.16 (2A)] Council shall furnish its comments within thirty days [S.16 (2B)] after the IEER, the Director General may request to submit EIAR [S.16 (2C)]. Therefore, the short tiles of the section state that any environmental impact assessment on the said coastal area should be approved by the Department of Coast Conservation. Furthermore, Waste/Damage or Detriment is addressed in Section 25 by saying that the Director should give directions for prevention or intrusion of waste or foreign matter into the Coastal Zone (S.25). The Director General shall have the power to issue directions to any person engaged in, or likely to engage in, any development activity which is causing or is likely to cause damage or detriment to the Coastal Zone or to the resources therein, regarding the measures to be taken in order to prevent or abate such damage or detriment, and it shall be the duty of such person to comply with such directions [S.26A (1)].

15.2.3 Sri Lanka Customs Ordinance

Customs Ordinance of Sri Lanka is one of the oldest customs administrations in the world. Sri Lanka Customs was established in 1806 and Ordinance was gazetted as No. 17 of 1869. Customs Ordinance was consecutively amended till 2003 and 14 major arts are established with new regulations and directorates by strengthening the ordinance.

Short title of Part IV states: No goods to be landed nor bulk broken before report. Times and places of landing, and care of officers. Goods not reported or entered, forfeited. Penalty. Except coin, bullion, cattle, passengers. Goods not reported or entered, forfeited. Penalty.

Customs Ordinances, which were established in colonial era, have not addressed the accidents we have faced currently and that loophole has been strengthened by the Prevention Directorate. The scope of the Preventive Directorate basically focuses on the enforcement aspect by way of prevention of smuggling, commercial frauds, and other offenses while safeguarding the socioeconomic, cultural, and environmental

Management and Policy Recommendations

interests of the country by being operative on broader protection routines at all the sea and airports (Sri Lanka Customs).

Furthermore, Preventive Directorate has six major functions that emphasize the act.

1. Preventing smuggling, commercial frauds, and drug offences.
2. Detections, seizures, investigations and prosecutions.
3. Safeguarding socio economic, cultural, ecological and environmental interests of the country, and enforcement of related laws and regulations.
   Preventive Directorate has been able to safeguard socio economic, cultural, ecological and environmental interests of the country through its operations.
      In addition to the regulations under the Customs Ordinance, officers of the Preventive Directorate ensure the compliance with other laws and regulations such as—
   - Arms & Ammunitions Act
   - Flora & Fauna Act
   - Import & Export Control Act
   - Cosmetic, Devices & Drugs Act
   - Telecommunication Regulatory Commission Act
   - Food and Drug Act
   - Exchange Control Act
   - Intellectual Property Act
4. Surveillance of Colombo seaport, other seaports, Free Trade Zones and Bandaranaike International Airport
5. Control of vessels movements and border operations
6. Disposal of goods forfeited by Sri Lanka Customs

Central Disposal Unit is entrusted with the disposal of seized and forfeited goods of various Directorates and Units of the whole Department. The provisions at the Customs Ordinance authorize the disposal of forfeited goods by public auction, and the Tender Sales Procedure established under the Financial Regulations is followed at such auctions.

According to this the waste generated by the MV X-Press Pearl ship incident is already brought through Custom Preventive Directorate and safely stored at a storage facility owned by the private party, but the ownership of this waste has not been confirmed and that causes another problem with the disposal of said waste. Therefore, the waste generated from MV X-Press Pearl ship incident is still in stored in a waste yard located in Wattala, Sri Lanka, causing huge costs and risks to the country.

15.2.4 National Environmental Act, No. 47 of 1980 and Its Further Amendments

This act functions as an umbrella-type cover to the environment. Three amendments, namely Act No. 56 of 1988 and Act No. 53 of 2000, along with regulations such as the

Solid Waste Management Regulation, Hazardous Waste and Chemical Management, and Environmental Pollution Prevention should be worked together on MV X-Press Pearl ship incident. The main purpose of this act is to make provision for the Protection and Management of the Environment and for Matters connected therewith or incidental thereto. "Environment" means the physical factors of the surroundings of human beings including the land, soil, water, atmosphere, climate, sound, odours, tastes and the biological factors of animals and plants of every description (NEA. 1980, S.33). License is required to be obtained under this Act (referred to as "prescribed activities"), which are activities which involve or result in discharging, depositing, or emitting waste into the environment causing pollution (NEA, Section 23 A1). Therefore, the waste disposal activities are prescribed in regulations published under the Gazette Notification No. 1534/18 dated February 01, 2008 (Re: License for discharge, emission or disposal of waste/scheduled waste management). According to the regulation, no person shall generate, collect, transport, store, recover, recycle, or dispose of waste or establish any site or facility for the disposal of any waste specified in Schedule VIII (hereinafter referred to as "scheduled waste") except under the Authority of a license issued by the Authority and in accordance with such standards and other criteria as may be specified by the Authority. All the waste that shows the characteristics of hazardous and noxious is prescribed under Schedule VIII on the same regulation. Therefore, those wastes generated from the MV X-Press Pearl ship incident should be categorized under this state regulation and most of the had been gone under scheduled waste.

Also there is a separate regulation for nonhazardous waste or general waste management in Sri Lanka, directives coming under Sections 129, 130, 131 in Municipal Council Ordinance, No. 16 of 1947; Sections 118, 119, 120 in Urban Council Ordinance No. 61 of 1939; and Sections 93, 94, 95 in Pradeshiya Sabha Act No. 15 of 1987. Waste includes any matter prescribed to be waste and any matter, whether liquid, solid, gaseous, or radioactive, which is discharged, emitted, or deposited in the environment in such volume, constituency, or manner as to cause an alteration of the environment (NEA 1980, Section 33).

Solid Waste Management Policy Technical Guideline 2004, published by Central Environmental Authority, designates state policy of waste management to ensure environmental accountability and social responsibility of all stakeholders, actively involve individuals and all institutions in integrated waste management practice, maximize resource recovery to minimize the amount of waste disposal, and minimize adverse environmental and health impact. According to the National Policy of the country, all waste which was generated from this ship incident should be disposed of adhering to those National Policies. Even though the state authorities have legal powers to intervene in waste management in this regard, it could be considered a default of the authorities.

Accordingly, Marine Environment Protection Authority and Central Environmental Authority have high responsibility for waste management and maritime accident management on any maritime-based accident or incident that takes place on Sri Lankan territorial sea, if the generated waste is disposed of within the country.

15.3 IDENTIFICATION OF WASTE TYPES GENERATED BY MV X-PRESS PEARL SHIP INCIDENT

Waste management was the most important part of the postaccident recovery process; hence it is important to determine actual waste types with their quantities to find out the way of final disposal. Waste generated by MV X-Press Pearl ship incident has been categorized into three main groups for easy management (UNEP Report 2021):

1. Nonhazardous waste
2. Hazardous and noxious waste
3. Chemicals/metals and scraps.

Also hazardous wastes can be divided into three broad categories: radioactive, infective, and chemical. Management of chemical wastes is a complex combination of treatment and storage (Carden, 1985).

Waste that has not been contaminated is categorized under nonhazardous waste and chemically or physically contaminated materials or cargo types are identified as hazardous wastes (HWM regulation 2008, CEA). Chemicals which had not been mixed or unreleased from the undamaged containers had to be categorized under hazardous waste and all other contaminated or chemicals, materials which are inside the damaged containers should be categorized under hazardous wastes.

It's been around 12 months since the MV X-Press Pearl ship was sunk, with nearly 1400 cargo containers, and most of them carried chemicals and industrial raw materials. That includes 7.985.1–8000 MT of chemicals, 3000 MT of several types of metals, 12,000 MT of polymers, 2700 MT of food items, and 5300 MT of other materials including paper items, tires, furniture, and pharmaceutical packaging. Also 26.7 MT electronic items were included inside the containers (cargo manifesto). Accordingly, that ship can be identified as a chemical carrying vessel, and sinking of that type of ship created a huge damage to the marine environment as well as the livelihood of the people like those involved in fishery and harbor activities. Risk assessment of sunken ships is necessary since they continuously have possibility for further oil spills or can cause other marine accidents (Hyuek-Jin et al, 2005). MV X-Press Pearl ship sunk at the main ship route of the busiest commercial harbor of the country. Therefore, it is important that we build a comprehensive management system for ensuring safety from sunken ships and for preserving a clean marine environment as well as develop related technology and facilities (Hyuek-Jin et al, 2005). Following analysis of cargo manifest was done by the Expert Committee appointed by MEPA to carry on the compensation and damage analysis process.

As per the report, most of the cargo containers had been carrying chemicals, dangerous goods, and some environmentally hazardous substances; moreover, containers labeled as environmentally hazardous had not been disclosed, at least to the so-called committee for damage assessment and waste management. Therefore, this ship accident created a contradictory situation in compensation process.

Waste can be further categorized into three main groups based on its generation and location of collection, as follows:

1. Shoreline waste
2. Underwater containers and material inside containers with debris
3. Shipwreck and containers remaining on cargo hold of the wreck.

Shoreline wastes, underwater container materials, including their contents, and the shipwreck itself can be categorized based on their chemical properties and hazardousness (HWM Guideline 2009). Those wastes are totally extraterrestrial, and country does not have sufficient facilities to manage them. Therefore, it was highly recommended to reexport or resend the waste through the Basel Convention, to a country with proper existing facilities; hence, all the costs of these activities should be borne by the ship owner. Basel Convention is the international agreement on control of transboundary movements of hazardous wastes and their disposal. Sri Lanka ratified the Basel Convention in 1992.

15.4 POLICY RECOMMENDATIONS FOR MANAGEMENT OF NONHAZARDOUS WASTE

After the laboratory analysis clarified that the wastes had not been contaminated by any hazardous substances, those identified were categorized under nonhazardous waste as per the CEA regulation of Hazardous Waste Management. All those identified wastes should be managed as follows,

All required approvals should be obtained from the relevant institutions, especially relevant local authorities who have the powers to handle Municipal Solid Wastes under the Municipal Council Ordinance No. 16 of 1947 – Sections 129, 130; Urban Council Ordinance No. 61 of 1939 – Sections 118, 119, 120; and Pradeshiya Sabha Act No. 15 of 1987, Sections 93, 94, 95. Any party who has planned to dump the waste within the Local Authority area should have to obtain prior approval from the relevant Local Authorities (LA) and waste handler should adhere to prevent, avoid, and minimize the environmental impact and pollution.

Basically these wastes are divided into two categories according to the National Guideline of Hazardous Waste Management, published in 2009, as follows:

1. Recyclable/recoverable waste, e.g., industrial raw materials, tires, furniture, etc.
2. Disposable waste, e.g., residues, debris, ash, etc.

During the handling of waste, all the waste materials were transported by using suitable enclosed vehicles to the storage or handling site. Upon arrival at the designated storage facility, the collected and transported waste was properly organized and stored according to its waste type in areas with ample space, as depicted in Figure 15.1. Furthermore, handling of collected waste had to be carried out only by skilled officers/workers wearing personal protective equipment (PPE).

Management and Policy Recommendations

FIGURE 15.1 Plastic pellets collected during the early stage of incident using manual collection method.

Different disposal options should be considered to get rid of this waste in order to minimize the possible environmental and health impacts further, as follows.

- Recyclable, recoverable waste should be handed over to recycling/recovery facilities having a valid EPL.
- Nonhazardous wastes can be disposed of through sanitary landfill facility at Dompe area operating under CEA and Dompe LA.
- As another option for disposing of nonhazardous burnable waste, it can be transported to Western Power Company (Pvt) Ltd. waste-to-energy plant operating at Kerawalapitiya.

15.5 POLICY RECOMMENDATIONS FOR MANAGEMENT HAZARDOUS WASTE

According to the National Environmental Act 47 of 1980, hazardous wastes are listed on Schedule VIII in Extraordinary Gazette notification 1534/18 of 2008. There are two main parts listed based on the generation of sources: nonspecific and specific. Hazardous wastes can be divided into three broad categories: radioactive, infective, and chemical (Hazardous Waste Guideline, 2009). Disposal of hazardous waste is not a simple process and a material becomes waste when it is discarded without expecting to be compensated for its inherent value. These wastes may pose a potential hazard to human health or the environment (soil, air, water) when improperly treated, stored, transported, or disposed of or managed (Misra, 2004). The sinking of the MV X-Press on 2 June raised alarm as to whether the bunker oil aboard the vessel would spill into the sea, especially as a sheen could be observed soon after the major fire outbreak (UNEP 2021). Huge oil spills were not observed and they could have caught fire onboard (CCD Report), but comprehensive report from UNEP said that a continuous release of oil had been flowing from the ship for nearly 1 month (8 June to 4 July). Furthermore a 'sheen' (code 1) of 0.04–0.3 μm thickness can be seen to be surrounding the main oil slick. Oil spills are also an important source of volatile organic compounds (VOCs) such as hexane, heptane, octane, nonane, benzene-toluene-ethylbenzene-xylene isomers (BTEX), and other lighter substituted benzene compounds (Sammarco et al., 2013).

Sri Lanka, which is a signatory to the Basel Convention on the transboundary movements of hazardous waste, has taken a number of important steps in line with its commitments to develop and implement all signatory countries, including the legal provisions required for proper risk management. The focal point of Basel Convention in Sri Lanka is Central Environmental Authority and they are the most responsible and accountable party for management of hazardous waste in the country. Unfortunately, the report released by the UN Environmental Advisory Mission did not even mention the role of the CEA or their contribution to it. According to the Extraordinary Gazette notification 1534/18 of 2008, any waste material included with their names or their source of generation in Schedule VIII is considered hazardous waste. In addition to wastes that exhibit hazardous characteristics according to the "Guideline for the Management of Scheduled Waste in Sri Lanka", wastes listed in the National Environmental (Protection & Quality) Regulation, Extraordinary Gazette No. 01 of 2008, published by the Central Environmental Authority, are also considered hazardous wastes.

15.5.1 Justification of Hazardous Waste Identification

Approximately 1,602 MT of waste were collected and safely stored at the aforementioned location. The waste collection was carried out by government agencies, primarily the Sri Lanka Navy. However, these agencies lack the facilities to independently determine the hazardousness of the collected waste beyond chemical contamination analysis. Hence, all collected wastes were stored under one storage. First batch of samples was sent to CEFAS, Pakefield Road, Lowestoft, Suffolk NR33, OHT on September 7, 2021 and the second batch was sent on November 10, 2021 to the same laboratory. According to the analytical reports hazardous substances contaminated waste should be disposed of through thermal destruction facility with valid SWML

Furthermore, buried plastic pellets/nurdles in the Sarakkuwa area in Gampaha District are washed and cleaned by using blue machine and manpower. Samples analysis was carried out by Prof. Meththika Vithanage in these areas with buried plastic pellets/nurdles and soil samples, and some biota in similar areas were chemically analyzed. According to the analysis, it has been confirmed that the plastic pellets/nurdles and soil of the surrounding shore of the area are contaminated by Mo, Li, and Zn. Molybdic oxide, Li-ion batteries, and some metal ingots were carried by the ship and those containers were placed in cargo hole under the deck, according to the given bay plan of ship. According to this chemical analysis, containers which were placed in cargo holes were also damaged and already dispersed.

Waste samples from waste storage were collected and sent to the CEFAS by MEPA. But still we have not received the test results at the time of writing this document. If the test results show that there is any contamination with chemicals, especially with heavy metals, collected total waste should be considered hazardous waste due to the occurrence of cross-contamination.

15.5.2 Identification of Shoreline Hazardous Wastes

According to the finalized analyzed cargo list of MV X-Press Pearl Ship (Table 15.1), 7000 MT of hazardous waste was identified. After the ship accident, containers which were placed on deck caught fire, were damaged, and were dropped down to the sea. MV X-Press was carrying 1486 cargo containers, and bay plan of the ship shows that nearly 1000 containers were placed on the deck, and most of them contained polymers, plastic pallets, and chemicals. Also nearly 81 containers were carrying 15 different types of hazardous substances under the International Maritime Dangerous Goods Code (IMDG code). Most of the hazardous and noxious substances carrying containers were located on the deck and they caught fire and burned totally or partially. Figure 15.2 shows most harmful and hazardous half-burned plastic pallets that were identified as toxic; some of them could have been evaporated and deformed. This made it challenging for authorities to determine the quantity of pollutants, requiring waste analysis to be performed according to the cargo manifesto provided by the ship's owner.

During the fire, some of the burned containers were dropped to the sea and some were washed ashore; also deformed plastic polymers and burned debris had been washed ashore with tidal waves. Figures 15.4, 15.5, and 15.6 are submitted as evidence of this disaster and all photographs were taken by author of this chapter during the field inspections. In the meantime, the sea around the country is very rough, and most of the dropped stuff might be washed away to the deep seas. There were reports of nurdles being found along the south coast and modeling indicated that the containers with the nurdles fell into the ocean prior to May 25, 2021. It would not be possible for the nurdles to travel to (say) Hikkaduwa in the elapsed time. Our model results indicated that the southward movement of the nurdles continued as far as Dondra Head until May 31, 2021 and then went offshore. There was also a northward movement as far as Chilaw (Pattiaratchi, 2021). Nearly 2000 MT of waste with contaminated sand was collected by Sri Lankan Government Authorities and safely stored in the Pamunugama area in a safe manner (MEPA 2021), which is illustrated by Figure 15.1 in this chapter.

Shoreline cleanup operations were initiated by MEPA with the support of stakeholder agencies and 1602 MT of wastes were collected as of November 17, 2021 (Table 15.2). The collected waste during the shoreline cleanup activities was temporarily stored in 45 containers in Wattala in early days of the incident. After that, waste collected from shoreline areas was stored at the warehouse in Pamungugama. Temporarily stored waste in containers was transported to the warehouse and all waste material collected in shoreline is stored at Pamunugama warehouse. Currently they are stored temporarily in Wattala area.

TABLE 15.1
Analyzed Cargo Manifesto Used in Expert Subcommittee of Waste Management

Analysis of Cargo Manifest

	Chemicals: 7,985.1 MT		Metals: 3,081.3 MT		Plastics and polymers: 11,939.2 MT		Food items: 2,781.5 MT Other 5,341.2 MT	
	Name of the Chemical	Amount	Name of the Metal	Amount	Name of Polymer Type	Amount	Name of Other Items	
1	Inorganic Chemicals:	495.3 MT	Aluminum stuffs:	2202.2 MT	Epoxyresins:97X1.8MT			
2	Caustic Soda:	1126.9 MT	Copper stuffs:	474.6 MT	Synthetic resins:	177.3 MT	Paper/Wastepaper items:	1705.9 MT
3	Chemical products:	160.2 MT	Iron and steel stuffs:	74.8 MT	HDPE:	747.8 MT	Used Items (Vehicle parts, Goods, Clothes, etc.):	263.9 MT
4	Nitric Acid:	28.7 MT	Lead:	187 MT	LLDPE:	245.4 MT	Cartons:	105.1 MT
5	Perfumery products:	8 MT	Metal scraps:	142.7 MT	LDPE:	574 MT	Fabrics, wadding, thread	1297.1 MT
6	Assorted perfumes:	16.8 MT					Tyres (New):	22.8 MT
7	Quicklime lumps:	1196.4 MT			Packages of PS pellets	31.9 MT	Furniture:	11.1 MT
8	Sodium Methylate Solution:	57.3 MT			Bare foam pig:	1.9 MT	Waterproof materials:	23.4 MT
9	Methanol:	235.6 MT			Packaging materials:	22.8 MT	Pharmaceutical packing:	16.5 MT
10	Bright Yellow Sulphur:	562.4 MT			Polycarbonate	60.9 MT	Carpets:	46.2 MT
11	Molybdic Oxide:	48 MT			Plastic pellets	81.3 MT	Aseptic pack:	29.5 MT
12	Env. Hazardous subs.:	10.6 MT			Polymeric beads:	19.3 MT	Personal cargo, effects	310 MT
13	Pharmaceutical stuffs:	7.5 MT					Scoured Goat hair:	22 MT
14	Paints:	21.2 MT			Polymers of propylene:	57.3 MT	Crushed Stone:	22.5 MT
15	Colours for ceramic ware:	15 MT			PVC film:	20.4 MT	Fire Protective Equipment:	109.3 MT

Management and Policy Recommendations 351

16	Shampoo.	16.3 MT	ALKYD resin:	19.5 MT	Automotive, vehicle parts (New):	126.3 MT
17	Cement conforming:	1154.8 MT	Unsaturated Polyester resin:	21 MT	Wagon Models:	28.9 MT
18	Modified Asphalt:	205.2 MT	Vinyl polymer:	18.6 MT	Exhaust stack anchor bolts, nuts,	16.1 MT
19	Silicon sealant:	12.5MT	Vinyl Acetate	46.3MT	CONSOL Cargo	43.6MT
20	Engine coolant and grease:	16.9 MT	Polybutadiene:	92.7 MT	Mobile and Stationary Accessories:	26.7 MT
21	Liquid Paraffin:	130.2 MT			Empty:	1114.3 MT
22	Brake fluid:	34.7 MT				
23	Lubricating oil:	387.8 MT				
24	BaseOil:	47.6MT				
25	Pails of lubricants:	145.9 MT				

FIGURE 15.2 A substantial accumulation of sand, nurdles, and burned plastic.

FIGURE 15.3 Burned plastic at Wellawatta Beach.

15.5.2.1 Pamunugama warehouse waster categories
See Figures 15.1 to 15.9.

15.5.3 ONSITE STORAGE OF HAZARDOUS WASTES

Central Environmental Authority is the focal point of Hazardous Waste Management in Sri Lanka. Authority has published a Guideline for the Management of Scheduled Waste in Sri Lanka in accordance with the National Environmental (Protection & Quality) Regulation, Extraordinary Gazette No. 01 of 2008, published by Central Environmental Authority; hazardous wastes are listed in Schedule VIII. All the recommendations for hazardous waste management have been based on the said guideline and Schedule Waste Management Regulation No. 1534/18 of 2008.

The approved waste management plan and environmental management plan related to the retrieval of containers and debris from the sea should be submitted to

Management and Policy Recommendations 353

FIGURE 15.4 Plastic pellets dispersal at Moratuwa Beach.

FIGURE 15.5 Collected waste, plastic pellets at Bambalapitiya Beach.

FIGURE 15.6 Plastic pellets dispersal within the groynes at Bambalapitiya Beach.

FIGURE 15.7 Dead fish with plastic ingested at Bambalapitiya Beach on June 1, 2021.

FIGURE 15.8 Dead turtle at Wellawatta Beach on June 16, 2021.

Management and Policy Recommendations

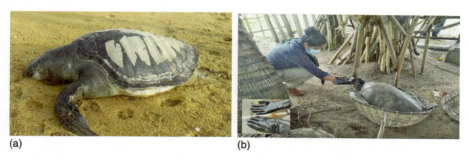

FIGURE 15.9 (a) Dead turtle at Angulana Beach on July 14, 2021. (b) Author looking at injured turtles at the rehabilitation center at Atthidiya.

the MEPA by the waste handling facilitator. Waste management plan and environmental management plan related to the onshore activities (i.e., transportation, sorting, storage and disposal) should be submitted directly to CEA and have to have prior approval. Furthermore, management plans should have been based on the risk analysis carried out by the external consultant, but unfortunately risk analysis was not carried out during the necessary period. Without proper environmental management plan, authorities should not have been allowed to bring those offshore wastes into the country. Therefore, especially CEA has possibility not to grant approvals to bring them inland. However, permission was granted and offshore wastes were transferred and stored at the approved storage facility (Figures 15.1, 15.2). Those scheduled wastes should not be stored in open ground, and they must be stored in closed containments or contained areas under specifically designed shelters and paved or concreted floor areas (HWM Guideline, CEA 2009).

Also those hazardous waste containers should be marked with the appropriate labeling approved by CEA and identified hazardous wastes should be stored specifically segregated for flammable, reactive, and noncompatible hazardous waste, and containers of such types of wastes should not be stored along with the other wastes, especially, with nonidentified wastes; moreover, hazardous waste should not store damaged, leaking, or deteriorated containers at the onsite storage. At the storage facility, stacking and shelving should provide sufficiently wide gangways in order to minimize the risk of mechanical damage at the waste handling. Also stacking heights must be limited to the maximum tolerable limit to avoid damages and accidents (HWM Guideline, CEA 2009). The types and quantities of waste currently in storage are detailed in Table 15.5 of this chapter.

Storage premises should be fenced properly, and access must be limited and allowed only for authorized personnel. It is very important to maintain hazardous waste stock report during the operation period and it should be submitted to the Central Environmental Authority. Any recovery or processing activity should not be carried out at the onsite premises. If the management of storage facility desires to carry out recovery activity, they should obtain Environmental Protection License (EPL) other than the Scheduled Waste Management Licence (SWML) from CEA (NEA, 1980).

TABLE 15.2
Types and Quantity of Waste Collected During the Shoreline Cleanup Operation

Details Waste Store at Pamunugama Warehouse

Solid Waste

	Type	Details	No. of Jambo Packs	Weight (MT)
1	15 bags in one pack	High sand, 1 pack 400 kg	1097	439
2	20 bags in one pack	High sand, 1 pack 500 kg	395	198
3	25 bags in one pack	Low sand, 1 pack 360 kg	65	23.4
4	25 bags in one pack	No sand, 1 pack, 240 kg	50	12
4	Burned plastic large parts	Unpack	0	20.4
5	Blue machine	sand removed plastics	38	10.3
6	Pallet bags 134	Undamaged 25 kg bags	0	3.3
7	Iron parts of beach washed containers (approx.)	Two containers	0	11
8	Waste received from Wattala storage			889
	Total		1645	1606.4

Wastewater

Item	Capacity (L)	Amount	Total Volume (L)
Barrel	200	38	7600
IBC Tank	1000	76	76,000
Total			83,600

(Source: MEPA)

15.5.4 TRANSPORT OF HAZARDOUS WASTE FROM ONSITE STORAGE TO FINAL DISPOSAL FACILITY

Transport of hazardous waste is one of the prescribed activities to obtain SWML from CEA; hence, transport of those hazardous wastes should be carried out with a SWML holding facility. Hazardous waste should be packed properly to minimize the potential risk at the transport. During the preparation of the package, significant chemical wastes should not be put together, and it is important to be careful as to select packaging materials that do not react with the contents of the package (HWM Guideline, CEA 2009). Before commencing the transportation activities, storage owner or responsible company who stores the hazardous waste should be reported to CEA through declaration forms, which are published by CEA under Extraordinary Gazette Notification 1534/18 of 2008.

15.5.5 Disposal of Hazardous Waste

Total hazardous wastes which were collected from this ship incident should be disposed of through a SWML holding facility.

There are many different waste disposal technologies available in the world. But Sri Lanka has only one technology for hazardous waste disposal, which is incineration under extremely high temperatures (1200–1300°C) called thermal destruction. Sri Lanka has only one facility to conduct thermal destruction and it is owned by a private organization. Also, there is only one SWML holding facility for total hazardous waste disposal in the country and they cater to more than 1000 island-wide industries for final disposal of industrial hazardous waste with very limited capacity.

Hazardous waste management in the country is regulated by Hazardous Waste Management and Chemical Management Unit in Central Environmental Authority, and they are the focal point of Basel, Rotterdam, and Stockholm Conventions. Hence management of scheduled waste/hazardous waste, including storing, transporting, recycling, recovering, and final disposal, should be regulated and supervised by said unit of CEA. Therefore, handling of generated wastes from this ship incident should be monitored by CEA, and hazardous wastes which cannot be disposed of within the country should be exported to a foreign facility following the procedures of Basal Convention. All the costs incurred in this regard should be borne by the ship owner.

15.5.6 Offshore Hazardous Waste

It has been about 12 months since the MV X-Press Pearl ship sunk with more than 1400 cargo containers. That includes chemicals, plastics, polymers, pharmaceuticals, food items, electrical and electronic devices, fabric materials, tires, and some environmentally hazardous substances, the names of which have not been disclosed. When the ship caught fire, it caused the destruction of containers which were located on deck, including polymers and plastic nurdles dropped off to the sea. Only two container wrecks were reported (at the time of writing this document) that washed onto the shore, and others remained undersea. Those scattered cargo containers might be damaged by the rough sea conditions, or they could remain still. Therefore, it is important to make proper map for scattered containers before the removal activities commence. During this period, a significant number of turtles, including the turtle shown in Figure 15.8, were found along the shore, some exhibiting physical injuries such as skin damage. Additionally, marine mammal carcasses and dead fish were also observed. Figure 15.7 of this chapter shows evidence that fish died due to the ingestion of plastic pallets.

Containers which are stuck with the shipwreck should be closely monitored before removal. Those containers contain some undisclosed environmentally hazardous substances and they can cause significant irreversible damage to the environment during the removal activities. Therefore, it is highly recommended to closely monitor the relevant organizations to avoid such environmental damage. All the relevant government organizations must be ensured to carry out these container and wreck removal activities in an environmentally sound manner.

15.5.6.1 Identification of Offshore Hazardous Waste

According to the risk assessment given by ITOPF, there were 15 dangerous cargo types carried by this MV X-Press Pearl ship. Those 15 types of dangerous cargo are contained in nearly 85 cargo containers. According to the given bay plan of ship, nearly 65 of them were located in hatch/cargo holes under the deck level. Hence they still might be stuck in the shipwreck and they may still not have been damaged. Cargo containers that were placed on or above the deck level might be dropped down to the sea and only 8 wrecked containers were washed off to the shore. Others may be scattered on the sea bed (UNEP 2021).

Finalized identification list of cargo is prepared by MEPA with quantities and it is rearranged and categorized by the waste management subcommittee as follows, based on the waste-categorizing mechanism introduced by CEA:

1. Hazardous substances
2. Chemicals
3. Nonhazardous substances

15.5.6.2 Analysis of Remaining Goods and Chemicals of MV X-Press Pearl Ship After the Accident

Estimations were very hard with the unknown total amount of waste generation due to this accident. Table 15.3 shows the rough estimation of offshore remains of hazardous substances and their quantities carried by the ship at the time of the accident. Some of the onboard cargo caught fire and totally burned out and evaporated. Table 15.4 illustrates the analysis of substances that can be evaporated, dissolved, or burned (not found after the accident) and their quantities carried by MV X-Press Pearl Ship at the time of accident.

Some of the containers were dropped down to the sea and washed away with high tide and current. Apart from that, some amount of containers were scattered on sea bed near the shipwreck. The water-soluble cargo, as mentioned previously, had been dissolved in seawater and fully dispersed. Plastic pellets/nurdles and some other materials were washed into the sea and spread over the different areas. But some cargo types, for example, solids or some clothing materials, still remain at the bottom of sea, and they should be removed from the containers and disposed of in a safe way to avoid terrestrial pollution. Table 15.5 shows the estimation of offshore remains of nonhazardous substances and their quantities carried by MV X-Press Pearl ship at the time of the accident.

The total quantity of hazardous substances is nearly 11,000 MT (except cross-contamination); that of nonhazardous substances is 3900 MT, but they could be cross-contaminated; and the quantity of chemicals is 16,300 MT, and more than 3000 MT of several types of metals other than those in the wrecked containers might have been offshore debris (Table 15.6). Without a proper map of scattered containers, it is hard to determine the hazardousness or nonhazardousness of the debris. Therefore, it is highly recommended not to bring that offshore debris to the country because of the limited facilities for disposal.

As per the information given by MEPA, offshore container removal activities will commence in mid-November 2021. It is highly recommended to take necessary

TABLE 15.3
List of Estimated Offshore Remains of Hazardous Substances and Their Quantities Carried by MV X-Press Pearl Ship at the Time of Accident

	Item Name	Estimated Quantity MT (Metric Tons)
1	Plastic and polymers (contaminated/burned)	
2	Ship oil	
3	Petroleum oils (Brake oil, Base Oil, Pails of Lubricants, Engine coolant and Grease)	604
4	Epoxy Resins	9700
5	Lithium ion batteries (burned)	
6	Pb	187
7	Environmental hazardous substances	11
8	PVC (burned)	
9	Paint	21
10	Modified asphalt	205.2
11	Console cargo with IMC.0	43.6
12	Extinguishers	
13	E-wastes (Electronic and electrical devices)	27 + Personal Cargo
14	Pharmaceuticals	7.5
15	Burned tires	
16	Silicon sealant	12.5
17	Environmentally hazardous substances	10.6
17	Polybutadiene	93
18	Other resin	40.5

(Source: Declared cargo manifest by the MV X Press Pearl Ship 2021/10/26)

actions to manage this hazardous waste through a foreign salvation company without bringing it to the country because disposal of more than 11,000 MT of hazardous waste is not possible here with very limited resources. As we mentioned previously, the country has only one private facility for hazardous waste management and Sri Lanka does not have an engineered landfill to manage nonburnable hazardous waste. Furthermore, it is highly recommended not to bring this offshore hazardous waste to the land due to the lack of resources and facilities.

If, somehow, waste has to be brought to the terrestrial land of Sri Lanka, subsequent recommendations should be followed for the management of offshore hazardous waste.

15.5.6.3 Recommendations for Onsite Storage Facility of Collected Hazardous Waste After Bringing to the Land

The recommendations outlined in section 15.5.3 of this chapter were adhered to during the establishment of a hazardous waste storage facility for the collected hazardous waste. Additionally, the following recommendations were proposed.

TABLE 15.4
Estimated List of Substances that Could Have Evaporated, Dissolved, or Burned (not Found after the Accident) and Their Quantities Carried by MV X-Press Pearl Ship at the Time of Accident

	Item Name	Estimated Quantity MT (Metric Tons)
1	Nitric acid	29
2	Perfumery products	8
3	Assorted perfumes	17
4	Methanol	236
5	Molybdic Oxide	48
6	Caustic Soda	1127
7	Urea	1844
8	Polymers	12,000
9	Sulfur	562
10	Inorganic chemicals	495
11	Sodium methylate solution	57
12	Chemical products	160
13	Quick lime products (CaO)	1200
14	Pharmaceutical stuff	7.5
15	Paints	21
16	Colors for Ceramic wave	15
17	Waterproof materials	23.4
18	Sulfur	562
19	Inorganic chemicals	495

(Source: Declared cargo manifest by the MV X Press Pearl Ship October 26, 2021)

Onsite storage facility should be in an isolated and not a congested area and it should be near where the said activity is taking place. Furthermore, this storage facility should obtain Environmental Recommendation (ER) from Hazardous Waste and Chemical Management Unit, CEA, before the activity commences.

The onsite storage area should be fenced properly and signed as "Danger and Limited Access" with lockable gates. Also the safety measures mentioned earlier, in section 15.5.3., should be adhered to.

Containers need to have clear and informative labels to provide easy comprehension of the materials they contain. Also, floor space including gangways between the waste containers, shelves, and stacks should be kept clear and uncluttered for easy inspection and good ventilation.

While storing at onsite storage, containers should be placed in a safe manner. Cross-contamination should be avoided at the temporary storage as well as at the offsite storage.

Waste declaration and record keeping is mandatory. Those records should be submitted to the relevant government authorities including CEA.

TABLE 15.5
List of Estimated Offshore Remains of Nonhazardous Substances and Their Quantities Carried by MV X-Press Pearl Ship at the Time of Accident

	Item Name	Estimated Quantity MT (Metric Tons)
1	Undamaged bulk of plastic and polymers (LDPE, LLDPE, PE, PP, HDPE, PVC, PET)	
2	Fabric, Wadding, Thread	1300
3	Furniture	11
4	Personal Cargo	310
5	Food	
6	Carpets	46
7	Pharmaceutical Packaging	16.5
8	Shampoo (noncontaminated)	
9	Cosmetics (noncontaminated)	
10	Used Items	265
11	Paper/Wastepaper	1700
12	Cartons	14.3
13	Tires (unburned)	
14	Vehicle Parts	126
15	Crushed Stone	22.5

(Source: Declared cargo manifest by the MV X Press Pearl Ship October 26, 2021)

15.5.6.4 Recommendations for Transportation of the Offshore Hazardous Waste after Bringing to the Land

Recommendations given in section 15.5.4 should be followed.

15.5.6.5 Recommendation for Offsite Waste Storage for Offshore Hazardous Waste after Bringing to the Land

Maintaining hazardous waste storage is one of the prescribed activities to obtain SWML and Environmental Protection License (EPL) from CEA. Therefore, offsite storage of hazardous waste should be managed with SWML and EPL holding facility

The storage area of the facility should be kept under a shelter, and floor area should be concreted or paved with any other suitable material to avoid soil contamination by any hazardous substances. Also proper ventilation and temperature control systems should be operated within the storage area according to the waste types. According to the cargo manifesto, some cargo containers containing volatile solvents or other low vapor pressure chemicals should be stored separately with adequate problems from direct exposure to sunlight. Any damaged, deteriorated containments should be removed immediately and contents of the containers should be transferred into compatible containers before final disposal.

TABLE 15.6
List of Estimated Offshore Remains of Metals/Scraps/Ingots and Their Quantities Carried by MV X-Press Pearl Ship at the Time of Accident

	Item Name	Estimated Quantity Metric Tons (MT)
1	Aluminum stuff	2200
2	Copper stuff	475
3	Iron and steel	75
4	Lead	187
5	Metal scrap	143
6	Wrecked containers	

(Source: Declared cargo manifest By the MV X Press Pearl Ship October 26, 2021)

The offsite storage should be located in a noncongested area and the perimeter of the facility could be fenced with minimum of 2 m high chain-link fencing and single nonemergency access to the facility with a gatehouse having 24 hours' security cover. Storage facility should have a containment system, and they should be designed and operated following the CEA guideline of scheduled waste management. Guidelines and instructions will be provided by CEA.

Storage areas should be specifically segregated for flammable, reactive, and noncompatible hazardous waste, and containers of such type of waste should not be stored along with the other wastes, especially with nonhazardous wastes. Also, operator of the storage facility must take precautions to prevent accidents of ignition or reactions. If there are any accidents, reactions, or spillages, it should be informed to the relevant authorities including CEA.

Record keeping is one of the most important parts of offshore debris removal activity. Operators of the storage facility should keep updated records of receiving and dispatching waste and it should be submitted to relevant authorities including CEA. Furthermore, operators of storage facility should follow declaration system introduced by Hazardous Waste and Chemical Management Unit, CEA, for SWML.

15.5.6.6 Recommendations of the Offshore Hazardous Waste Disposal

Hazardous wastes in the recovered cargo containers should be disposed of through a facility that holds a valid SWML and EPL as a hazardous waste disposal facility.

An appropriate technical process and recovery mechanism or proposal for them should be submitted to the CEA for prior approval. Moreover, contaminated sand/soil recovery mechanism should be submitted with a detailed technical proposal to the CEA for prior approval. Also, waste declaration and record keeping is mandatory, as mentioned previously.

After the thermal destruction, destruction certificates should be submitted to the relevant authorities. Also the disposal activities must be supervised by the CEA and other relevant authorities. Waste material balance of reception and destruction must be calculated by the authorities and should be submitted to the CEA for final approval.

15.6 POLICY RECOMMENDATIONS FOR CHEMICALS/METALS/SCRAP MANAGEMENT

15.6.1 Chemicals

According to the chemical cargo list (Table 15.1), most chemicals are highly inflammable and they could be evaporated or burned out (Table 15.4).

15.6.1.2 Identification of Chemicals

MV X-Press Pearl ship was identified as a chemical-carrying ship. That is proven by cargo manifestos received from ship agent, Sri Lanka Customs and Ports Authority, Sri Lanka.

As per the data given by the MEPA, nitric acid and caustic soda caught on fire. According to the video footage submitted by the Sri Lanka Navy, there was heavy fire and explosion occurred on board before the ship sank. Therefore, we can assume that ship oil and heavy oil might have burned out. Also substances which were unstable at high temperatures such as perfumery products, assorted perfume products, methanol, and liquid paraffin may have evaporated.

List of chemicals that might have remained in the sunken ship is shown in Table 15.5. and undamaged or sealed containers with noncontaminated chemicals can be considered nonhazardous waste and it should follow nonhazardous waste disposal methods. After confirming by chemical analysis for contamination, chemicals that are contaminated should be considered hazardous waste, the disposal of which should follow the hazardous waste disposal methods and recommendations given in section 4.0 onward.

15.6.2 Metal Ingots

Several types of metals were identified at the cargo manifesto and most of them were carried as ingots. Chemical analysis should be carried out to check the hazardousness and contamination. If the analysis reveals that they are nonhazardous or noncontaminated they can be sent to the channel of nonhazardous waste management. Otherwise they have to follow the hazardous waste management procedures and recommendations.

15.7 WRECK MANAGEMENT OF SUNKEN SHIP

Shipwreck removal has still not commenced, and the following recommendations should be followed by the responsible parties of wreck removal.

It is recommended to remove all the containers placed in cargo holes in the sunken ship before managing the shipwreck. According to the given bay plan there are nearly 65 dangerous cargo-carrying containers placed in cargo holes and they would not be damaged during operation, since they may contain highly hazardous substances to the environment. It is highly recommended that shipwreck or debris not be brought to the Sri Lankan lands due to the lack of recovery measures within the country. If this waste is somehow brought to Sri Lanka, necessary prior approvals should be obtained from the CEA. Comprehensive proposal for each activity including risk assessment should be submitted to the CEA to obtain prior approvals.

Contents of removed containers should be categorized based on the chemical characteristics, cross-contaminations, and chemical analysis into hazardous and nonhazardous. Non-hazardous waste can be managed using the same methods as previously employed for non-hazardous waste, while hazardous waste should be managed in accordance with the provided recommendations.

After removing all containers, ship should be brought to the dry docks of the ship-building yard in Sri Lanka and ship-breaking activities can be carried out only within the dry docks. Also, the party engaging in those activities should submit the proposal and agreements with dry docks facility holding company with a valid Environmental Protection License (EPL) and SWML that has obtained prior approvals from CEA as well as their continuous monitoring and guidance.

15.8 CONCLUSION

MV X-Press Pearl ship caught fire on May 20, 2021 that continued for more than five days with a major explosion, causing huge damage to the ship as well as the marine environment of Sri Lanka. After seven days of fire, the ship had sunk 9.5 nautical miles away from the Port of Colombo. This chapter considers four major Acts and regulations of the country, emphasizes the strengths and losses of policy recommendations, and tries to evaluate the gaps between the actual scenario and written legal framework. Generated waste types are categorized as hazardous, nonhazardous, and chemical/metal, and disposal channels were recommended as per the legal framework of the country as well as the international lawsuits.

Marine Pollution Prevention Act, Coast Conservation Act, Coast Conservation and Coastal Resource Management Act, Sri Lanka Customs Ordinance, and National Environmental Act are discussed as principal legal frameworks in this regard. Accordingly, Marine Pollution Prevention Act is the latest lawsuit that introduced "Polluter Pays Principal" to the Sri Lankan context for the first time. National Environmental Act is one of the upgraded lawsuits that gives an umbrella-type cover to the total environment of the country. NEA has been amended several times and new regulations are introduced directly based on International Conventions and Treaties, such as Hazardous Waste Management and Chemical Management regulations.

Though the country has strong and updated lawsuits, when these types of sudden disasters occur, authorities face difficulties when coming to apply the lawsuits practically and when integrating them due to the lack of experts who have practical experience in these types of severe disasters, especially in maritime incidents. Gaps and conflicts between practical consequences and theoretical phenomena were experienced while working on the ground during this time period. In Sri Lankan context, hazardous waste management is practiced in an international standard manner in CEA accordance with NEA. However, most environment-related government authorities and line agencies do not prioritize it. It is doubtful whether the country has been able to restore the damaged environment or obtain fair compensation for the recovery of this tragic marine destruction. This was the latest hit that the country faced due to the ignorance of the most important parts of environmental management. Nature always prefers equilibrium and she is recovering on her own.

REFERENCES

Basel Convention on the Control of Transboundary Movements of Hazardous Wastes and their Disposal, Basel, 22 March 1989 Sri Lanka Customs Ordinance Nos. 17 of 1869 and its further amendments.

Basnayake, B.F.A., Visvanathan, C. (2014). Solid waste management in Sri Lanka. In: Pariatamby, A., Tanaka, M. (eds) *Municipal Solid Waste Management in Asia and the Pacific Islands*. Environmental Science and Engineering. Springer, Singapore. https://doi.org/10.1007/978-981-4451-73-4_15.

Carden, J.L. Jr. (1985). *Hazardous Waste Management*. United States: Springer-Verlag New York Inc.

Coast Conservation Act, No. 57 of 1981.

Coast Conservation (Amendment) Act, No. 64 of 1988.

Coast Conservation (Amendment) Act, No. 49 of 2011.

Coast Conservation and Coastal Resource Management Act.

Farhad Nadim, Amvrossios C. Bagtzoglou, and Jamshid Iranmahboob. (2008). Coastal management in the Persian Gulf region within the framework of the ROPME programme of action. *Ocean & Coastal Management*, Volume 51, Issue 7, Pages 556–565, ISSN 0964-5691, https://doi.org/10.1016/j.ocecoaman.2008.04.007 (https://www.sciencedirect.com/science/article/pii/S0964569108000483)

Giusti, L. (2009). A review of waste management practices and their impact on human health. *Waste Management*, Volume 29, Issue 8, Pages 2227–2239. ISSN 0956-053X, https://doi.org/10.1016/j.wasman.2009.03.028 (www.sciencedirect.com/science/article/pii/S0956053X09001275)

Guideline for the Management of Scheduled Waste in Sri Lanka" in accordance to the National Environmental (Protection & Quality) Regulation, Extraordinary Gazette No.01 of 2008.

Hazardous Waste Management Guideline, 2009, Central Environmental Authority.

Hyuek-Jin, Choi, Kim, Hongtae, Seung-Hyun, Lee, Chang-Gu, Kang, and Lew Jae-Moon. "Development of Risk-based Information Systems for Management of Sunken Ships." Paper presented at the Fifteenth International Offshore and Polar Engineering Conference, Seoul, Korea, June 2005.

IMO – the International Maritime Organization – is the United Nations specialized agency with responsibility for the safety and security of shipping and the prevention of marine pollution by ships. June 2000 (https://www.imo.org/).

Jens Auer and Antony Firth (2007). The 'Gresham Ship': an interim report on a 16th-century wreck from Princes Channel, Thames Estuary, Post-Medieval Archaeology.

Knut Breivik, James M. Armitage, Frank Wania, Andrew J. Sweetman, and Kevin C. Jones (2016). *Environmental Science & Technology* Volume 50, Issue 2, Pages 798–805.

Manahan, S.E. (1990). Hazardous Waste Chemistry, Toxicology, and Treatment. CRC Press.

Marine Pollution Prevention Act, No.35 of 2008.

Misra, V. and S.D. Pandey. Hazardous waste, impact on health and environment for development of better waste management strategies in future in India. *Environment International*, 2005, Volume 31, Issue 3, Pages 417–431.

National Environmental Act, No. 47 of 1980.

National Environmental (Amendment) Act, No. 56 of 1988.
National Environmental (Amendment) Act, No. 53 of 2000.
National Environmental Act (No. 47 of 1980) – Sect 33. (http://www.commonlii.org/lk/legis/num_act/nea47o1980294/s33.html)
Regulations Published under the Gazette Notification No. 1534/18 dated 01.02.2008.
Order Published under the Gazette Notification No. 1533/16 dated 25.01.2008, Environmental Protection License Prescribed Activities.
Order Published under the Gazette Notification No. 1534/18 dated 01.02.2008, National Environmental Protection & Quality Regulations.
Sammarco, P.W., Kolian, S.R., Warby, R.A.F., Bouldin, J.L., Subra, W.A., Porter, S.a., 2013. Distribution and concentrations of petroleum hydrocarbons associated with the BP/Deepwater Horizon Oil Spill, Gulf of Mexico. Mar. Pollut. Bull. 73, 129–143.
Virendra Misra, S.D. Pandey. (2005). Hazardous waste, impact on health and environment for development of better waste management strategies in future in India. *Environment International*, Volume 31, Issue 3, www.sciencedirect.com/science/article/pii/S0160412004001448.
Victoria Tornero, Georg Hanke (2016). Chemical contaminants entering the marine environment from sea-based sources: A review with a focus on European seas. *Marine Pollution Bulletin,* Volume 112, Issues 1–2, Pages 17–38.
Xin, C., Wang, J., Wang, Z. et al. (2002). Reverse logistics research of municipal hazardous waste: a literature review. *Environment, Development and Sustainability*, Volume 24, Pages 1495–1531. https://doi.org/10.1007/s10668-021-01526-
X-Press Pearl Maritime Disaster, Sri Lanka Report of the UN Environmental Advisory Mission, July 2021. A UN report called the incident in May 2021 the "single largest plastic spill" in history, with about 1,680 tonnes of nurdles released into the ocean.

16 X-Press Pearl Maritime Disaster
A Framework for Valuation of Environmental Damages

U.A.D.P. Gunawardena, J.M.M. Udugama, and G.A.N.D. Ganepola

16.1 INTRODUCTION

The burned X-Press Pearl caused significant damage particularly on certain coastal parts of the country and is expected to pose negative impacts on the marine resources which are yet to be assessed. The cargo ship claims to have been carrying chemicals, mainly hazardous and noxious substances, among many other materials. The ocean currents have carried the debris southward and northward, creating more damage. As the ship has sunk, hundreds of tons of oil may have leaked into the sea, resulting in irreversible impact on marine life. Currently the debris, mostly comprising tiny plastic pellets, spread mainly along Sri Lanka's western coastline.

According to marine biologists the microplastics can never be removed completely. Hazardous material on the ship such as nitric acid caustic soda solid, sodium methoxide solution, cosmetics, methanol, and vinyl acetate once mixed with seawater causes reactions that could alter the pH level of the seawater, causing the death of marine wildlife. These externalities can cause an inestimable threat to the surrounding biodiversity including birds and mammals in the area. Dead marine animals, including turtles, started appearing on the beaches which are likely to have succumbed to exposure to toxic chemicals from the ship.

According to experts, smaller fish and organisms such as phytoplankton and coral, which provide food for a wide range of smaller sea creatures, could have died from the spills which will adversely affect the marine food chain needed to sustain a balanced ecosystem, creating long-term irreversible damages. Social impacts are already visible. Apart from tourism and ancillary services, the main occupation for many coastal dwellers is fishing, which was banned temporarily post the incident.

Oil spills, waste, marine debris (all solid waste materials) or marine litter (plastics and other man-made objects), and microplastics (component of marine debris less than 5 mm in size), liquid and hazardous waste can be expected to cause one of the worst marine and coast pollution in history in the medium and long run as well. Apart from the irreversible losses to the environment and its natural resources, this can be expected to result in lost benefits to society and economy as a whole.

16.1.1 Marine and Coastal Ecosystem Services Affected

Coastal and marine ecosystems provide a variety of ecosystem services and can be described under the frameworks of both total economic value and ecosystem service values established under the Millennium Ecosystem Assessment framework. These include direct consumable food products such as fish and other plants and animals. Nonconsumptive direct uses include recreation, education, and scientific research–related benefits. Mostly coastal ecosystems provide indirect functions such as regulatory services including storm protection, flood control, shoreline stabilization, and waste assimilation functions. Importance of these functions will be more felt with the climate change impacts especially as a potential insurance for coastal properties. Marine ecosystems have complex interactions among their components in supporting complex food webs, nutrient cycling, and climate regulation, among others.

In addition, coastal and marine resources have cultural values associated with them. They are often sources of knowledge-related values such as bequest and existence values, and cultural and heritage values. These values may be felt beyond the local and national boundaries; for example, marine charismatic organisms such as blue whales and turtles are valued by global communities and are being protected by global conventions.

Almost all of these value types have been affected by the ship accident and associated impacts. The most visible impacts included those resulting from the immediate fishing ban, damages to fishing vessels, potential contamination of fish that has reduced consumption, etc. The incident incurred irrevocable damage to the marine ecological system, mariculture, and various coastal businesses, potentially food security of the people and to the larger economy.

In addition, the coastal habitats were mostly affected by the flood of large-, small-, micro-, and nano-level plastic waste that resulted from the ship accident. The beach itself was fully covered by plastic material, and even after a year, the area is getting continuous influx of plastic pellets. The coastal and inland communities were largely deprived of the use of beach for a variety of uses that they had been accustomed to. The lack of such activities implies large loss to the recreational experiences for the people in the country and hence economic losses. The tourism sector of the country will experience a direct impact due to the pollution of the coastal and beach fronts.

According to experts, smaller fish and organisms such as phytoplankton and coral, which provide food for a wide range of smaller sea creatures, will be negatively affected and will have long-term irreversible damage to the marine food chains. Impacts toward the cultural services are quite intense. The damage to turtles, for example, has been felt beyond the coastal region and perceived highly by the entire nation. The impact will be felt in the future, by the neighboring countries, resulting in significant impact on the level of cultural services.

In addition, the air pollutants generated from the fire of the ship have created significant damage to the ocean ecosystem as the pollutants have fallen on the ocean again which are most often identified as persistent organic pollutants and possible carcinogens. The impacts will be felt along wider spatial and longer temporal dimensions.

… A Framework for Valuation of Environmental Damages

16.1.2 Impacts on Terrestrial Ecosystems

The ship contained a considerable amount of chemicals that had been released into the atmosphere as vapor or combustion products that cause harmful damage to humans, biota, and environment. The air pollutants generated from the fire are most often identified as persistent organic pollutants and possible carcinogens. The damages to the population over which the pollutants have been deposited could be significant. Since the MV X-Press pearl ship fire lasted for more than seven days with significant emission of air pollutants, this could potentially have affected ambient air quality, particularly in the surrounding area. Pollutants such as SO_2, NO_2, CO_2, and PM are the prominent pollutants generated due to the fire which dispersed in the atmosphere, deteriorating the quality of ambient air. Levels of deterioration are based on the emission rate of the pollutant, pollution load (the time and process efficiency), meteorological condition of the atmosphere, the distance from the source, etc. (Tatano *et al*, 2021).

16.1.3 Natural Re source Damage Assessment (NRDA)

NRDA implies placing the 'correct value' on types of goods and services provided by the ecosystems which are usually not recognized by the market. It is important to identify the damage as the first step. Ship accidents caused thermal damages with fire and explosion which has impaired the functioning of the variety of ocean ecosystems and marine life. Physical damage to the coastal areas and ocean-related ecosystems including ocean bed was caused by the debris generated by the burned materials, spilled plastic nurdles, and scattered containers and due to the physical presence of the shipwreck itself. There was a variety of chemicals that were released into the ocean waters and coastal areas and there are complex reactions that could have taken place among the chemicals as well as due to the chemical and thermal interactions. Firefighting chemicals are known to have hazardous effects on the ecosystems. Variety of oils found in the ship causes whole range of damages to the marine and coastal ecosystems. Each of these impacts has its current and future damages which may span over several decades.

An extensive matrix would be available as candidates for valuation when the abovementioned impacts of the ship accident are arranged against the variety of values found within each and every ecosystem and their individual components (Table 16.1). Variety of methods are available to value each of the valuation candidates including fresh valuation studies and benefit transfer approaches. The next section elaborates detail on a variety of examples which lead to the development of a comprehensive environmental valuation framework for determining the damage cost due to the ship accident.

Each of the cells in the above table could be elaborated for different components of each ecosystem and their variety of functions. When just one component of an ocean ecosystem such as turtle is considered, it consists of a variety of life stages. For example, adult reproducing female turtles, juveniles, turtle eggs, and adult males, each associated with a variety of economic values (Table 16.2). The ship accident has

TABLE 16.1
Damages to Marine and Coastal Ecosystems, Their Components and Functions Due to Different Environmental Impacts

Marine and Coastal Ecosystems and Their Components and Functions (MCECF)	Damages Due to Individual Chemicals (Short and Long Term)	Damages Due to Mixtures of Chemicals (Short and Long Term)	Damages Due to Physical Presence of Nurdles (Micro- and Nanoparticles)	Damages Due to Chemicals Adsorbed to Nurdles	Air Pollution	Thermal Damage	Other Physical Damages
Estuaries	$	$	$	$	$	$	$
Lagoons	$	$	$	$	$	$	$
Coral reefs	$	$	$	$	$	$	$
Mangroves	$	$	$	$	$	$	$
Open ocean	$	$	$	$	$	$	$
Deep-sea ocean	$	$	$	$	$	$	$
Sand beaches	$	$	$	$	$	$	$
Seagrasses	$	$	$	$	$	$	$
Other Ecosystems	$	$	$	$	$	$	$

TABLE 16.2
Values of Lost Turtles

Component	Value Type (Losses)	Valuation Method
Adult female turtles	Direct nonconsumptive use values – viewing value of turtle nesting sites in Sri Lanka	Market prices
	Educational values (For both local communities and visitors from other countries)	Travel cost method/choice modeling
Adult male turtles	Viewing value – occasionally as a substitute for whales	Market prices
All turtles – male/female	Indirect use values – role of turtle in the ecosystem (local and global values)	Choice modeling approach
All turtles – male/female	Nonuse values (Bequest value and Existence values) Values for local communities and values of these turtles for rest of the world (regional and global communities)	Contingent valuation method Benefit Transfer Approach

caused the damage mostly to adult reproducing female turtles which may have negative implications for other life forms in the long run.

Total loss is the summation of what is actually lost and cost of restoration and monitoring till the full restoration or 'before the event' status is achieved. Costs of restoration have to consider chemical, thermal, and plastic-related damages relevant for both present and future contexts.

16.1.4 Impacts on Humans

The entire event has created several impacts on humans from various fronts. During the initial phases, firefighting may have created certain undesirable impacts on firefighters. Pollutants generated during the firefighting may have led to health impacts. For example, exposure to polycyclic aromatic hydrocarbons (PAHs) may lead to genotoxicity in peripheral blood mononuclear cells. The burning of the ship has brought large quantities of air pollutants to the inland areas. Fires are a major source of PAH and persistent organic pollutants (POPs) such as dioxins and furans (Blomqvist, Persson and Simonson, 2007) which have serious toxicological impacts on natural ecosystems and in humans because of their widespread nature in various environmental compartments. A variety of chemicals, plastics, and other materials added to ocean may have significant impact on edible fish stocks and finally on human health. In addition, losses due to the impacts on nonuse values and cultural and other similar values are significant components of damages. The losses and damages to the marine and coastal ecosystems and their functions may result in unexpected thresholds and nonlinear responses which may result in large negative consequences on humankind.

16.1.5 THE VALUATION CONTEXT

The valuation of economic damage requires identification of the impacts and selection of the candidate items for the valuation exercise. During the identification stage, the complex behavior of hazardous and noxious substances (HNSs) in the marine environment needs to be identified. HNS may act as evaporators, sinkers, floaters, or dissolvers and could show intrinsic qualities (such as toxicity, flammability, corrosiveness, and reactivity with other substances or autoreactivity). Such behaviors of HNS could be modeled in order to generate environmental and socioeconomic vulnerability maps dedicated to HNS. However, the lack of technical capacity for such exercises may seriously weaken the valuation attempts.

The most obvious impacts could be attempted first: for example, the scattered waste on the beaches has impacted the tourism industry, polluted waters may have been taken in by the salterns, the consumers may have reduced fish consumption, and air pollutants may have impacted human health. Impacts on regulatory functions, however, could only be attempted with more detailed information related to impairments of functions along spatial and temporal scales. The nonuse value components could be attempted with most obvious impacts on charismatic megafauna such as turtles, whales, and dolphins (Figure 16.1).

16.2 ECONOMIC VALUATION OF IMPACTS

16.2.1 ECONOMIC VALUATION OF VESSEL SPILLS

Oil spills in the coastal ecosystem harm the marine ecosystem and human activities. The disrupted environment of the waters by the oil spills causes environmental and social-economic losses, specifically impacting the decline of fishery products and the community's living standard. Several major and minor oil spills have occurred around the world in the past few decades, e.g. Exxon Valdez 1989, Gulf War 1991, Prestige 2002, Hebei Spirit 2007, Deepwater Horizon 2010, MV MSC Chitra 2010, and Sanchi 2018. Many of these incidents have witnessed major impacts on the marine ecosystem (Peterson, 2003; Kim, 2014; Boufadel, 2014; French-McCay et al., 2021). The incidents have damaged the marine ecosystem and decreased the demand for fish and other seafood as a result of marine pollution caused. Pollution caused by different sources rigidly affects the commercial use of marine ecosystems and most of the world's largest fish industries are debased (Islam and Tanaka, 2004). Alvernia (2021) evaluated the fishermen's economic losses caused by the oil spills using the fishermen's income before and after the contamination based on production and the cost of fish.

Loureiro (2009) used a Contingent Valuation Method (CVM) to estimate environmental use and passive use losses due to the Prestige oil spill took place. On its way to the bottom of the sea, the tanker spilled more than 60,000 MT of oil, polluting the coastline (Loureiro, 2009). This was the first study to employ CVM to study an

A Framework for Valuation of Environmental Damages

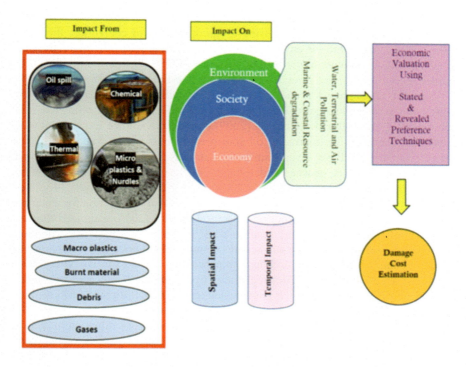

FIGURE 16.1 A framework for economic assessment of the damage.

oil spill in Europe. The parametric willingness to pay to avoid a similar oil spill in future was estimated using a Logit model. This value was estimated as €40.51 per household.

Morris (2002) reported the results of a large-scale contingent valuation study conducted after the Exxon Valdez oil spill to assess the harm caused by it. Among the issues considered are the design features of the CVM survey, estimation of household willingness to pay to prevent another Exxon Valdez–type oil spill, and issues related to validity and reliability of the estimates obtained. Furthermore, on the social aspect, cases of divorce have occurred mainly because the source of income for the family was affected by the oil spill and also mistrust among community members mainly due to the anger directed at those who take cleaning jobs from the polluting companies (Nelson and Grubesic, 2020). Bengtsson's (2013) CVM performed on the Exxon Valdez oil spill estimated nonuse values lost from a typical oil spill. The study found a total lost value of US$2.8 billion as the lower bound on aggregated passive use losses. Subsequent developments in econometric applications have advanced the estimation of nonparametric models and more flexible parametric models of the distribution of willingness to pay in contingent valuation. Estimates using these approaches have amounted to passive use losses of up to US$7.19 billion (Richardson and Brugnone, 2018).

According to Campos (2002) in November 2000 the oil spill of the ship Vergina II occurred due to the collision with the docks, spilling 86 m^3 of oil, causing a damage of

US$15,450 just in lost oil. Besides the environmental damage, up to US$26,287,603, other costs due to the incident, referring to the operations of contention, the repair of the damages caused to the environment, cleaning of the spill, and the recovery of oil, generating a total cost of the spill in the value of US$4737 per barrel have been counted, except the applied fines. CVM was used to estimate the cost.

Ahtiainen (2007) used CVM to estimate the willingness to pay (WTP) for improving the oil spill response capacity in the Gulf of Finland. The estimated WTP amounted to €112 million in total. Liu (2009) employed a Choice Experiment (CE) to estimate the WTP for improving oil spill preparedness in the Wadden Sea. The survey estimated WTP for several attributes that may be affected by a future spill including coastal waters, beaches, Eider ducks, and the oil collection ratio (collected/spilled). The study estimated a mean WTP of €29 per household. Given 39 million households in Germany, the study estimated a total WTP of €1.1 billion for a number of improvements in oil spill response capacity. Loureiro (2006) estimated the cost of the Prestige oil spill along the Spanish coast in 2002. As a basis for valuation, they used market prices of lost catches in commercial fisheries and other seafood industries such as fish farming and the fish processing sector. Van Biervliet (2011) used CVM to estimate the WTP for preventing hypothetical oil spill scenarios along the Belgian Coast. The total WTP is estimated to be €120 million to €606 million, depending on the size and the frequency of the oil spill scenario.

Liu and Wirtz (2006) also developed a series of economic evaluation models in calculating the impacts of incidental oil spills. Their model consisted of two principal steps, namely: (i) integrating the lost services with a unit value of injured natural resources, which was either measured by economic valuation methods, and (ii) measuring the lost services of injured natural resources. Liu and Wirtz (2006) also examined the relationship between the total oil spill cost and its admissible claims, leading to two major findings: (i) admissible claims do not cover the overall costs of the oil spill as research costs are neglected and (ii) admissible claims cannot be fully compensated in the case of large spills.

Reineking (2002) studied the economic loss of a burning cargo Pallas stranded 2 nautical miles off the island Amrum in Schleswig-Holstein, Germany. Incident caused 250 tons of fuel oil to be released, causing numerous environmental and economic damages. More than 7300 dead Eider ducks were found and approximately 14 million DM had been used for clean-up and salvage. For the same incident, economic assessment was coupled with three-dimensional oil spill simulation model, which also describes direct response measures. For the economic assessment, not only socioeconomic losses but also damaged natural resources resulting from oil pollution were also taken into account. The results show that total oil spill costs range from €1.28 million to €41.27 million, highly depending on spill size, weather conditions, and ecological importance of the area polluted by oil (Liu and Wirtz, 2009).

16.2.2 Economic Valuation of Plastic and Other Pollutants

According to Lee (2015) costs of marine litter and estimated economic damage can be considered as lost benefits, which are the benefits of clean-up and prevention to society. The aggregate damages of marine litter to maritime industry include the impact of microplastics. The estimated potential aggregated costs resulting from the consequence of adverse marine litter effects on marine and maritime industries are identified at over £54 million, while the costs of clean-up and prevention costs are over £34 million per annum. Coastal tourism together with bathing water quality was also vulnerable to beach litter, and the potential costs show up to be £16 million per year. An economic cost valuation of the biophysical reactions of organisms from ingesting microplastics was estimated using an ECoMip (economic cost of microplastic) model in order to capture uncertainty in biological actors and responses. Limitations in such calculations include the difficulty to address the food chain effect as marine organisms and sea creatures at every level of the food web ingest microplastics due to high levels of complexity and data unavailability. Also this stochastic approach included direct and indirect use-value components within a framework of a total economic valuation model in order to tackle the uncertain effects of marine contaminants and pollutants of microplastics and plastics on marine organisms.

Cai and Li (2011) analyzed the total economic value of marine ecosystem adjacent to Pearl River estuary of China and its economic losses from current pollution. The methods used for ecosystem valuation fall into four basic types: (i) direct market valuation, (ii) indirect market valuation, (iii) contingent valuation, and (iv) group valuation. The econometrics approaches indicated that total economic value of the studied marine ecosystem amounted to US$30.5 billion annually, among which the value of industrial utilization was the most important. The pollution situation caused an economic loss of US$5.04 billion annually, which accounted for 16.5% of the total economic value of the marine ecosystem.

Zambrano-Monserrate and Ruano (2020) experimented with a significant case of plastic pollution occurring in the Galapagos Islands which are threatened by plastic pollution. They quantified the cost of environmental damage generated by plastic waste in the Galapagos Islands using CVM. The WTP of Ecuadorian families to reduce plastic pollution was between US$4.90 and US$14.51 per year, with a median of US$7.65.

Ofiara and Brown (1999) examined recent marine pollution events associated with recreational activities in New Jersey and, in addition, assessment of economic losses from long-term contamination of fish from the New York Bight to New Jersey. Aggregate economic losses of the pollution and wash-up events to New Jersey were conservatively estimated to range from US$379.1 million to US$1597.8 million. In addition, the Hudson River Estuary and New York Bight have been exposed to long-term contamination by toxic substances (PCB, DDT) which have been detected in finfish from these waters. Effects of eating contaminated fish can result in sizeable economic losses and can provide justification for public policy regarding toxicants in seafood and in the marine environment.

Wielgus (2004) analyzed payments for damages to coral reef resources and established the basis for legal claims against parties responsible for damage caused by boat and ship groundings at Eilat, Israeli Red Sea. Previous economic and ecological studies and a set of working hypotheses have been used as a basis for estimating the required compensation for damages to the coral reef.

Ytreberg and Åstr (2021) evaluated factors responsible for a range of different pressures affecting the marine environment, air quality, and human welfare. The results for this Baltic Sea case that employed a cost-benefit analysis showed the total annual damage costs of Baltic Sea shipping to be €2.9 billion in 2010. The damage costs due to impacts on marine eutrophication and marine ecotoxicity were in the same range as the total damage costs associated with reduced air quality and climate change.

McArthur and Osland (2013) employed an impact-pathway approach to monetize costs from emissions from ships at berth in the Port of Bergen in Norway. Sensitivity analysis was conducted by varying the underlying assumptions. The cost of these emissions was estimated at between €10 million and €21.5 million per year.

16.2.3 Economic Valuation of the Impacts of Ship Accidents

Calculating the social cost of an oil spill requires considering a more comprehensive set of damages than the limited assessments carried out for compensation purposes (Garza-Gil, Prada-Blanco and Vázquez-Rodríguez, 2006). Dichmont (2016) analyzed the process of damage valuation and compensation oils to discuss the main factors explaining the observed divergence between three categories of numbers [(i) estimates by experts; (ii) compensation claims; and (iii) compensation eventually paid to claimants], and their implications in terms of the allocation of the costs of pollution. The cost of oil spills is also a key figure in debates on the development of preventive measures limiting the risks of pollution, and this at three levels.

Although the focus was initially on economic losses (Bonnieux and Rainelli 2003), the social impact of pollution events has been examined in a number of studies, and progress has been made in our understanding of the magnitude of the total costs associated with the release of toxic or hazardous substances into the sea, both from a theoretical and an applied point of view. The Amoco Cadiz oil spill raised the first questions about the valuation extent and procedures (Bonnieux and Rainelli, 1993; Grigalunas *et al.*, 1986; Grigalunas *et al.*, 1998; Thébaud *et al.*, 2005). Since then, some studies have adopted a more general framework that includes collective nonmarketed losses (Cohen A. N., Carloton J. T. and Fountain M. C., 1995) in the Exxon Valdez case (Leis and Carson-Ewart, 2003).

The Hebei Spirit Oil Spill (HSOS) was the biggest incident in Korea's coastal water and had a substantial effect on the general public and the Korean government (Ryu *et al.*, 2010). Emergency response operations for removal of bulk oil and subsequent secondary response had been conducted until the following year, and approximately 20% of the spilled oil was recovered in total (Jung *et al.*, 2012). A rapid assessment of shoreline oil contamination was performed to prioritize secondary clean-up operations and to make decisions for recreational beach reopening, supported by a modified fluorometric on-site analysis of pore water in the affected area (Theodotou, 2018).

Kim *et al.* (2014) highlighted that there was a considerable gap between the amount of economic loss claimed by residents and the actual compensation received;

approximately KRW4.3 trillion (US$4 billion) in total was requested for compensation, but only KRW478 billion (US$440 million) was approved and awarded. A parametric WTP estimation indicates that respondents in the sample are willing to pay about €40.51 per household to avoid vessel oil spills in Spain.

16.2.4 Impact of Vessel Spills on Seafood Consumption

According to the estimates of FAO, about 1 billion people in the world rely on fish as their primary source of animal protein. Not only as an important food source, but the fish industry also plays a major role in the income of over 36 million people all over the world who are directly employed in fishing (Eskola *et al.*, 2020). Certain incidents that occurred around the world damaged the marine ecosystem and decreased the demand for fish and other seafood irregularly as a result of marine pollution caused by the incidents. Most of the marine pollution is accidental and caused by oil spills. Not only the highly visible oil covers but also the light components severely affect the health of marine ecosystems (Clark, Millet and Marshall, 2017).

Unlike the other causes of marine pollution such as incineration of garbage, engine disposals, gas releases, and polluted water, accidental oil spills have a considerable impact on fish consumption patterns in the world. Pollution caused by different sources rigidly affects the commercial use of marine ecosystems and most of the world's largest fish industries (Islam and Tanaka, 2004). The study of Allen, Ewel, and Jack (2001) highlighted that, with time, fishing became more industrialized and is now criticized as it may lead to environmental problems and be a threat to the marine ecosystem.

Shkëlqim Sinanaj (2020) studied the impact Hebei Spirit tanker oil spill had on fishing and tourism. Study showed sales reductions of fish due to the oil spills heavily affecting the marine-based economy. Consumers feared to purchase fish products, thinking that fish had been contaminated. The assessment of economic damages from the prestige oil spill by Garza-Gil, Prada-Blanco, and Vázquez-Rodríguez (2006) revealed that the oil spill generated an income loss due to the closed fishing activities in the coastal area as well as the sales decrements.

Chang (2014) also commented on the economic impact of oil spills. According to the study, historic oil spills caused those incidents to experience economic losses as a result of declined market demand for fish products. Hebbar (2016) also pointed out that those changes in consumer demand following a major spill from any of the tankers in the coastal area in India may result in economic decline. The research conducted by Wenaty A (2018) highlighted that health concern is one of the main factors that affect consumers' preference when consuming fish. Therefore, consumers' demand for fish may be reduced due to oil spills in the ocean.

Oken and Essington (2016) explored fish consumption among pregnant women after the national mercury advisory and the key findings of the study revealed that there was a decline in dark meat fish, canned tuna, and white meat fish consumption from approximately 1.4 servings per month in the period from 2000 to April 2021. Shkelqim Sinanaj (2020) also commented on the mercury incident and found that some target consumers significantly reduced fish consumption. This literature is evidence for the fact that consumers are more concerned about health and food safety when consuming fish products.

16.2.5 THERMAL IMPACT OF VESSEL ACCIDENTS

Direct and indirect thermal impacts from a vessel accident on the ecosystem can be considered one of the main types of pollution and damage. According to the study by Mumby et al. (2001) thermal impacts may kill off corals and temperature-sensitive organisms.

Fire is the second major cause of ship accidents (Kwiecińska, 2015). Some of the basic causes of ship fires are damage to the electrical items, mechanical instruments, and ship's hull and damage occurring during maintenance ignition of cargo, etc. (Dumax and Rozan, 2021).

16.2.6 AIR POLLUTION DUE TO SHIP ACCIDENTS

Pollutants such as volatile organic compounds (VOCs), primary particulate matter, CO_2, SO_2, and NO_x will increase global warming and acid rain formation. Some ships which have chlorofluorocarbon cool systems may also lead to air pollution (Kwiecińska, 2015). According to the report of Miola and Ciuffo (2011), CO_2 has the highest emission factor among marine emissions. Iduk and Samson (2015) investigated the effects and solutions of marine pollution from ships in Nigerian waterways. The study finds that sulfur in the air creates acid rain which could damage crops and buildings. When inhaled sulfur is known to cause respiratory problems and even increase the risk of a heart attack.

16.2.7 LONG-TERM IMPACTS OF SHIP ACCIDENTS

According to Chen (2022), the long-term impacts of ship accidents have the potential to create hazards to human health, harm living resources and marine life, damage amenities, or interfere with other legitimate uses of the sea. Accidents caused by crude oil spills can cause short- and long-term impacts to the environment and humans (Jung et al., 2011).

Jang et al. (2014) studied the environmental and ecological effects and recoveries five years after the Hebei Spirit oil spill in Taean, Korea. Although several years have passed since the incident it appears that compensation and recovery efforts are far from being satisfactory and the affected communities are still suffering various adverse impacts incurred by the disaster (Kim et al., 2014).

The comprehensive review conducted by Gohlke et al. (2011) pointed out that several studies have outlined evidence of high polyaromatic hydrocarbon (PAH) concentrations in seafood harvested after oil spills. The study also explored the presence of metals such as Zn, Hg, Mn, As, Co, Cr, Cd, etc., which were found in seafood harvested from the areas surrounding the ship accident. Jung et al. (2011) also commented on heavy metal accumulations in the fish organs due to various marine pollution. However, not much is known about the long-term impacts. The unhealthy substances accumulated in fish may lead to toxicity to human health along the food chain rather than providing nutritional value. Although most seafood was shown to be uncontaminated by PAHs and untainted, local consumers proved very reluctant to accept these results (Heikes et al., 2002).

16.2.8 Impacts on Fishery Industry

Law and Hellou (1999) who studied impacts of Valdez oil spill suggest that there is some risk to consumers from eating fish and shellfish contaminated with oil. Immediately after an oil spill, the respective government closes all the fisheries and bans fishing in affected areas which will lead to huge losses in the fish industry and market shortage (Mathuru et al., 2012). Therefore, marine pollution may lead to fish food insecurity in the world. According to the study by de Oliveira Estevo et al. (2021), food security is strongly damaged by the oil spill in Brazilian coastal fishery communities. Hebbar (2016) explored the fisheries dimensions of marine oil spills and also highlighted that there was a high economic loss due to fishery restrictions because fishers could not meet the market demand. A similar finding was highlighted in the study of Sukumaran (2014), as he indicated that around 60 fish markets in Mumbai were empty and closed due to the lack of quality fish harvest resulting from the Chitra spill in India. These chemicals, mainly polyaromatic hydrocarbons (PAHs), are exposed to fish and other aquatic organisms.

Therefore, the pollution opens the way to the potential threat of consuming the fish for human consumers as well. The oil and tank discharge load a considerable amount of oil to the marine system (Law and Hellou, 1999). Crude oil contains a large number of toxic substances with varying components that can cause adverse effects on marine and terrestrial ecosystems including human health (Harwell and Gentile, 2006). In the samples of fish captured after the gas oil spill, 5 of the 11 PAHs found form part of the IARC list (BaA, BaP, BbF, Chr, DahA and Nap). Gurumoorthi (2021) investigated the fate of the MV Wakashio oil spill in the Indian Ocean Island of Mauritius by using the GNOME model in an attempt to identify the meteorology and oceanographic forces.

16.2.9 Cost to Beach Users

A major part of the natural resource damages settlement, according to Morris J. Berman oil spill incident, had an impact on public beach use due to reduced visits to the San Juan National Historic Site. Penn (2005) determined the restoration for the beach use and historic site use losses by estimating the dollar value of the lost services and then selecting the scale of restoration that has a cost equivalent to the lost value. Penn (2005) quantified the number of lost beach visits and the number of impacted site visits by using attendance information and then quantified the value loss associated with each type of affected trip by applying a value per affected trip from an appropriate literature study.

Melstrom (2019) presented the economic damages from a hypothetical worst-case oil spill at the Straits of Mackinac between Lakes Huron and Michigan. Potential economic damages to outdoor recreation, shipping, commercial fishing, energy, residential properties, and water supplies were quantified using the benefit transfer method. WTP for activities in the Great Lakes and estimated the damages to five recreational uses of the Great Lakes from the spill, including beach use, state park camping, visits to state parks, recreational fishing, and recreational boating. Projected loss from the worst-case scenario is calculated to be at least US$1.3 billion (Melstrom, 2019).

According to Bonnieux and Rainelli (2003) the residents were disturbed by the Amoco Cadiz accident, through their leisure activities, mainly fishing on foot, a very popular activity on this coastline. Benefit transfer has been used for substitute outdoor recreational activities.

16.2.10 Loss of Welfare Due to the Impact

Economic losses from a ship accident led to lost economic welfare due to pollution, closure, exclusion of users from using resources, and other related losses. Lost net economic welfare includes welfare loss to the consumer (consumer surplus) and to the producer (producer surplus). This total loss can be measured below.

Consider the typical demand-supply market equilibrium model (Figure 16.2). The behavior of buyers is represented by the demand curve 'D' and supply curve 'S'. The consumer surplus (CS) is represented by share area 'A' while the producer surplus is indicated by area 'B'. Economic effects are measured by calculating the changes in the CS and PS. The ship incident damage can lead to both supply side and demand side effects. For example, the temporary closure of the sea and pollution-led destruction of fish will affect the supply. The reduction in demand for fish due to fear of contamination will affect the demand side of the model. Both supply effects and demand effects lead to loss of total economic welfare.

The net economic losses from marine pollution are assessed by calculating the net economic values 'before' and 'after' the impact. For example, consider a case

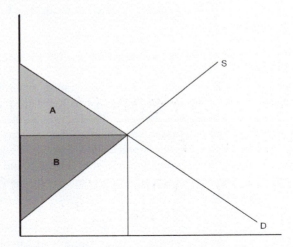

FIGURE 16.2 Market equilibrium – consumer and producer surplus.

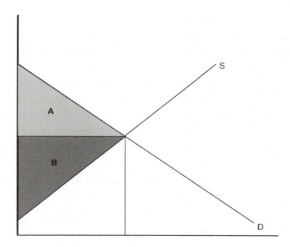

FIGURE 16.3 Change in consumer and producer surplus due to demand shift.

where the demand for fish resulted in an 'inward' shift in the demand curve D to D' (Figure 16.3). The change in the CS and the PS can now be calculated due to the change in demand to find the net change in the economic welfare.

Decision-making for valuation is a dynamic process that must consider both quantitative and qualitative aspects with the involvement of all key stakeholders, from citizens to policymakers, for evidence-based management of the situation. It must also be mentioned that spillover effects of the accident that might affect ancillary sectors must not be overlooked.

16.3 EVALUATION FOR COMPENSATION

The International Oil Pollution Compensation Fund (IOPCF) sets out guidelines for presenting claims for environmental damage. According to the guidelines, compensation is available for environmental damage, subject to criteria set out in the claims manual and provided claims are based on sound science. 'Pollution damage' in the manual is defined as means loss or damage caused outside the ship by contamination resulting from the escape or discharge of oil from the ship, wherever such escape or discharge may occur, provided that compensation for impairment of the environment other than loss of profit from such impairment shall be limited to costs of reasonable measures of reinstatement actually undertaken or to be undertaken.

Guidelines define three types of claims in relation to impairment of the environment: (i) claims for loss of profit, (ii) claims for the costs of postincident studies, and (iii) claims for the costs of reinstatement measures. The criterion that measures should not result in the degradation of other habitats or in adverse environmental or economic consequences calls for the application of net environmental benefit analysis.

However, the damage to the ecosystems and many types of ecosystem benefits are recognized as inadmissible by the claims manual. It has been also revealed that substantial heterogeneity exists in terms of performance of international conventions on oil spill compensation across nations. It should also be noted that certain accepted and tested models such as the 'Basic Oil Spill Cost Estimation Model (BOSCEM)' introduced by the United States Environmental Protection Agency (US EPA), designed to model oil spill response and damage cost, must be considered in estimation. The damages undoubtedly will persist for years to come. For example, habitats may never be restored to their 'pre' condition, and the value of a turtle's existence in the ocean ecosystem is lost for at least another 60 years on average into the future. The damages of microplastics and nanoplastics will persist for several hundred years. Chemical damages and subsequent secondary impacts could last for several decades. Damages to coral reefs may never recover back to their original status. These future costs therefore must be calculated with an appropriate discount rate for a determined number of years to estimate the long-term lost values. It is essential to use low or zero discount rates to ensure the rights of future generations.

National governance is a key factor in the proper functioning of international institutions. Countries with good governance were capable of mobilizing resources efficiently and coordinating activities to minimize and repair the damages resulting from accidents. In addition, such countries were able to ensure adequate compensation for the victims; to identify and hold accountable those responsible for the spill; and to effectively execute the judicial or extrajudicial mechanisms for financing the compensations. However, countries with poor governance showed weaknesses in enforcing the law, assisting victims, and minimizing the damages from the spills (Soto-Onate and Caballero, 2017).

16.4 CONCLUSION

The damage cost associated with X-Press Pearl ship accident is a clear case of externality. Damages to the nation's marine and coastal ecosystems, their components and functions would be felt for a long period of time. Many damages would be felt beyond the borders within Asia and the world at large. Estimating the total damage and getting the compensation will be a daunting task given the magnitude of the event and many other constraints in conducting a proper assessment. A comprehensive framework has been prepared, however, to assist the exercise. Environmental economists need to undertake multiple valuation studies with the assistance of ecologists, chemists, marine biologists, and sociologists till the full valuation is complete. Proper governance structure, strong institutions, and ethical principles are essential in getting compensation for the victims along the multiple spatial and temporal scales which includes current and future generations of both human and nonhuman species (both known and unknown) belonging to national, regional, and global territories.

REFERENCES

Ahtiainen, H. (2007). 'The willingness to pay for reducing the harm from future oil spills in the Gulf of Finland: An application of the contingent valuation method.' (Discussion papers / Department of Economics and Management, University of Helsinki; No. 18). Helsinki: University of Helsinki, Department of Economics and Management.

A Framework for Valuation of Environmental Damages

Allen, J. A., Ewel, K. C. and Jack, J. (2001) 'Patterns of natural and anthropogenic disturbance of the mangroves on the Pacific Island of Kosrae', *Wetlands Ecology and Management*, **9**(3), pp. 291–301.

Alvernia, P. *et al.* (2021) 'Studies of fishermen's economic loss due to oil spills', *IOP Conference Series: Earth and Environmental Science*, **802**(1).

Bengtsson, L. (2013) 'Treatment of liability, economic losses and environmental damage of ship source oil pollution', pp. 1–86.

van Biervliet, K., Le Roy, D. and Nunes, P. A. L. D. (2011) 'An Accidental Oil Spill Along the Belgian Coast: Results from a CV Study', *SSRN Electronic Journal*, (1).

Blomqvist, P., Persson, B. and Simonson, M. (2007) 'Fire emissions of organics into the atmosphere', *Fire Technology*, **43**(3), pp. 213–231.

Bonnieux, F. and Rainelli, P. (1993) 'Learning from the Amoco Cadiz oil spill: Damage valuation and court's ruling', *Organization & Environment*, **7**(3), pp. 169–188.

Bonnieux, F. and Rainelli, P. (2003) 'Lost Recreation and Amenities: The Erika Spill Perspectives', ... *Effects of the Prestige Spill, Saint-Jacques de ...*, pp. 1–22.

Boufadel, M. C. *et al.* (2014) 'Simulation of the landfall of the deepwater horizon oil on the shorelines of the Gulf of Mexico', *Environmental Science and Technology*, **48**(16), pp. 9496–9505.

Cai, M. and Li, K. (2011) 'Economic losses from marine pollution adjacent to Pearl River Estuary, China', *Procedia Engineering*, **18**, pp. 43–52.

Campos, J. J. F., Pereira, N. M. and Cavalcanti, R. N. (2002) 'Economic valuation of environmental damages at coastal areas: Methodologies and challenges', *Water Studies*, **11**, pp. 353–363.

Chang, S. E. *et al.* (2014) 'Consequences of oil spills: A review and framework for informing planning', *Ecology and Society*, **19**(2).

Chen, E. (2022) 'IMO 2012 Solution Notes', (October), pp. 1–9.

Chen, J. *et al.* (2018) 'Identifying critical factors of oil spill in the tanker shipping industry worldwide', *Journal of Cleaner Production*, **180**, pp. 1–10.

Clark, L. P., Millet, D. B. and Marshall, J. D. (2017) 'Changes in transportation-related air pollution exposures by race-ethnicity and socioeconomic status: Outdoor nitrogen dioxide in the United States in 2000 and 2010', *Environmental Health Perspectives*, **125**(9), pp. 1–10.

Cohen A. N., Carloton J. T. and Fountain M. C. (1995) 'Introduction, dispersal and potential impacts of the green crab Carcinus maenas in San Francisco Bay, California', *Marine Biology*, **122**, pp. 225–237.

Dichmont, C. M. *et al.* (2016) 'A generic method of engagement to elicit regional coastal management options', *Ocean and Coastal Management*, **124**, pp. 22–32.

Dumax, N. and Rozan, A. (2021) 'Valuation of the environmental benefits induced by a constructed wetland', *Wetlands Ecology and Management*, **29**(6), pp. 809–822.

Eskola, M. *et al.* (2020) 'Worldwide contamination of food-crops with mycotoxins: Validity of the widely cited "FAO estimate" of 25%', *Critical Reviews in Food Science and Nutrition*, **60**(16), pp. 2773–2789.

French-McCay, D. P. *et al.* (2021) 'Oil fate and mass balance for the Deepwater Horizon oil spill', *Marine Pollution Bulletin*, **171**, p. 112681.

Garza-Gil, M. D., Prada-Blanco, A. and Vázquez-Rodríguez, M. X. (2006) 'Estimating the short-term economic damages from the Prestige oil spill in the Galician fisheries and tourism', *Ecological Economics*, **58**(4), pp. 842–849.

Gohlke, J. M. *et al.* (2011) 'A review of seafood safety after the Deepwater Horizon blowout', *Environmental Health Perspectives*, **119**(8), pp. 1062–1069.

Grigalunas, T. A. et al. (1986) 'Estimating the Cost of Oil Spills: Lessons from the Amoco Cadiz Incident', *Marine Resource Economics*, **2**(3), pp. 239–262.

Grigalunas, T. A. et al. (1998) 'Liability for oil spill damages: Issues, methods, and examples', *Coastal Management*, **26**(2), pp. 61–77.

Gurumoorthi, K. et al. (2021) 'Fate of MV Wakashio oil spill off Mauritius coast through modelling and remote sensing observations', *Marine Pollution Bulletin*, **172**(January), p. 112892.

Harwell, M. A. and Gentile, J. H. (2006) ' Ecological significance of residual exposures and effects from the Exxon Valdez oil spill ', *Integrated Environmental Assessment and Management*, **2**(3), pp. 204–246.

Hebbar, A. (2016) 'Fisheries dimension of marine oil spills', *Journal of the Marine Biological Association of India*, **58**(2), pp. 81–88.

Heikes, B. G. et al. (2002) 'Atmospheric methanol budget and ocean implication', *Global Biogeochemical Cycles*, **16**(4), pp. 80-1-80–13.

Iduk, U. and Samson, N. (2015) 'Effects and Solutions of Marine Pollution from Ships in Nigerian Waterways', *International Journal of Scientific & Engineering Research*, **6**(9), pp. 81–90.

Islam, M. S. and Tanaka, M. (2004) 'Impacts of pollution on coastal and marine ecosystems including coastal and marine fisheries and approach for management: A review and synthesis', *Marine Pollution Bulletin*, **48**(7–8), pp. 624–649.

Jang, Y. C. et al. (2014) 'Estimation of lost tourism revenue in Geoje Island from the 2011 marine debris pollution event in South Korea', *Marine Pollution Bulletin*, **81**(1), pp. 49–54.

Jung, J. H. et al. (2011) 'Biomarker responses in pelagic and benthic fish over 1 year following the Hebei Spirit oil spill (Taean, Korea)', *Marine Pollution Bulletin*, **62**(8), pp. 1859–1866.

Jung, J. H. et al. (2012) 'Spatial variability of biochemical responses in resident fish after the M/V Hebei Spirit Oil Spill (Taean, Korea)', *Ocean Science Journal*, **47**(3), pp. 209–214.

Kim, D. et al. (2014) 'Social and ecological impacts of the Hebei Spirit oil spill on the west coast of Korea: Implications for compensation and recovery', *Ocean and Coastal Management*, **102**(PB), pp. 533–544.

Kwiecińska, B. (2015) 'Cause-and-effect analysis of ship fires using relations diagrams', *Zeszyty Naukowe / Akademia Morska w Szczecinie*, nr **44** (116(116), pp. 187–191.

Law, R. J. and Hellou, J. (1999) 'Contamination of fish and shellfish following oil spill incidents', *Environmental Geosciences*, **6**(2), pp. 90–98. d

Lee, J. (2015) 'Economic valuation of marine litter and microplastic pollution in the marine environment: An initial assessment of the case of the United Kingdom Growth Strategies of Sub-Saharan Africa View project Economic valuation of marine litter and microplastic pol', *Discussion Paper*, (March), pp. 1–16.

Leis, J. M. and Carson-Ewart, B. M. (2003) 'Orientation of pelagic larvae of coral-reef fishes in the ocean', *Marine Ecology Progress Series*, **252**, pp. 239–253.

Liu, X. et al. (2009) 'Willingness to pay among households to prevent coastal resources from polluting by oil spills: A pilot survey', *Marine Pollution Bulletin*, **58**(10), pp. 1514–1521.

Liu, X. and Wirtz, K. W. (2006) 'Total oil spill costs and compensations', *Maritime Policy and Management*, **33**(1), pp. 49–60.

Liu, X. and Wirtz, K. W. (2009) 'The economy of oil spills: Direct and indirect costs as a function of spill size', *Journal of Hazardous Materials*, **171**(1–3), pp. 471–477.

Loureiro, M. L. et al. (2006) 'Estimated costs and admissible claims linked to the Prestige oil spill', *Ecological Economics*, **59**(1), pp. 48–63.

Loureiro, M. L., Loomis, J. B. and Vázquez, M. X. (2009) 'Economic Valuation of Environmental Damages due to the Prestige Oil Spill in Spain', *Environmental and Resource Economics*, **44**(4), pp. 537–553.

Mathuru, A. S. *et al.* (2012) 'Chondroitin fragments are odorants that trigger fear behavior in fish', *Current Biology*, **22**(6), pp. 538–544.

McArthur, D. P. and Osland, L. (2013) 'Transportation Research Part D Ships in a city harbour: An economic valuation of atmospheric emissions', *Transportation Research Part D*, **21**, pp. 47–52.

Melstrom, R. T. *et al.* (2019) 'Economic damages from a worst-case oil spill in the Straits of Mackinac', *Journal of Great Lakes Research*, **45**(6), pp. 1130–1141.

Miola, A. and Ciuffo, B. (2011) 'Estimating air emissions from ships: Meta-analysis of modelling approaches and available data sources', *Atmospheric Environment*, **45**(13), pp. 2242–2251.

Morris, R. (2002) 'Community radio FM broadcast planning', *ABU Technical Review*, **(202)**, pp. 17–22.

Mumby, P. J. *et al.* (2001) 'A bird's-eye view of the health of coral reefs', *Nature*, **413**(6851), p. 36.

Nelson, J. R. and Grubesic, T. H. (2020) 'Oil spill modeling: Mapping the knowledge domain', *Progress in Physical Geography*, **44**(1), pp. 120–136.

Ofiara, D. D. and Brown, B. (1999) 'Assessment of economic losses to recreational activities from 1988 marine pollution events and assessment of economic losses from long-term contamination of fish within the New York Bight to New Jersey', *Marine Pollution Bulletin*, **38**(11), pp. 990–1004.

Oken, K. L. and Essington, T. E. (2016) 'Recovery Within Marine Protected Areas', *ICES Journal of Marine Science*, **73**, pp. 2267–2277.

de Oliveira Estevo, M. *et al.* (2021) 'Immediate social and economic impacts of a major oil spill on Brazilian coastal fishing communities', *Marine Pollution Bulletin*, **164**(January).

Penn, T. *et al.* (2005) 'Morris J. Berman oil spill: Natural resource damages settlement and restoration scale', *2005 International Oil Spill Conference, IOSC 2005*, pp. 9222–9227.

Peterson, C. H. *et al.* (2003) 'Long-Term Ecosystem Response to the Exxon Valdez Oil Spill', *Science*, **302**(5653), pp. 2082–2086.

Reineking, B. (2002) 'The Wadden Sea Designated as a PSSA History', *Wadden Sea Newsletter*, **(2)**, pp. 10–12.

Richardson, R. and Brugnone, N. (2018) 'Oil Spill Economics: Estimates of the Economic Damages of an Oil Spill in the Straits of Mackinac in Michigan', *Michigan State University*, (May).

Ryu, J. H. *et al.* (2010) 'Characteristics and triage of a maritime disaster: An accidental passenger ship collision in Korea', *European Journal of Emergency Medicine*, **17**(3), pp. 177–180.

Shkëlqim Sinanaj (2020a) 'The Impact of Shipping Accidents on Marine Environment in Albanian Seas', *Journal of Shipping and Ocean Engineering*, **10**(1), pp. 10–23.

Shkëlqim Sinanaj (2020b) 'The Impact of Shipping Accidents on Marine Environment in Albanian Seas', *Journal of Shipping and Ocean Engineering*, **10**(1), pp. 27–32.

Sukumaran, S. *et al.* (2014) 'Impact of "Chitra" Oil Spill on Tidal Pool Macrobenthic Communities of a Tropical Rocky Shore (Mumbai, India)', *Estuaries and Coasts*, **37**(6), pp. 1415–1431.

Soto-Onate, D., & Caballero, G. (2017). Oil spills, governance and institutional performance: The 1992 regime of liability and compensation for oil pollution damage. *Journal of Cleaner Production*, 166, 299–311.

Tatano, H., Collins, A. E. and Kyōto Daigaku. Bōsai Kenkyūjo (2021) *Proceedings of the 3rd Global Summit of Research Institutes for Disaster Risk Reduction.*

Thébaud, O. *et al.* (2005) 'The cost of oil pollution at sea: an analysis of the process of damage valuation and compensation following oil spills', *Economic, Social and Environmental Effects of the Prestige Oil Spill de Compostella, Santiago*, pp. 187–219.

Theodotou, S. (2018) 'Corporate Liability and Compensation Following the Deepwater Horizon Oil Spill: Is There a Need for an International Regime?', *Groningen Journal of International Law*, **6**(1), p. 161.

Wenaty, A., Mabiki, F., Chove, B. and Mdegela, R. (2018) 'Fish Consumers Preferences, Quantities of Fish Consumed and Factors Affecting Fish Eating Habits: A Case of Lake Victoria in Tanzania', *International Journal of Fisheries and Aquatic Studies*, **6**(6), pp. 247–252.

Wielgus, J. (2004) 'No Title', (August), pp. 1–31.

Ytreberg, E. and Åstr, S. (2021) 'Valuating environmental impacts from ship emissions – The marine perspective', **282**(August 2020).

Zambrano-Monserrate, M. A. and Ruano, M. A. (2020) 'Estimating the damage cost of plastic waste in Galapagos Islands: A contingent valuation approach', *Marine Policy*, **117**(December 2019), p. 103933.

17 Lessons from the Pearl

Deshai Botheju

17.1 INTRODUCTION

The X-Press Pearl disaster (XPPD) exposed a substantial swath of weaknesses in the existing methodologies, infrastructure, decision making, management, and response to maritime disasters, at the local, regional, and international levels. This chapter aims to recognize and discuss relevant issues, as well as to propose suitable amendments and improvements. The ultimate ambition is to reduce the likelihood of future maritime accidents of similar nature and to enhance the general preparedness against any societal disasters originating from technological events. XPPD is demanding changes to the current domain of international maritime safety doctrines and operational procedures. The International Maritime Organization (IMO) must seriously evaluate the facts surrounding this accident and absorb relevant lessons into their future regulations and guidelines.

17.2 MARITIME TRANSPORT OF DANGEROUS GOODS

The initial cause of the accident is suspected to be an acid leak from one cargo container carrying nitric acid (Botheju, 2021). Such leaks often result from poor securing of the primary containment (i.e. the acid holding container in this case) within the shipping container, or due to inappropriate handling of the shipping containers. The International Maritime Dangerous Goods Code (IMDG code by IMO, 2018) clearly indicates the requirements applicable for the sea transport of nitric acid (ref. IMDG code, Part 3 – Dangerous Goods List, special provisions and exceptions. See under UN no. 1796). Whether these requirements had been properly followed in XPPD case remains an unanswered question.

Recent observations have revealed that significant amounts of hazardous goods are transported without correct declarations, and with no or poor securing inside their shipping containers (Kallada, 2020). The common excuse given by the originators of such hazardous cargo is the excessive cost for additional securing, which is claimed to be beyond their affordability. Such excuses are unacceptable and high attention must now be drawn to ensure that the relevant international regulatory standards are always maintained. Reference is also made to the CTU Code (IMO/ILO/UNECE

Code of Practice for Packing of Crago Transport Units, 2014), providing a detailed guidance on the recommended practice with respect to packing of cargo units.

Further, it seems that noncompliance is sometimes becoming the norm, just because in the majority of cases these noncompliances do not end up in accidents or noteworthy incidents. However, in the rare occasions when they do end up in accidents, the consequences can be very severe, just like in the case of XPPD and other recent maritime accidents including the major fires onboard the container ships Yantian Express (Manaadiar, 2020) and KMTC Hong Kong (Voytenko, 2019).

The difficulty of enforcing every IMO regulation or procedural code (like IMDG code, CTU code, etc.) with respect to every cargo is a practically difficult task. Such action will be almost impossible considering the tremendous cargo volumes and shipments handled in the world today. Increased spot checks and more severe penalties should be used to demote the high number of noncompliances related to the transport of dangerous goods. The lack of manpower and relaxed work procedures in some ports during the COVID pandemic time might have also played a role, further increasing the number of noncompliance cases during recent times.

17.3 RESPONSIBILITY OF PORTS

At least two ports (Hamad in Qatar and Hazira in India) had refused to unload the leaking nitric acid container when XPP captain requested "re-work" on the concerned container. The lack of facilities to handle rework on dangerous cargos was given as the reason. It looked more like the relevant ports authorities did not want to be involved in additional work that they are not bound to do (or not responsible for). One can still interpret this as an irresponsible act from these authorities. IMO should ensure that ports authorities around the globe are held more responsible in future to provide positive responses to such requests involving dangerous cargo.

On the other hand, it is not clear if the XPP captain had properly emphasized the severity of this leaking container event at the abovementioned ports. Understanding the exact nature of the communication that took place between XPP and those ports is important before one could hold any of those parties accountable.

17.4 ONBOARD DECISION MAKING

The decisions made onboard XPP during a ten-day period preceding the accident had played a critical role in deciding the fate of the ship. The captain and the ship company were responsible for making these critical decisions. Some of the decisions could have been made by captain alone or together with the other senior staff members onboard. However, it is perceived that many of the critical decisions must have been made by captain after consulting his shipping company superiors. The way captain reported the events unfolding onboard the ship may have of course influenced the opinions of the company offices on land. In any case, it is quite apparent that this decision-making process had not properly captured the actual gravity of the situation. It is a question indeed whether any of the stakeholders involved in the decision making at that stage had understood the potential consequences of that acid leak event. Note that the IMDG code has a clear reference to possible hazards of leaking nitric acid onboard ships.

The crew members of the vessels transporting dangerous cargo, and particularly the key personnel involved in major decisions during an emergency, must be well acquainted with the regulations and safety procedures applicable to the goods that they are transporting. IMO and other international regulatory bodies must increase their attention on the industry's awareness of relevant standards and codes, beyond just developing the standards. Further, the critical decision makers must be made acquainted with the relevant risk assessment and risk mitigation principles.

17.5 HUMAN FACTORS IN CRITICAL DECISION MAKING

Human factors constitute a vital part of decision making during an emergency situation. In the case of XPPD, the actions of the ship captain obviously reflected the human factor scenario that was influencing him at that time. For example, an experienced captain would make different decisions compared to a less experienced one. Also, the level of stress tolerance decides the effectiveness of decision making under stressful conditions. Further, the surrounding company culture influences the decisions. A company having a good (positive) safety culture can encourage a decision maker to make timely and bold decisions that can avert accidents from happening, irrespective of potential negative economic consequences. In a company having a negative safety culture, persons are reluctant to take such decisions over fear of bad repercussions for his/her own career. They would rather try to "play safe" by prioritizing economic aspects rather than paying attention to potential safety consequences.

17.6 ONBOARD HANDLING OF THE EVENT

The response actions taken onboard XPP after observing the acid leak (until the ship arrived in Colombo) had not been made public. Based on the eventual result, it is fair to believe that not many effective response actions had been carried out.

In an acid leak scenario, it is prudent to respond with a large amount of water. This will not only dilute the leaking acid but also help to cool and slow down the exothermic chemical reactions involving nitic acid and various organic substances (or metals) in the surrounding environment. Therefore, the leaking acid container and the surrounding area should have been deluged with a lot of water. In contrast, using a minor amount of water can give rise to the opposite results. However, the practical possibility of carrying out such water spraying at the location of this acid container may possibly have been limited due to the limited access availability among the cargo containers.

The feasibility of disposing of leaking acid container in the sea can be seen as a controversial idea. Then again, such action could have resulted in far less damage to the environment than the eventual catastrophe that unfolded due to the XPPD. Note that even according to MARPOL convention (IMO, 1973), such dumping of containers in the sea is allowed under dire situations. Nevertheless, the possibility of operating the onboard cranes in the sea (for removing the acid container) depends on the prevailing weather conditions and other technical and operational restrictions. At least, such action should have been thoroughly evaluated during the 10-day voyage before the accident unfolded in Colombo.

IMO should clearly indicate various options in case of a dangerous chemical leak onboard ships. Even though many of these are already included in various codes and regulations, a summarized incident action manual can be truly helpful in case of an emergency. Alternatively, a maritime emergency expert center must be established to provide real-time advice and guidance during an ongoing incident.

17.7 ARRIVING AT COLOMBO

It was clearly noted that even arriving at Colombo, XPP did not indicate any emergency or urgent situation onboard the ship. The ship was directed to the outer harbor anchorage until the scheduled berthing slot next day. It was later revealed that XPP only sent an "email" communication to its shipping agent in Colombo asking for a "re-work" on this leaking acid container. This clearly demonstrates that the ship did not consider this leaking acid container a seriously hazardous situation. This lack of understanding of the impending danger could explain many of the previous actions (or the absence of actions) during the voyage.

Once again, it is high time to raise concerns about the prevailing risk awareness among the seafaring community. IMO and various regulatory authorities must also take initiatives to raise awareness and improve safety culture within the maritime sector.

17.8 ONBOARD FIRE RESPONSE

A day after arriving at Colombo, fire (smoke) detection was alarmed onboard the ship. This was then communicated to the Colombo Ports Authority. The ship had then activated their CO_2-based fixed firefighting system within one cargo compartment and subsequently notified the port that the fire had been successfully extinguished. After a while the automatic fire alarms were activated again and the ship had decided to release all remaining CO_2-based cargo fire protection systems, thus exhausting all of them at once. The important point here is to understand that the actual epicenter could still have been outside of these (below deck) cargo compartments (i.e. a fire above the cargo deck) while CO_2 extinguishing system release was carried out inside the cargo compartments. Note that the leaking nitric acid container was located above the cargo deck. It was likely that seeping acid went straight through the rubber sealing of the cargo compartment hatches and reacted with some organic substances stored below deck to release nitrogen dioxide gas (yellow-brown smoke; see Figure 17.1) to initiate the smoke-based fire detectors. Alternatively, the smoke alone could have penetrated through the hatch cover to trigger smoke alarms within the cargo compartment. In any case, releasing the cargo firefighting system (CO_2) inside the cargo compartment (below cargo deck) may not have been effective in curtailing the origin of the fire, which was likely to be above the cargo deck. Further, CO_2-based fixed firefighting systems have the inherent weakness of requiring an airtight surrounding to perform effectively. Also these systems do not provide any cooling effect which is essential for curtailing exothermic chemical reactions–based fires (such as the reactions between acids and organic matter).

Lessons from the Pearl

FIGURE 17.1 XPP after arriving in Colombo on 20 May 2021. Notice the yellow-brown smoke (NO_2 gas) emanating from a location at the forward section of the ship. (Photo credit: Wikimedia Commons, "Isuruhetti".)

The basis is that this ship had been designed as an ordinary container ship and did not have specialized firefighting systems or procedures typically needed to handle risks posed by a large consignment of hazardous cargo. In contrast, oil tankers and other specialized vessels transporting flammable and explosive cargo do have extensive firefighting systems such as fire monitors (water cannons), deluge systems, sprinkler systems, fire hoses (at strategic locations), water mist, and inert gas fixed firefighting systems. Further, such ships are designed with enough space and layout arrangements to rapidly direct firefighting systems to the epicenter of hazardous cargo compartments or tanks. The crew members of such vessels also possess sufficient training and risk awareness to handle relevant emergency scenarios.

International maritime regulators must closely look into XPPD-type cases where ordinary container ships are used to transport large amounts of dangerous goods without sufficient safety barriers such as emergency equipment, layout design, trained crew, or safety culture to match the actual risk picture.

17.9 INITIAL ASSESSMENT BY COLOMBO AUTHORITIES

On 20 May 2021 evening, after the two reported fire (smoke) detection events, Colombo Ports Authority personnel went onboard XPP to observe the situation of the ship. They had observed a large cloud of yellow-brown smoke emanating from the forward section of the ship. They had apparently then concluded that, since no fire was visible, there was no imminent fire hazard, and therefore no further action was taken. In fact, the large cloud of nitrogen dioxide (yellow-brown smoke) was a clear sign of an ongoing exothermic chemical reaction involving nitric acid. At least at this point, deluging the epicenter (i.e. the leaking nitric acid container) with large amounts of water must have been initiated. This could have been possibly done with the firefighting tugboats available in the port. Alternatively, attempts should have been made to remove the leaking acid container from the ship as an emergency measure.

The lack of a clear strategy to assess the nature of the ongoing hazardous event and the absence of a preplanned emergency response procedure to trigger during this kind of scenario was apparent.

17.10 INITIAL FIRE RESPONSE

On the night of 20 May, a large fire suddenly erupted onboard XPP at a location close to the leaking acid container. The ongoing exothermic chemical reaction must have now been developed into a full-blown fire event. The Colombo Ports Authority's fire response team was immediately notified, and they responded to the fire with their tugboats carrying water cannons. By next morning, the visible fire was gone but smoke prevailed. For next few days, eruption of the fire occurred repetitively but the fire response team was able to extinguish it every time with water cannons onboard the fire response vessels. It is clear that the origin of the fire, i.e. the exothermic chemical reaction, had been deeply embedded within the cargo containers, and the water jets and sprays were only able to temporarily subside the visible flames. This is a typical nature of fires originating from exothermic chemical reactions. Until the reaction is suppressed, or the reactants are removed, the fire will not normally be fully extinguished just by spraying water. Water use can only be effective if used in large enough quantities to completely submerge the reacting substances. In XPPD, that was not likely to be the case. The water cannons onboard the tugboats were not apparently close enough to spray water directly to the epicenter, or not enough in quantities to submerge the reacting masses. The only effective response action at this stage should have been to remove the cargo containers surrounding the epicenter and then divert more water to the epicenter. It was later reported that a local marine service company may have offered their equipment and know-how to remove containers from the burning ship and to help extinguish the fire. The shipping company might not have accepted this offer. Note also that a specialized foreign expert team had been brought onboard by the shipping company by this time in order to assist and advise the ongoing firefighting operation.

17.11 EXPLOSION AND EVACUATION

With the weather conditions deteriorating, and the exothermic reactions continuing deep inside the fire epicenter, by 24th May 2021 the fire started to escalate. There was a clear danger of explosions due to the presence of various hazardous cargo including marine fuel oil, vinyl acetate, methanol, polymer resins, and aluminum remelting byproducts. Despite this risk, the crew members and the additional firefighting team had not been evacuated from the ship yet. Neither the ship owner nor the local authorities took required steps to order an evacuation. On 25 May night, a sizable explosion rocked the vessel and at least one seaman received injuries needing hospitalization. Only then was the evacuation carried out. This is a strong sign of incorrect decision making and poor safety culture, leading to actions resulting in potential loss of human life.

After the first explosion event, even firefighting tugboats had to relocate themselves further away from the ship, thus reducing their effectiveness, before they had to even completely retreat from the burning ship later. Reportedly more explosions had occurred on the ship subsequently.

17.12 SOCIETAL AND MEDIA RESPONSE

The local communities, the society in general, and the local media displayed a great deal of confusion during the accident. The proper risk communication techniques were not observed. The news broadcasts were often agitating and blaming rather than aiming at providing the actual risk picture. Social media posts and discussion forums were mostly focused on accusing somebody rather than fact finding. Establishing a proper media center for reporting the latest and unbiased accident status by a responsible disaster management body is essential in this kind of situation. The correct risk communication principles must be utilized to provide a justifiable risk perception to society. When no trustworthy Authority is available to present the correct risk picture, society and the media promote their own distorted risk perceptions based on assumptions, aided by various outside experts and "citizen scientists" who are expressing their opinions without facts or the context to back them.

The local community members were seen scavenging and taking away various hazardous debris washed ashore after the first explosion on XPP (see Figure 17.2). This was a totally irresponsible and dangerous act, which was also broadcast internationally. This act alone showed the lack of proper risk communication and proper implementation of a local emergency response plan by the relevant authorities. The debris zones should have been immediately restricted to public access, and the large pieces of debris must have been mapped and removed to a safe location for investigation.

17.13 DISASTER MANAGEMENT BY LOCAL AUTHORITIES

The accident response had to be coordinated via a multitude of different government agencies, causing communication and implementation delays along the command chain. The necessity of a single apex Authority who can command directions during a large-scale industrial accident was evident. It is highly recommended to establish an authoritative body in every country to manage, respond, and investigate industrial accidents. The existing disaster response institutions in many developing countries lack the necessary facilities, know-how, and experience for responding to highly dynamic and complicated industrial accidents like the XPPD.

FIGURE 17.2 Scavenging of debris washed ashore after the first explosion on XPP. (Photo credit: UTV news.)

17.14 FIRE RESPONSE DURING THE EXTENDED ACCIDENT PERIOD

After escalation of the accident on 25 May 2021, and when the fire was spread all along the vessel, it was clear that the continued firefighting was not going to make any significant difference, except for perhaps helping some minor cooling on the ship hull. On the other hand, letting the hazardous substances burn at a higher temperature could have been more favorable than subsiding the fire which then led to burning substances at low temperatures, thus generating partially combusted products such as polychlorinated dibenzo-p-dioxins, polychlorinated dibenzofurans, and other volatile organic compounds which were more hazardous and persistent in the environment (see Figure 17.3). Further, firefighting water that accumulated in the ship's bottom created a highly hazardous chemical cocktail which eventually released to the sea when the ship sank. The extended use of firefighting foam chemicals also increased the chemical release load on the environment. Some use of dry chemical powders by helicopters was also noted during this period. Looking at the fully spread nature of the fire and the limited impacts (if at all) of a few drops of dry chemicals, this use was probably unnecessary and useless.

17.15 TAIL-END OF THE FIRE EVENT

At the end of the fire event by the 1st June 2021, when the fire extinguished mostly itself after burning through the ship, there was a brief window (before sinking of the ship) in which it was possible for the shipping company representatives and

Lessons from the Pearl

FIGURE 17.3 Large plumes of black smoke generated during the fire onboard XPP, caused by partial combustion. (Photo credit: SL Airforce Media.)

the salvage company agents to go onboard the burned ship. This was an excellent opportunity for collecting samples and documenting the whole aftermath, possibly revealing crucial evidence related to the accident. The local authorities should have used this golden opportunity to gather evidence for accident investigation and for other legal proceedings. It is not known if this opportunity had been properly used. Apparently no sample gathering had been done during this brief window of opportunity. Such sampling could have revealed the true nature and the extent of pollution eventually unfolded. Also, proper observations (including photographs and videos) could have revealed the status of the remaining cargo and the hazardous substances, thereby providing a correct estimate of the pollution.

17.16 SINKING OF THE SHIP

On the 2nd June 2021, it had been decided to tow the ship away from the accident location toward the deep sea. This was a decision made without proper engineering judgment. After an intense fire reaching almost the steel melting temperature (see Figure 17.4), the ship hull and the structure must have been so weak that any hard movement would almost certainly end up breaking the ship apart. Any towing or relocating of the ship could only have been done with significant reinforcements and extra supports and should only have been carried out by a qualified salvage company with heavy equipment for the operation. Also, the removal of all possible remaining containers should have been the priority action. This not only would have reduced the pollution load in case of eventual sinking but also would have helped to lighten up the weight, thus improving the ship's ability to stay afloat.

Shortly after starting the haphazard towing of the ship, a large section of the front hull had come loose and within hours the ship sank to the bottom of the sea. Due to the shallow depth of the sea (about 21 m), the living quarter structure and part of the cargo deck was kept above the surface (see Figure 17.4).

FIGURE 17.4 The partially sunken XPP, sitting on the seabed (approx. 21 m depth). Notice the cargo handling crane at the forward section of the ship sagged due to fire temperature. (Photo credit: SL Airforce Media.)

17.17 POSTSINKING WRECK HANDLING

Sinking of XPP resulted in total release of its remaining cargo into the sea, thereby resulting in uncontrolled and unquantified pollution damage to the natural environment. There were various telltale signs of released chemical substances including oily matters. Several samplings had been carried out at different locations around the shipwreck during the period following the sinking event. However, a systematic and real-time understanding of the extent of pollution zone and its dynamics was never realized.

17.17.1 Wreck Recovery

The wreck recovery started after more than 6 months from the sinking incident. A salvage company employed by the ship owner carries out this work. In reality, the wreck recovery phase involves a large risk potential that is comparable to the sinking incident.

The shipwreck removal process disturbs the settlement status of the ship and its damaged cargo. Therefore, a comprehensive contingency plan should have been made to prepare for unexpected events that may unfold during the debris removal process. For example, sudden leaks of marine fuel oil may occur. Therefore, the whole shipwreck perimeter (with enough distance to work on) must be encircled with oil-collecting floating booms. Oil-absorbing pads and oil dispersant chemicals must be readily available in case a need arises to use them. Moreover, other hazardous chemical substances can get disturbed and might start to undergo accelerated leaking. Preparation for such events should have been planned for (such as neutralization of acids or bases). Facilities for collecting/storing damaged liquid containers must be available on site (such as a dedicated vessel with secured cargo holds). Assistance from a group of experts must be constantly available during the whole operation to ensure correct guidance is available as rapidly as possible if an unexpected event occurs. Further, it is essential to ensure the safety of divers working during the debris removal operation. Therefore, any underwater leaking of chemicals or any potential gas emissions should have been identified prior to extensive diving operations. Water samples collected at the depths should have been analyzed for hazardous chemicals. Portable gas meters that can identify toxic and flammable gases should be available on site.

There is no clear evidence of comprehensive contingency planning (as briefly explained above) with respect to the XPP shipwreck removal process. Further, involvement of the regulatory authorities in this process seemed to be minimal. This is an abnormal and risky way of handling a hazardous shipwreck. In addition, the whole wreck recovery methodology must have been presented and approved by the local authorities. For example, a piecemeal approach, where the ship is broken down into smaller pieces before removal, can end up being a highly polluting process. Floating of the whole ship (or large parts of it – based on the technical feasibility) should have been the preferable option.

17.18 CONCLUSIONS AND RECOMMENDATIONS

X-Press Pearl disaster represents a marine accident of a unique nature that has led to one of the most extensive marine pollution anywhere in the globe. Therefore, the lessons learned from this accident must affect some of the existing rules and guidelines applicable to international maritime trade. Any necessity of introducing new requirements must be assessed based on a critical review of the XPPD.

The highlighted attention points include:

- Enforcement and quality control mechanisms for dangerous goods transport requirements.
- Reviewing of ports' responsibility with respect to rework requests involving dangerous goods.
- Attention to human factor and safety culture aspects with respect to critical decision making onboard ships during emergencies or developing incidents.
- Enhancing and ensuring awareness among seafarers regarding the existing codes and regulations with respect to the transport of dangerous goods.
- Critically reviewing the existing design requirements and procedures with respect to the transport of dangerous goods in regular container ships.

While the IMO and other relevant institutions and regulatory authorities must carefully evaluate the abovementioned aspects with respect to the international regulations and codes, the individual countries must initiate their own assessments to recognize measures for safeguarding their coasts and the waters from these types of events in future. To reduce similar risks in future, the following key measures are proposed.

Countries must establish a single point Authority to plan, respond, investigate, and manage aftermaths related to maritime and land-based catastrophes involving complicated chemical processes and technical scenarios. The X-Press Pearl accident showcased the current weaknesses of existing management structures involving multiple parties that are difficult to coordinate and manage during a crisis situation. Alternatively, the above-proposed management body can be a part of an Accident Prevention and Investigation Authority that may be entrusted with a wider responsibility of handling various other industrial incidents and accidents (such as transport accidents, aviation accidents, factory and refinery accidents, etc.). The structures of the US Chemical Safety and Hazard Investigation Board and Norway's Petroleum Safety Authority are some of the good examples to follow.

Apparently, there is a vast void when it comes to the legal basis for claiming environmental damage costs involving international parties. Countries must pass suitable parliamentary acts with proper attention given to the US OPA (United States Oil Pollution Act), in which NRDA (Natural Resource Damage Assessment) models are embedded.

Countries must enhance their capabilities and capacities to respond to similar events. Strategic technical facilities such as improved marine firefighting equipment, underwater ROVs, underwater survey facilities, coastal surveillance aircraft, and deep sea response vessels must be recognized and acquired. Sufficient stocks of oil spill dispersants, oil collecting booms, oil absorbing pads, firefighting agents, etc. must be stockpiled.

Countries must establish a national pool of experts related to various disciplines and subdisciplines. Receiving expert opinion at urgent moments is a critically important aspect that can decide the eventual outcome of an industrial accident while it is still evolving. Information on resident and expatriate scientists and engineers must be collected and the individuals must be contacted to establish such a resource pool.

REFERENCES

Botheju, D. (2021). X-Press Pearl: What went wrong? The scientific basis. Available online: www.ft.lk/opinion/X-Press-Pearl-What-went-wrong-The-scientific-basis/14-719287.

IMO/ILO/UNECE. (2014). Code of Practice for Packaging of Cargo Transport Units (CTU Code).

International Maritime Organization (IMO, 1973). International Convention for the Prevention of Pollution from Ships (MARPOL).

International Maritime Organization (IMO, 2018). International Maritime Dangerous Goods Code, Volume I & II.

Kallada, S. (2020). What are the major reasons for misdeclared or undeclared dangerous goods. Available online on shippingandfreightresource.com.

Manaadiar, H. (2020). Misdeclared cargo and its investigation route-Yantian Express fire. Available online on shippingandfreightresource.com.

Voytenko, M. (2019). Major fire on KMTC container ship in Thailand. Available online on maritimebulletin.net.

Index

A

accident 1, 4, 13, 15–31, 52, 60, 62, 69, 70, 91, 109, 114, 121–5, 127, 130, 131, 134–7, 139, 144–9, 160, 175, 177, 180–4, 190–7, 202, 203, 205, 215, 216, 218–21, 223–9, 230, 232, 234, 236, 238, 240, 242, 244–65, 273, 276, 278, 279, 281, 283–7, 298–301, 310, 338–65, 376–82, 387–90, 393, 394, 397, 398
acetyl cedrane 60
acid container 2, 5–7, 12, 47, 48, 58, 59, 65, 388–90, 392
afloat 52, 61, 395
aft of the vessel 11
air pollution 25, 46, 58, 61, 81–2, 198, 278, 279, 370, 378
Alaska 21
algal blooms 52, 228, 242
allergic reactions 27
aluminium re-smelting by-products 47
Amaranayake 67
Amararathna 173
Amarathunga 134
ammonia 9
anchor 1, 8–9, 56, 278
aquatic life 27, 160, 198, 259
aquatic organisms 27, 252, 254, 379
Aquatic Resources 25, 69, 71, 226, 229, 231, 232
aquatic toxicity 52
ashore 18
authorities 4, 5, 8, 12, 23, 56, 57, 69, 70, 72–4, 77, 82, 115, 121–5, 130, 131, 178, 184, 204, 276, 289, 290, 306, 339–41, 344, 346, 348, 349, 352, 355, 357, 360–364, 388, 390–5, 397, 398
authority 23, 82, 115, 116, 121–6, 130, 131, 178, 184, 275–6, 289, 339–41, 344, 346, 348, 349, 352, 355, 360, 362–4, 388, 390, 391–3, 395, 397, 398
awareness 4, 72–5, 81, 104, 111, 122, 124–5, 147, 178, 389, 390, 391, 397
Ayeshya 215

B

Bandara 338
bathymetry 23
beach 18, 25–7, 29, 30, 53, 54, 56, 62, 68, 72–7, 79, 82, 84, 87–91, 95–98, 100, 104–107, 109, 112, 114–16, 121, 124, 125, 128, 131, 135, 136, 138–44, 146–9, 156, 160, 173–4, 176–9, 182–4, 191, 196, 197, 202, 204, 215, 216, 218, 246–7, 250, 260–2, 264, 273–6, 278, 279, 282, 285, 287, 290, 297–310, 339, 341, 352, 353–6, 367, 368, 370, 372, 374–6, 379–80
beads 10
benthic 31
berthing 7, 390
bioaccumulating 52
bioconcentration 53
biota 23, 200, 232, 257, 264, 319, 324, 325, 348, 369
birds 27, 196, 220, 273–90, 320, 367
blue economy 22
BOL 40
booms 24, 29, 72–5, 79, 104, 105, 108, 115, 122, 396, 398
Botheju 56–65, 387–98
boundary cooling 7
breathing difficulties 27
BTEX 53, 348
bunker oil 49, 68, 76, 79, 82, 103, 121, 348
burning ship 62, 71, 81–3, 89, 105, 111, 115, 123, 137, 181, 196, 392, 393

C

cadmium 52, 201–2, 378
cancer 27, 200, 202
capabilities 9, 70, 398
captain 7, 13, 27–8, 388–9
carbon dioxide firefighting system 7
carcasses 10, 54, 56, 62, 125, 147, 177, 179, 182, 184, 191, 196, 197, 199, 200, 204, 205, 215–16, 218, 233, 246–65, 281, 282, 287, 357
cargo 1, 2, 5, 7, 9, 10, 11, 13–15, 18, 19, 22–6, 29, 30, 40–54, 56, 58–61, 64, 65, 68, 84, 85, 87, 88, 111, 115, 116, 135, 141, 146, 147, 160, 190, 191, 200, 202, 215, 216, 221, 228, 244, 245, 247, 250, 273, 278, 279, 289, 298, 315, 318, 338, 339, 345, 346, 349–51, 357–63, 367, 374, 378, 387–96
cargo compartments 390, 391
Cargo Hold 7–8, 11
caustic 40, 43, 62, 68, 121, 190, 195, 202, 215, 249, 256, 278, 350, 360, 363, 367
chemical carriers 13
chemical cocktail 49–53, 61, 394
chemical dispersion 23
chemical hazard management 62
chemical spills 7, 19, 21–3, 27, 29, 31, 100, 103–4, 107, 111, 114, 116, 121, 123, 124, 131, 195, 278, 289

399

Index

chemicals 1, 7, 15–31, 40, 41, 43–5, 48, 49, 52–4, 61, 68, 72, 73, 79–80, 100–2, 111, 115, 121, 123, 124, 127, 129, 131, 144, 146, 148, 162, 164, 166, 182–4, 191, 195, 200–2, 215, 221, 228, 245, 263, 274–5, 278, 281, 288–91, 300, 307, 318, 338, 339, 345, 349, 350, 357, 358, 360, 361, 363, 367, 369–71, 379, 394
Chief Engineer 7–8
chromium 52
chronology 7, 18–19
chronology of events 7–12
clamshell grapple 64
Coast Guard 10, 18, 25, 69, 71, 84, 215
coastal 18, 20, 23, 25, 31, 53, 68, 71–5, 77, 79–82, 84, 87, 88, 104–9, 111, 115, 121–4, 130, 131, 134–49, 155, 160, 162, 164, 167, 175, 176, 180, 182–4, 190, 191, 193, 197, 200, 203–5, 215–65, 273–90, 297–300, 308, 315, 338–42, 367–72, 374–7, 379, 382, 398
Coastal Management 20
coastal waters 18, 22–3, 72, 73, 104, 146, 175, 178, 182, 193, 195, 197, 203–5, 215, 224, 232, 262, 280, 281, 290, 315, 374, 376
coastline 10, 18, 53, 56, 61, 106, 131, 135, 137–9, 145, 148–9, 259, 160, 218, 219, 279, 285, 288, 367, 372–3, 380
Colombo 1, 2, 4, 5, 7, 8, 10, 12, 13, 18, 29, 40, 56–8, 68, 75, 77, 80, 83, 84, 87, 89, 103, 105, 107, 109, 114, 135, 138, 178, 195, 215, 278, 282–5, 315, 338, 343, 364, 389
combustibles 45–6
communication 5–6, 12, 20, 42, 122–4, 130–1, 175, 177, 190, 204, 249, 320, 322, 343, 388, 390, 393
consequences 2–7, 13, 21, 65, 114, 121, 123–5, 145, 148, 190, 191, 275, 300, 317, 364, 372, 381, 388, 389
container ship 18, 19, 56, 135, 190, 276, 289, 315–30, 388, 391, 397
containers 1, 2, 4, 5, 10, 18, 24, 40, 41, 46–9, 52, 53, 56, 58, 60, 63, 64, 68, 69, 84, 100, 102, 103, 121, 135, 137, 148, 155, 158, 159, 160, 182, 190, 191, 204, 215, 221, 224, 245, 247, 278, 279, 290, 315, 330, 338, 339, 345, 346, 349, 352, 355–8, 360–4, 369, 387, 389, 392, 395, 396
contingency plan 63, 109, 116, 122, 396, 397
convention 25, 29, 31, 348
cooling 46, 48, 58, 390, 394
coral reefs 27, 127, 145, 161, 375
COVID-19 9–10
crane boom 60
crew 4, 17
critical decisions 28, 388–9
crude oil 19

CTU code 4, 12, 387–8
Cumaranatunga 190, 215, 216, 242

D

Dalpathadu 215, 224
dangerous goods 1, 4, 13, 29, 40, 49, 81, 121, 160, 190, 345, 349, 387–8, 391, 397
dead fish 10
debris 10, 25, 27, 29, 49, 54, 56, 63, 64, 74, 79, 84, 96–9, 102, 103, 115, 135, 137, 138, 144, 148, 155, 160–2, 176, 182, 183, 204–5, 215–16, 221, 244, 245, 273, 278, 281, 287–90, 298, 304–6, 315, 320, 338, 346, 349, 352, 362, 367, 369, 393, 394, 396, 397
debris removal 49, 63, 64, 102, 304, 362, 396, 397
decision making 22, 76–9, 87, 317–19, 387–9, 393, 397
deep sea 1, 57, 73, 106, 109, 121, 142, 177, 182, 191, 224, 287, 349, 395, 398
Deepwater Horizon 21–3, 176, 197, 274–5
deluge systems 391
depth 2, 11–13, 31, 57, 63, 68, 137, 142, 143, 193, 197, 248, 290, 300, 303–5, 395–7
design 17, 24, 41, 49, 190, 197, 301–4, 308, 318, 325, 344, 346, 355, 362, 373, 382, 391, 397
detection 196, 325, 343, 390, 391
dibenzofurans 394
dilution 50, 51, 61
dioxins 62, 81, 279, 325, 327, 371, 394
disaster 2, 12, 13, 18, 26–9, 40, 41, 46, 53, 54, 65, 67–117, 121, 123, 130, 134–49, 155–7, 182, 184, 190–206, 215–65, 274–6, 278–9, 297–310, 315, 318–24, 326, 328–30, 339–41, 367–82, 393–4
dispersant 29, 72, 74, 79, 104, 111, 396, 398
dispersion 23, 31, 49, 61, 137, 148, 251
dissolved oxygen 23, 62
divers 59, 63, 64, 224, 397
diving 176, 228, 230, 397
dolphins 27
dry chemical 4, 56, 394

E

ecological catastrophe 53
economic impacts 21, 25, 26, 31, 218, 377
ecosystem 19, 20, 24, 27, 31, 40, 52, 53, 65, 69, 71, 72, 74, 75, 97, 108, 144–7, 182, 193, 195, 200–3, 216, 218, 220, 221, 223, 226, 230, 244, 245, 281, 289, 297, 300, 307, 319, 321, 322, 325, 328–30, 367–72, 375, 377–9, 381, 382
Egodage 40
Ekanayake 173, 174, 177, 178, 180, 182–3
emergency 7, 69, 70, 76, 104, 376, 389–93

Index

emissions 22, 24, 46, 61, 68, 82, 100, 134, 147, 195, 199, 308, 344, 369, 376, 397
enforcement 397
engine room 9, 11, 49, 57
engineering judgment 395
environment 15, 16, 18, 20, 22, 24, 25, 29, 31, 40, 46, 52, 60, 68–71, 84, 102, 121, 122, 134–49, 155–65, 184, 195, 199, 215, 216, 219–21, 226, 228, 230, 244, 247, 259, 264, 273, 276, 279, 281, 282, 289, 297–301, 307–9, 315, 316, 328, 329, 338–40, 344, 345, 348, 357, 363, 364–367, 369, 372, 374–6, 378, 381
environmental assessment 20
environmental catastrophe 1
environmental damage 29
Environmental impacts 27, 31
epicenter 84, 390–2
epoxy resins 25, 42, 43, 46, 47, 58, 135, 137, 159, 160, 191, 259, 298, 359
escalate 56, 64, 394
escalated 10
estimate 51, 182, 216–17, 274–5, 282–3, 309, 341, 359, 360, 372–7, 380, 382, 395
evacuated 9–10, 56, 393
evidence 2, 4, 5, 7, 8, 10, 24, 52, 56, 60, 104, 114, 138, 139, 141, 147, 148, 156, 185, 200, 247, 250, 257, 261, 262, 264, 275, 303, 316–18, 320, 321, 326, 328, 329, 349, 378, 381, 395, 397
e-Waste 41, 42
exothermic 46–52, 57, 58, 65, 389, 390, 392–3
exothermically 46, 47, 58
experts 62, 68, 82, 114, 290, 322–3, 330, 364, 367, 368, 376, 393, 397, 398
explosion/explosions 2, 9, 10, 11, 17, 26, 40, 42, 46–9, 53, 56–8, 60, 65, 68, 69, 70, 102–6, 108, 114, 115, 137, 140, 148, 156, 190, 191, 195, 204, 205, 215, 216, 245, 247–9, 315, 316, 318, 338, 363, 364, 369, 392–4
explosive substance 47
extinguished 1, 8, 11, 49, 56–7, 60, 390, 392, 394
Exxon Valdez 21, 26, 27, 274, 275, 372, 373, 376

F

feeder vessel 1
Fernando 125, 215
fire 1, 2, 4, 5, 7–11, 13–15, 17, 18, 25, 31, 40, 42, 44–50, 52–4, 56–62, 64, 65, 68–70, 81, 82, 103, 104, 122, 123, 130, 134–7, 141, 155, 160, 179, 181, 190, 198–200, 205, 206, 215, 217, 223, 228, 244, 247–50, 274, 275, 278, 298, 315, 316, 318, 338, 339, 341, 348, 349, 350, 357, 358, 363, 364, 368, 369, 371, 378, 388, 390–6, 398
fire alarms 390
fire cannons 56

fire hoses 391
fire protection 44, 350
firefighters 9
firefighting 5, 7–10, 47, 49, 52, 56, 58–60, 64, 68–70, 369, 371, 390–4, 398
firefighting foam 394
firefighting vessels 10
firefighting water 394
fish 10, 20, 26, 27, 43, 48, 54, 56, 62, 79, 82, 122, 124, 125, 127–31, 145–7, 157, 162, 175, 191, 193, 195, 202, 203, 215–65, 273, 277–82, 287, 288, 300, 319, 321, 325–8, 330, 339, 357, 367, 368, 371, 372, 374, 375, 377–81
fishermen 27, 29, 122–5, 127, 130, 131, 177–8, 372
fishing community 40, 73–5, 81, 87
fixed firefighting system 5, 390
flammable 10, 13, 40, 47, 51, 56, 58–60, 195, 355, 362, 363, 391, 397
flammable substances 10
food chain 27, 31, 53, 54, 61, 146, 193, 195, 201, 223, 226, 228, 242, 245, 250, 252, 300, 321, 328, 367, 368, 375, 379
forward section 11, 47, 56, 57, 391, 392, 396
fuel 9
fuel oil 12, 24, 48–9, 60, 64, 68, 70, 215, 250, 256, 274, 279, 374, 393, 396
fuel tanks 42, 49, 57, 60
fume 7–8, 46, 68
furans 62, 81, 279, 371

G

Gamage 67
Ganepola 367
gas cloud explosion 60
Gayathry 215
government agencies 393
guidance 322–9, 364, 388, 390, 397
Gunawardena 367
Gunawardhana 155
Guruge 67

H

Hamad 1, 4, 57, 388
Haputhantri 215
hatch cover 3, 5, 390
hatches 48, 390
hazardous 15
hazardous and nox-ious substances 15–16, 42
hazardous material 42, 191, 195
hazardous shipwreck 397
hazardous situation 390
hazardous substances 54, 64, 155, 259, 315, 340, 345–9, 357–9, 361, 363, 376, 395

Hazira 1, 4, 5, 57, 68, 388
HDPE 24–5, 43, 46, 53, 58, 59, 84, 135, 136, 157–9, 288, 350
heavy equipment 395
heavy fuel 12, 48, 60, 64, 256, 274
heavy metals 27, 31, 42, 52, 53, 61, 121, 131, 145, 161–2, 195, 200–2, 228, 252, 265, 278, 279, 327, 349, 378
HFO 12, 48–9
HNS 15, 16, 19, 24, 42, 289, 338, 372
hull 11, 13, 17, 52, 57, 62–5, 378, 394, 395
human error 17, 22, 31, 190, 329
human factor 389, 397
human health 24–7, 30, 31, 53, 54, 61, 65, 82, 148, 252, 254, 256, 258, 259, 338, 347–8, 371, 372, 378
human life 393
Human Machine Interface 17
hydrocarbons 23, 24, 29, 42, 48, 49, 53, 131, 157, 199–200, 256, 257, 273, 274, 278, 279, 288, 324, 325, 371, 378, 379
hydrodynamic 63

I

IMDG code 349, 387–9
IMO 2, 24, 25, 309–10, 338, 387–90, 397
impact zone 61
incident control 65
incidents 18, 19, 24, 27, 30, 31, 111–112, 115, 123–4, 131, 137, 147, 198, 205, 218, 223, 228, 244–6, 248–50, 265, 273, 341, 364, 372, 377, 388, 397, 398
incomplete combustion 40, 199
industrial 15, 121, 160, 303, 339, 345, 346, 357, 375, 393, 398
infrastructure 122, 324, 328, 387
injuries 9
inspection 11
intense heat 10
International Maritime Organization 24, 310, 387
investigation 4, 11, 13, 18, 114, 264, 323, 341, 393, 395
ion 5, 46, 47, 58, 183, 253, 254, 307
ITOPF 2, 5, 7, 67, 74, 88, 91, 358

J

Jabel Ali 1, 2, 4, 5
James 134, 139, 141, 144, 176, 178, 278, 279, 289
Jayasekara 40

K

KMTC Hong Kong 388
Kumara 127, 134, 224, 297

L

Lasitha Cumaranathunge 13
LDPE 18, 25, 43, 53, 59, 84, 135, 136, 157–60, 164, 165, 278, 350, 361
leak 4, 5, 7, 13, 42, 46, 48, 57, 58, 135, 200, 279, 387–90
legal 25, 29–30, 82, 114, 117, 185, 317, 339–41, 344, 348, 364, 375, 395, 398
legislation 15
lessons 13, 115, 289, 316, 387, 389, 391, 393
Li-ion battery 348–9
lithium 5, 13, 42, 46–8, 58, 202, 215, 228, 245, 254–5, 278, 359
living quarter 62, 57, 395
Liyanage 173, 190, 197
local community 54, 393
long-term health 27, 131
low-density polyethylene 10
lube oils 45, 60, 64
lubricants 49, 274, 351

M

Maes 67
management 17, 18, 20, 22, 23, 25, 62, 65, 67, 71–4, 76, 78–9, 84, 100, 115–17, 134, 178, 184, 297, 316, 317, 319, 322, 323, 338–41, 344–52, 355, 357–9, 362–4, 381, 387, 393, 397–8
mangrove forests 27, 143, 279, 280
Mangroves 27, 143, 279, 370
marine accidents 15
marine animals 52, 111–12, 114, 116, 147, 220, 257, 303, 367
marine biology 62
marine diesel 49, 60
marine environment 15, 16, 18, 22, 25, 54, 60, 67, 69, 70, 74, 84, 134–5, 138, 141, 144, 146, 155–7, 160, 161, 164, 195, 204, 205, 215, 220, 221, 224, 226, 228, 229, 242, 244, 247, 264, 273, 276, 279, 281, 287, 289, 290, 297–301, 306–9, 328, 329, 338, 340, 341, 344, 345, 364, 372, 375, 376
marine life 15, 16, 27, 30, 52–4, 60, 107, 114, 116, 147, 157, 161, 195, 200, 219–20, 273, 301, 302, 338, 367, 369, 378
marine oil spill 23, 274, 379
marine organisms 10, 16
marine safety 24
maritime accident 1, 15–23, 27, 28, 123–4, 134, 139, 190, 195, 275, 287, 298, 344, 388
maritime emergency 390
maritime regulators 391
maritime safety 17, 20, 22, 23, 72, 387
MARPOL 25, 29, 389

Index

marshes 26, 27
media 9, 10, 57, 60, 78, 81, 111, 121–4, 130, 131, 166, 307, 325, 393, 395, 396
MEPA (Marine Environment Protection Authority) 3, 9, 10, 41, 67, 69–71, 73–7, 80, 84, 87, 88, 95, 100, 102, 104, 107, 109, 111, 114, 289, 290, 340, 341, 345, 349, 356, 358, 363
mercury 27, 52, 201, 278, 377
methanol 41, 44, 46, 47, 58–60, 68, 121, 200, 215, 278, 350, 360, 363, 367, 393
methodologies 308, 397
methyl nitrate 47, 58
microplastic 16, 23, 25, 27, 29, 41, 53, 62, 134, 136–7, 141–8, 155, 157–62, 165–7, 183, 191, 196, 203, 264, 297–310, 367, 375, 382
molybdic oxide 52
monsoon 1, 40, 53, 56, 71, 82, 102, 115, 137, 142, 147, 160, 191, 197, 200, 205, 215, 216, 220, 221, 223, 226, 232, 248, 249, 279, 281, 287, 290, 299, 300

N

nanoplastics 53, 54, 141, 157, 164, 205, 258, 263, 264, 300, 382
negligence 12, 17, 22
nickel 52, 201
Niroshan 273
nitric acid 2, 5–7, 12–13, 17, 18, 41, 42, 46–50, 57–62, 65, 68–9, 111, 121, 135, 155, 164, 190, 191, 195, 200, 202, 215, 249, 255, 278, 350, 360, 363, 367, 387–90, 392
nitrogen dioxide 46, 58, 81, 111, 390, 392
Norway 18
nurdles 10, 18, 25, 26, 29, 41, 53, 54, 68, 83–6, 88–98, 100, 102, 114–16, 121, 124, 125, 128, 131, 134–49, 156, 157, 160, 164, 165, 167, 182–4, 191, 195, 202–3, 215, 221, 228, 245–7, 261, 262, 264, 276, 278, 288–90, 298–300, 304–6, 310, 315, 339, 348, 349, 357, 358, 369, 370

O

ocean current 16, 23, 27, 30–1, 54, 72, 82, 83, 115, 137, 148, 160, 182, 220–2, 298, 367
oceanography 62
offshore oil and gas 22, 30, 31
oil sheen 49, 111
oil slick 12, 49, 71–5, 77, 104, 105, 107–9, 111, 348
oil spill 15, 18, 19, 21–4, 26–31, 64, 67–79, 82, 84, 91, 102–9, 111, 115–16, 122, 131, 176, 191, 196, 197, 199, 200, 251, 256, 273–5, 278, 279, 287, 289, 345, 348
operating procedure 17

orange smoke 7
organic substance 7
organic substances 57, 68, 389, 390
organics 46, 53
overboard 10
oxidizer 13, 46, 57
oxidizing 7

P

pellets 10, 18, 21, 24–6, 30, 31, 56, 68, 88, 115, 121, 123–7, 134–9, 143, 147, 148, 155, 157, 159, 160, 176, 191, 195–6, 202, 205, 221, 247, 249, 250, 260–4, 275, 276, 278, 281, 285–6, 288, 289, 298, 299, 338, 339, 347, 348, 350, 353, 354, 358, 367, 368
Perera 68, 107, 124, 134, 138, 155, 297
perfumery products 60, 363
persistence 49, 53, 157, 316
persistent 61, 62, 121, 124, 200, 202, 204, 206, 208, 244, 300, 368, 369, 371, 394
piecemeal 62, 63, 65, 397
piecemeal strategy 62–3
Piyatilleke 40
plan 5, 16, 67, 70, 71, 73–7, 79, 84, 88, 102–5, 109, 116, 122, 123, 131
planktonic 16
plastic 1, 10, 15–31, 40, 41, 43–5, 53–4, 56, 61, 62, 68, 69, 84–98, 100, 102, 114–16, 121, 124, 125, 127, 128, 131, 134–49, 155–67, 176, 182–4, 191, 194–6, 199, 202–5, 215, 216, 221, 228, 245–7, 249, 250, 257, 258, 260, 261–5, 273–8, 281, 285, 288–90, 297–310, 315, 320, 330, 338–9, 347–50, 352–4, 356, 382
plastic beads 10
plastic pellets 18
plastic spills 15–16, 18–19, 21–31, 53, 134, 135, 138, 144, 147, 148, 202, 288
pollutants 24, 29, 40, 42, 47, 53, 54, 62, 81, 82, 84, 141–2, 155–7, 162, 164, 191, 193, 195, 199, 201, 202, 204, 221, 223–6, 228, 230, 242, 244, 252, 254, 256, 258, 260, 265, 279, 281, 282, 287–90, 307, 326, 340, 341, 349, 368, 369, 371, 372, 374, 375, 378
pollution 15, 19, 20, 22, 24–7, 29–31, 42, 46, 53, 54, 58, 61–5, 67–70, 81, 82, 88, 89, 124, 136, 137, 139, 143, 148, 155, 156, 160, 162, 191, 195, 198, 200, 203, 204, 228, 230, 273, 278, 279, 281, 300, 304, 308, 310, 330, 340, 341, 344, 358
Pollution Prevention 15
pollution zone 396
polychlorinated dibenzofurans 61
polychlorinated dibenzo-p-dioxins 61
polymer beads 10

polymers 40, 157, 159, 160, 162, 164–7, 202, 298, 307, 308, 345, 349, 350, 357, 359–61, 393
pool fire 49, 56
Port of Colombo 1, 2, 68, 343, 364
ports authority 390, 391, 392
Prasad 190
primary containment 387
Priyadarshani 215
PVC 44, 61, 135, 157, 159, 162, 302, 350, 359, 361
pyrophoric 64

R

Ranasinghe 67, 316
Ratnayake 155
reactions 30, 46, 48–52, 57, 58, 61, 65, 159, 164, 165, 200, 259, 362, 367, 369, 375, 389, 390, 392
recovery period 31
Reddy 134
regulatory authorities 397
reinforcements 161, 395
research, 20
resins 10, 25, 42, 43, 47, 58, 134, 135, 137, 159, 160, 191, 196, 259, 298, 350
response 20, 21, 26, 28–31, 67–79, 82, 91, 100, 102–5, 107, 109, 111, 114–17, 122, 123, 177, 252, 253, 257, 264, 310, 316, 317, 321, 325–30, 341, 371–2, 374–6, 382, 387–90, 392–4, 398
re-work 388, 390
reworked 5
risk 13, 15, 19, 20, 22, 23, 27, 30, 63, 65, 69, 75–7, 105, 107, 109, 123, 145, 146, 148, 160, 163, 197, 228, 257, 258, 289, 308–9, 316, 318, 319, 321, 324, 339, 343, 345, 348, 355, 356, 358, 364, 376, 378, 379, 389–91, 393, 396, 397
risk mitigation 389
risk perceptions 393
root causes 16–17
ROV 49
rubber sealing 390
Rubesinghe 26, 27, 80, 82, 87, 97, 111, 121, 131, 182, 275, 276, 278, 279, 288, 289, 315
rules 29, 289, 397
runaway reaction 46, 58

S

safer location 10
safety barriers 391
safety culture 389–91, 396, 397
safety precaution 9
salvage company 395, 396
salvage vendor 62, 65
salvors 1, 11, 18, 57

samples 48, 54, 96, 102, 124, 139, 143, 144, 182, 184, 201–3, 226, 230, 232, 242, 264, 265, 273, 290, 305, 324, 325, 328, 339, 348, 349, 377, 379, 395, 397
sampling 100, 102, 114, 116, 142, 143, 228, 232, 243, 260, 264, 282, 301, 303, 318, 322, 324, 395, 396
sank 1, 13, 26, 40, 49, 53, 121, 155, 273, 363, 394, 395
satellite 12, 111, 175, 178–9
scrap metals 61
sea turtles 10
seabed 11, 12, 27, 54, 63, 64, 219, 290, 308, 396
seafarers 397
seafood 27, 61, 126, 131, 147, 157, 301, 372, 374, 375, 377–9
sediment 22, 23, 160, 167, 193, 195, 196, 200, 205, 227, 245, 249, 250, 252, 253, 256, 257, 259, 265, 301–3, 309, 319, 324, 325, 328
Seneviratne 273
Sewwandi 25, 29, 134, 136, 137, 139–41, 145, 148, 155, 157, 160, 164, 278, 279, 288, 289, 297–300, 304, 315
sharks 27, 195
sheerleg barges 63–5
ship bottom 60, 64
ship owner 17, 49, 343, 346, 349, 357, 393, 396
shipping company 82, 114, 388, 392, 394–5
shipwreck 11, 27, 42, 52–4, 62–5, 102, 135, 138–41, 143, 144, 146–8, 195, 216, 217, 223, 228, 230, 232, 242, 244, 278, 288, 298, 299, 338, 346, 357, 358, 363, 369, 396, 397
shoreline 10, 56, 62, 67, 73, 84, 87, 100, 114–17, 121–2, 124, 126–7, 138, 145, 273, 279, 281, 288–90, 346, 349, 356, 368, 376
side scan sonar 49
sinking 1, 11, 22, 49, 57, 62, 102, 105, 107–9, 190, 196, 197, 200, 216, 217, 248, 250, 288, 307, 308, 315, 345, 348, 394–6
SLPA 13, 67, 69, 71, 73, 74, 77, 102, 104, 109
smoke 13, 15, 41, 46, 57, 58, 61, 62, 81, 111, 191, 195, 198–9, 206, 215, 217, 248, 287, 390–2, 395
smoke plumes 11
social impacts 25, 26
socio-economic 19–22, 24, 65, 343
sodium hydroxide 52, 61, 62, 278
sodium methylate 41, 60
sodium methylated solution 41
sonar 63
spreading of the fire 9
stakeholders 20, 30, 31, 82, 322–3, 344, 349, 381, 388
steel 13, 43, 60, 61, 352, 362, 395
Stolt Rotterdam 13

structure 13, 48, 54, 57, 122, 155, 162, 164, 165, 223, 250, 277, 324, 328, 382, 387, 395, 398
sulphur 24, 40, 44, 52
surveillance 72, 73, 79, 104, 398
survey 49, 63, 67, 81, 84, 85, 88, 104, 111, 138, 177–8, 218, 219, 230, 232, 280, 282, 288, 316, 328–9, 373, 374, 398

T

tanker 19
temperature 7, 27, 50–2, 60, 62, 141, 148, 162, 164, 176, 183, 184, 197, 219–20, 228, 248, 287, 298, 357, 361, 363, 378, 394–6
thallium 52
thermal explosion 46, 58
Thushara 40
Tonnage 2
tourism 27, 54, 176, 316, 341, 367, 368, 372, 375, 377
tow 11, 18, 65, 82, 395
towed 1, 62, 65, 155
towing 11, 18, 395
toxic 27, 29, 46, 47, 50, 52, 53, 58, 59, 61, 62, 79, 82, 111, 114, 116, 121, 123, 124, 131, 141, 147, 155, 157, 160, 162, 182, 195, 199–202, 228, 248, 250, 252, 253, 254, 257, 259, 276, 279, 281, 298, 300, 315, 321, 324, 349, 367, 375, 376, 379, 397
toxicities 16
training 13, 17, 22, 114, 116, 184, 190, 391
transhipment 22
tugboats 8, 56, 392, 393
turbidity 53, 165
turtles 10, 18, 26, 27, 52, 54, 56, 62, 68, 112–14, 125, 145, 147, 148, 157, 173–85, 196, 216, 273, 278, 287, 301, 339, 354, 357, 367–9, 371, 382

U

Udugama 367
underwater 22, 63–5, 289, 346, 397, 398
United Nations 25, 132, 144, 310
urea 40, 43, 52, 190, 195, 215, 228, 360

V

Vassanadumrongdee 297
ventilation 49, 60, 61, 360, 361
vinyl acetate 41, 135, 288
visible flames 8, 392
Vithanage 134, 297, 348
VLSFO 24
volatile 50, 58, 162, 348, 361, 378, 394

W

waste 20, 24, 27, 54, 59, 72, 73, 74, 78, 82–5, 87, 88, 89–97, 100–2, 104, 114–16, 134, 157, 161, 166, 167, 248, 289, 297, 298, 301, 303, 305–7, 309, 338–40, 342–7
water cannons 10
water jets 392
water mist 391
water spraying 392
wave 16–18, 23, 49, 52, 53, 61–3, 82, 84, 87, 122, 123, 137–9, 141–5, 147, 162, 164, 216, 249, 278, 279, 290, 298–300, 304, 309, 349, 360
weaknesses 338, 382, 387, 397–8
weather conditions 22, 56, 62, 124, 232, 308, 374, 389, 392–3
Weerasekera 215
weight 18, 60, 63, 88, 159, 164, 165, 201, 260, 278, 288, 316, 317, 321, 395
Wells 197, 302, 304, 315, 316, 323, 324, 330
whales 18, 26, 27, 112, 147, 191–5, 197–9, 201, 203–4, 287, 368, 371, 372
Wimalasiri 107, 215
wind 9
Withanage 68, 121, 125, 131
wreck removal 1, 200, 363

Y

Yantian Express 388

Z

Zhang 15, 18, 19, 61–2, 147, 164, 165, 167, 297, 300, 315, 317, 324, 327